COMPLEX ANALYSIS

(For Undergraduate Students of Mathematics, Physics and Engineering)

P. DURAIPANDIAN

KAYALAL PACHAIYAPPA

S. CHAND
PUBLISHING

S Chand And Company Limited

(ISO 9001 Certified Company)

S Chand And Company Limited

(ISO 9001 Certified Company)

S. CHAND PUBLISHING

Head Office: D-92, Sector–2, Noida – 201301, U.P. (India), Ph. 91-120-4682700

Registered Office: A-27, 2nd Floor, Mohan Co-operative Industrial Estate, New Delhi – 110 044, Phone: 011-49731800

www.**schandpublishing.com**; e-mail: **info@schandpublishing.com**

Marketing Offices:

Chennai	:	Ph: 23632120; chennai@schandpublishing.com
Guwahati	:	Ph: 2738811, 2735640; guwahati@schandpublishing.com
Hyderabad	:	Ph: 40186018; hyderabad@schandpublishing.com
Jalandhar	:	Ph: 4645630; jalandhar@schandpublishing.com
Kolkata	:	Ph: 23357458, 23353914; kolkata@schandpublishing.com
Lucknow	:	Ph: 4003633; lucknow@schandpublishing.com
Mumbai	:	Ph: 25000297; mumbai@schandpublishing.com
Patna	:	Ph: 4011400; patna@schandpublishing.com

First Edition 2014
Reprints 2017, 2018, 2019, 2020, 2022 (Twice), 2023

Reprint 2025

ISBN: 978-93-837-4646-0 **Product Code:** H6CAY61MATH10ENAA14R

PRINTED IN INDIA

By Vikas Publishing House Private Limited, Plot 20/4, Site-IV, Industrial Area Sahibabad, Ghaziabad – 201 010 and Published by S Chand And Company Limited, A-27, 2nd Floor, Mohan Co-operative Industrial Estate, New Delhi – 110 044.

PREFACE

This book provides a logical development of the subject from its elementary root. The book is meant for the undergraduate students of Mathematics, Physics and Engineering. Complex Analysis is a subject developed from an imaginary number but it is beautifully applied in many Engineering disciplines.

The book is an introductory treatise. However, it does not suffer from lack of rigour. Our aim is to provide the reader an elementary, thorough and systematic introduction of the subject. In doing so, a stress is made on important basic ideas. For example, the idea of the complex infinity is brought out through the Stereographic Projection and also through the mapping $w = 1 / z$ in a lucid manner. This presentation enhances the understanding of the physical meaning of the complex infinity and its neighbourhood.

The text is supported by a large number of solved sums including questions asked in university examinations. These sums pertain to

- evaluation of the integrals of analytic functions along simple closed curves, using Cauchy's theorem and Cauchy's integral formulae for derivatives.
- construction of analytic function when its counterpart, namely, real or imaginary part is given.
- expansion of analytic functions as Taylor's series or Laurent's series.

Even a first reading of the book will give the reader a sufficient mastery over the subject. The feature will commend the book to the students.

We are grateful to the Professors who have given us their valuable suggestions during preparation of the manuscript. We look forward for further suggestions from them.

We are thankful to the management and the editorial team of S.Chand & Company Pvt. Ltd. for help, support and guidance in publication of this book.

Thanks to all those who have helped us directly or indirectly in preparation of this project.

P. Duraipandian
Kayalal Pachaiyappa

CONTENTS

CHAPTER

1

COMPLEX NUMBER SYSTEM

1.1. COMPLEX NUMBER SYSTEM

The square of real number is never negative. As such, no real number satisfies the equation $x^2 = -1$. Equations of this type caused the introduction of a new type of numbers called complex numbers.

Complex numbers. All ordered pairs (α, β) of all real numbers α and β with equality among them being defined by that

(*i*) $(\alpha_1, \beta_1) = (\alpha_2, \beta_2)$ implies if and only if $\alpha_1 = \alpha_2$, $\beta_1 = \beta_2$ and with two binary operations + and . defined by

(*ii*) $(\alpha_1, \beta_1) + (\alpha_2, \beta_2) = (\alpha_1 + \alpha_2, \beta_1 + \beta_2)$,

(*iii*) $(\alpha_1, \beta_1) . (\alpha_2, \beta_2) = (\alpha_1 \alpha_2 - \beta_1 \beta_2, \alpha_1 \beta_2 + \alpha_2 \beta_1)$,

are called complex numbers. The operations are respectively called complex addition and complex multiplication. The set of all complex numbers is denoted by **C**.

1.2. FIELD OF COMPLEX NUMBERS

It can be easily seen that the set **C** of complex numbers is a field with the above two operations as additive and multiplicative operations. The element (0, 0) is the *additive identity* and the element (1, 0) is the multiplicative identity. The inverse of (α, β) under addition is $(-\alpha, -\beta)$. If (x, y) is the multiplicative inverse of (α, β), then

$$(\alpha, \beta) . (x, y) = (1, 0)$$

or $$(\alpha x - \beta y, \alpha y + \beta x) = (1, 0).$$

This leads to solving the equations $\alpha x - \beta y = 1$ and $\alpha y + \beta x = 0$ for x and y whose values are

$$x = \frac{\alpha}{\alpha^2 + \beta^2}, \quad y = \frac{-\beta}{\alpha^2 + \beta^2}$$

if $\alpha \neq 0$ and $\beta \neq 0$. So the multiplicative inverse of (α, β), when $(\alpha, \beta) \neq (0, 0)$, is

$$\left(\frac{\alpha}{\alpha^2 + \beta^2}, \frac{-\beta}{\alpha^2 + \beta^2}\right).$$

1.3. SCALAR MULTIPLICATION OF A COMPLEX NUMBER

The scalar multiplication of a complex number (α, β) by a real number λ, is defined by

$$\lambda (\alpha, \beta) = (\lambda \alpha, \lambda \beta).$$

Now the following results are evident:

(*i*) $(-1)(\alpha, \beta)$ or $-(\alpha, \beta) = (-\alpha, -\beta)$, (*ii*) $(r+s)(\alpha, \beta) = r(\alpha, \beta) + s(\alpha, \beta)$,

(*iii*) $(r\{(\alpha_1, \beta_1) + (\alpha_2, \beta_2)\} = r(\alpha_1, \beta_1) + r(\alpha_2, \beta_2)$,

where r, s are any two real numbers.

1.4. FIELD OF REAL NUMBERS

If **R** is the set of all real numbers, then it can be shown that the set of complex numbers

$$\{(\alpha, 0) \mid \alpha \,\varepsilon\, \mathbf{R}\}$$

is a sub-field of **C**. Further the results

$$\left.\begin{aligned}(\alpha, 0) + (\beta, 0) &= (\alpha+\beta, 0),\\ (\alpha, 0)\,.\,(\beta, 0) &= (\alpha\beta, 0)\end{aligned}\right\} \text{ for all } \alpha, \beta, \varepsilon\, \mathbf{R}$$

show that the complex numbers of the form $(\alpha, 0)$ have the same arithmetic properties as real numbers and the complex number $(\alpha, 0)$ may be identified with the real number α. Consequently we get that the real number system is embedded in the field of complex numbers and we can denote $(\alpha, 0)$, $(1, 0)$, $(0, 0)$ simply by α, 1, 0.

1.5. EXPRESSION OF THE NUMBER (α, β) IN THE FORM $\alpha + i\beta$

If we denote the complex number $(0, 1)$ by i, then

$$(\alpha, \beta) = (\alpha, 0) + (0, \beta) = \alpha(1, 0) + \beta(0, 1) = \alpha + \beta i.$$

It is important to note that

$$i\,.\,i = i^2 = (0, 1)\,.\,(0, 1) = (0-1, 0+0) = (-1, 0) = -1.$$

Thus we see that the equation $z^2 = -1$ has a solution in the complex field **C**. Also, addition and multiplication of complex numbers

$$(\alpha_1, \beta_1), (\alpha_2, \beta_2)$$

can be identified with the arithmetic addition and multiplication of the expressions

$$\alpha_1 + i\beta_1, \alpha_2 + i\beta_2$$

with the restriction that $i^2 = -1$. For, their arithmetic sum and product are

(*i*) $$\begin{aligned}\alpha_1 + i\beta_1 + \alpha_2 + i\beta_2 &= \alpha_1 + \alpha_2 + i(\beta_1 + \beta_2)\\ &= (\alpha_1 + \alpha_2, \beta_1 + \beta_2)\\ &= (\alpha_1, \beta_1) + (\alpha_2, \beta_2)\end{aligned}$$

(*ii*) $$\begin{aligned}(\alpha_1 + i\beta_1)(\alpha_2 + i\beta_2) &= \alpha_1\alpha_2 + i^2\beta_1\beta_2 + i(\alpha_1\beta_2 + \alpha_2\beta_1)\\ &= \alpha_1\alpha_2 - \beta_1\beta_2 + i(\alpha_1\beta_2 + \alpha_2\beta_1)\\ &= (\alpha_1\alpha_2 - \beta_1\beta_2, \alpha_1\beta_2 + \alpha_2\beta_1)\\ &= (\alpha_1, \beta_1)\,.\,(\alpha_2, \beta_2).\end{aligned}$$

Division. Given $z_1 \neq 0$ and z_2, the process of finding z_3 so so that

$$z_1 z_3 = z_2$$

is called division of z_2 by z_1. Here z_3 is written as

$$\frac{z_2}{z_1}.$$

1.6. CONJUGATION

Real and imaginary parts. In the complex number $z = \alpha + i\beta$, the real numbers α and β are respectively called the real and imaginary parts of z and are denoted by

$$\mathbf{Re}\, z, \quad \mathbf{Im}\, z.$$

Conjugation. The transformation which transforms a complex numbers $\alpha + i\beta$ into the complex number $\alpha - i\beta$ is called complex *conjugation*. Each of the complex numbers $\alpha + i\beta$ and $\alpha - i\beta$ is called the *conjugate* of the other. The conjugate of z is denoted by $\bar{z}$. Now it follows that

$$\textbf{Re}\, z = \frac{z + \bar{z}}{2}, \quad \textbf{Im}\, z = \frac{z - \bar{z}}{2i}$$

and that $z = \bar{z}$ implies that z is real.

Remark. Geometrically $\bar{z}$ is the reflection of z on the real axis.

Theorem 1.1. If z_1 and z_2 are any two complex numbers, then

$$\overline{(z_1 + z_2)} = \bar{z}_1 + \bar{z}_2,$$

$$\overline{(z_1 z_2)} = \bar{z}_1 \bar{z}_2.$$

Proof: If $z_1 = x_1 + iy_1, z_2 = x_2 + iy_2$, then

$$\begin{aligned}\overline{(z_1 + z_2)} &= \overline{(x_1 + iy_1) + (x_2 + iy_2)} = \overline{x_1 + x_2 + i(y_1 + y_2)} \\ &= x_1 + x_2 - i(y_1 + y_2) = x_1 - iy_1 + (x_2 - iy_2) \\ &= \bar{z}_1 + \bar{z}_2\end{aligned}$$

$$\begin{aligned}\overline{(z_1 z_2)} &= \overline{(x_1 + iy_1)(x_2 + iy_2)} = \overline{\{x_1 x_2 - y_1 y_2 + i(x_1 y_2 + x_2 y_1)\}} \\ &= x_1 x_2 - y_1 y_2 - i(x_1 y_2 + x_2 y_1) = (x_1 - iy_1)(x_2 - iy_2) \\ &= \bar{z}_1 \bar{z}_2.\end{aligned}$$

Remark 1. Extending the second result to n factors,

$$\overline{(zzzz \ldots z)} = \bar{z}\,\bar{z}\,\bar{z}\,\bar{z} \ldots \bar{z} \quad \text{or} \quad \overline{(z^n)} = \bar{z}^n.$$

Remark 2. From $\overline{(z_1/z_2)\, z_2} = \bar{z}_1, \overline{(z_1/z_2)}\, \bar{z}_2 = \bar{z}_1$ or

$$\overline{\left(\frac{z_1}{z_2}\right)} = \frac{\bar{z}_1}{\bar{z}_2}.$$

Theorem 1.2. If the coefficients in a polynomial equation are real, then its complex roots form pairs of conjugate complex numbers.

Proof: Let a be a root of the equation

$$c_0 + c_1 z + c_2 z^2 + \ldots + c_n z^n = 0.$$

Then $$c_0 + c_1 a + c_2 a^2 + \ldots + c_n a^n = 0.$$

Now conjugation gives

$$c_0 + c_1 \bar{a} + c_2 \bar{a}^2 + \ldots + c_n \bar{a}^n = 0$$

which shows that $\bar{a}$ also is a root.

1.7. ABSOLUTE VALUE OF A COMPLEX NUMBER

Absolute value (*modulus*) of a complex number $z = x + iy$ is defined to be the non-negative real number $\sqrt{x^2 + y^2}$ and it is denoted by the symbol

$$|z| \quad \text{or} \quad |x + iy|$$

Remark. Geometrically, $|z|$ is the distance of the complex point z from the origin.

Theorem 1.3. If z, z_1, z_2 are any three complex numbers, then

(*i*) $|z| = |\bar{z}|$ (*ii*) $z\bar{z} = |z|^2$ (*iii*) $|z_1 z_2| = |z_1||z_2|$

Proof: (*i*) $|z| = \sqrt{x^2 + y^2} = \sqrt{x^2 + (-y)^2} = |\bar{z}|$

(*ii*) $z\bar{z} = (x + iy)(x - iy) = x^2 + y^2 = |z|^2$.

(*iii*) $|z_1 z_2|^2 = (z_1 z_2)\overline{(z_1 z_2)} = (z_1 z_2)(\bar{z}_1 \bar{z}_2)$

$= (z_1 \bar{z}_1)(z_2 \bar{z}_2) = |z_1|^2 |z_2|^2.$

$\therefore \quad |z_1 z_2| = |z_1||z_2|.$

Corollary. Further $|(z_1/z_2) z_2| = |z_1|$ gives $|z_1/z_2||z_2| = |z_1|$.

or

$$\left|\frac{z_1}{z_2}\right| = \frac{|z_1|}{|z_2|}.$$

Remark: Extending the result $|z_1 z_2| = |z_1||z_2|$,

$$|z_1 z_2 z_3 \ldots z_n| = |z_1||z_2||z_3| \ldots |z_n|,$$
$$|z^n| = |z|^n.$$

1.8. INEQUALITIES IN TERMS OF MODULI

There is no order relation in the complex number system; that is, there is no meaning in saying that $z_1 > z_2$ unless z_1 and z_2 are real. Now we prove the following inequalities (order relations) in terms of the real parts, imaginary parts and the moduli.

Theorem 1.4. If z, z_1, z_2 are any three complex numbers, then

(*i*) $-|z| \le \mathbf{Re}\, z \le |z|, -|z| \le \mathbf{Im}\, z \le |z|$

(*ii*) $|z_1 + z_2| \le |z_1| + |z_2|$

(*iii*) $|z_1 - z_2| \ge ||z_1| - |z_2||$

of which the second and third are called *triangle inequalities*.

Proof: (*i*) Let $z = x + iy$. Then

$$|z| = \sqrt{x^2 + y^2}, \quad \mathbf{Re}\, z = x, \quad \mathbf{Im}\, z = y.$$

Now it is evident that, for all positive and negative values of x and y,

$$-\sqrt{x^2 + y^2} \le x \le \sqrt{x^2 + y^2}$$

$$-\sqrt{x^2 + y^2} \le y \le \sqrt{x^2 + y^2}$$

from which the first result follows.

(*ii*) $|z_1 + z_2|^2 = (z_1 + z_2)\overline{(z_1 + z_2)}$ $(\because |z|^2 = z\bar{z})$

$= (z_1 + z_2)(\bar{z}_1 + \bar{z}_2)$

$= z_1 \bar{z}_1 + z_1 \bar{z}_2 + \bar{z}_1 z_2 + z_2 \bar{z}_2$

$= |z_1|^2 + (z_1 \bar{z}_2 + \bar{z}_1 z_2) + |z_2|^2$

But $\bar{z}_1 z_2$ is conjugate to $z_1 \bar{z}_2$. So $z_1 \bar{z}_2 + \bar{z}_1 z_2 = 2\, \mathbf{Re}\, (z_1 \bar{z}_2)$. Thus

$$|z_1+z_2|^2 = |z_1|^2 + 2\,\mathbf{Re}\,(z_1\bar{z}_2) + |z_2|^2$$
$$\le |z_1|^2 + 2\,|z_1\bar{z}_2| + |z_2|^2 \qquad (\because \mathbf{Re}\,(z_1\bar{z}_2) \le |z_1\bar{z}_2|)$$
$$= |z_1|^2 + 2\,|z_1|\,|\bar{z}_2| + |z_2|^2$$
$$= |z_1|^2 + 2\,|z_1|\,|z_2| + |z_2|^2 = (|z_1| + |z_2|)^2.$$

$\therefore$ $\qquad |z_1+z_2| \le |z_1| + |z_2|.$

(*iii*) Since $(z_1 - z_2) + z_2 = z_1$ and $(z_2 - z_1) + z_1 = z_2$, by using $|z_1| + |z_2| \ge |z_1 + z_2|$,

$$|z_1 - z_2| + |z_2| \ge |z_1|,$$
$$|z_1 - z_2| + |z_1| \ge |z_2|.$$

These two inequalities give

$$|z_1 - z_2| \ge |z_1| - |z_2|$$
$$|z_1 - z_2| \ge |z_2| - |z_1|$$

$\therefore$ $\qquad |z_1 - z_2| \ge ||z_1| - |z_2||.$

Remark 1. The extention of the first triangle inequality is

$$|z_1 + z_2 + z_3 + \ldots + z_n| \le |z_1| + |z_2| + |z_3| \ldots |z_n|.$$

Remark 2. The triangle inequalities are established in the next chapter geometrically.

1.9. EXAMPLES

1.7. Absolute Value of a Complex Number

1. **Solved sum.** Show that every complex number z whose absolute value is 1, can be expressed in the form

$$z = (1 + it)/(1 - it),$$

where t is a real number.

✍ Since $|z| = 1$, z is of the form $\cos\theta + i\sin\theta$. Denoting $\tan\frac{1}{2}\theta$ by t,

$$z = \cos\theta + i\sin\theta = \frac{1-t^2}{1+t^2} + i\,\frac{2t}{1+t^2}$$

$$= \frac{(1+it)(1+it)}{(1+it)(1-it)} = \frac{1+it}{1-it}.$$

2. Show that

(*i*) $\mathbf{Re}\,(z_1\bar{z}_2) = \mathbf{Re}\,(\bar{z}_1 z_2) = \frac{1}{2}(z_1\bar{z}_2 + \bar{z}_1 z_2)$

(*ii*) $|z_1 + z_2|^2 = |z_1|^2 + 2\,\mathbf{Re}\,(z_1\bar{z}_2) + |z_2|^2$

(*iii*) $|z_1 - z_2|^2 = |z_1|^2 - 2\,\mathbf{Re}\,(z_1\bar{z}_2) + |z_2|^2$

(*iv*) $|z_1 + z_2|^2 + |z_1 - z_2|^2 = 2(|z_1|^2 + |z_2|^2)$

(*v*) $|1 - \bar{z}_1 z_2|^2 - |z_1 - z_2|^2 = (1 - |z_1|^2)(1 - |z_2|)^2.$

[*Hint* : (*iv*) $|z_1 + z_2|^2 = (z_1 + z_2)\,\overline{(z_1 + z_2)} = (z_1 + z_2)(\bar{z}_1 + \bar{z}_2)$]

3. **Solved sum.** If one of $|a|$ and $|b|$ is equal to 1, show that

$$\left|\frac{a-b}{1-\bar{a}b}\right| = 1.$$

✍ $$\left|\frac{a-b}{1-\bar{a}b}\right|^2 = \frac{|a-b|^2}{|1-\bar{a}b|^2} = \frac{(a-b)\,\overline{(a-b)}}{(1-\bar{a}b)\,\overline{(1-\bar{a}b)}}$$

$$= \frac{(a-b)(\bar{a}-\bar{b})}{(1-\bar{a}b)(1-a\bar{b})} = \frac{(a\bar{a}-a\bar{b}-\bar{a}b+b\bar{b})}{1-a\bar{b}-\bar{a}b+(a\bar{a})(b\bar{b})}$$

$$= \frac{|a|^2-a\bar{b}-\bar{a}b+|b|^2}{1-a\bar{b}-\bar{a}b+|a|^2|b|^2}$$

which reduces to 1 when either $|a| = 1$ or $|b| = 1$.

4. **Solved sum.** a, b, c are complex constants and if $|a| \neq |b|$, show that there is only one z satisfying the equation

$$az + b\bar{z} + c = 0.$$

✍ Let $z = x + iy$, $a = \alpha_1 + i\,\alpha_2$, $b = \beta_1 + i\,\beta_2$, $c = \gamma_1 + i\,\gamma_2$.

Then the equation is

$$(\alpha_1 + i\,\alpha_2)(x+iy) + (\beta_1 + i\,\beta_2)(x-iy) + (\gamma_1 + i\,\gamma_2) = 0$$

or $$(\alpha_1 x - \alpha_2 y + \beta_1 x + \beta_2 y + \gamma_1) + i(\alpha_1 y + \alpha_2 x - \beta_1 y + \beta_2 x + \gamma_2) = 0.$$

$\therefore$ $$(\alpha_1 + \beta_1)x + (-\alpha_2 + \beta_2)y + \gamma_1 = 0,$$
$$(\alpha_2 + \beta_2)x + (\alpha_1 - \beta_1)y + \gamma_2 = 0.$$

A unique solution for x, y exists if and only if

$$\frac{\alpha_1 + \beta_1}{\alpha_2 + \beta_2} \neq \frac{-\alpha_2 + \beta_2}{\alpha_1 - \beta_1}$$

which leads to the result that

$$\alpha_1^2 + \alpha_2^2 \neq \beta_1^2 + \beta_2^2 \quad \text{or} \quad |a| \neq |b|.$$

1.8. Inequalities in Terms of Absolute Values

5. **Solved sum.** Show that, if $|z| < \frac{1}{2}$, then

$$|(1+i)z^3 + iz| < \frac{3}{4}.$$

✍ L.H.S. $\leq |(1+i)z^3| + |iz|$ by triangle inequality

$$= |1+i||z|^3 + |i||z| = \sqrt{2}\,|z|^3 + |z|$$

$$< \sqrt{2}\left(\frac{1}{2}\right)^3 + \frac{1}{2} < 2\left(\frac{1}{2}\right)^3 + \frac{1}{2} = \frac{3}{4}.$$

6. **Solved sum.** Find z which satisfies $|z| - z = 1 + 2i$.

✍ Let $z = x + iy$. Then the given result is

$$\sqrt{x^2+y^2} - (x+iy) = 1 + 2i$$

$$\therefore \quad \sqrt{x^2 + y^2} - x = 1, -y = 2.$$

Using $x^2 + y^2 = (1 + x)^2$ and $y = 2$, $x = 3/2$. So $z = (3/2) + 2i$.

7. **Solved sum.** If $\lambda_1 + \lambda_2 + ... + \lambda_n = 1$ and if $\lambda_k \geq 0$, $|z_k| < 1$ for $k = 1, 2, ... n$, then show that

$$|\lambda_1 z_1 + \lambda_2 z_2 + ... + \lambda_n z_n| < 1.$$

✍

$$\begin{aligned} \text{L.H.S.} &\leq |\lambda_1 z_1| + |\lambda_2 z_2| + ... + |\lambda_n z_n| \\ &= \lambda_1 |z_1| + \lambda_2 |z_2| + ... + \lambda_n |z_n| \\ &< \lambda_1 (1) + \lambda_2 (1) + ... + \lambda_n (1) \\ &= \lambda_1 + \lambda_2 + ... + \lambda_n = 1. \end{aligned}$$

8. Show that the set of matrices of the form

$$\begin{pmatrix} \alpha & \beta \\ -\beta & \alpha \end{pmatrix}$$

with matrix addition and multiplication, where α, β are real, is isomorphic to the field of complex numbers.

CHAPTER 2

COMPLEX PLANE

2.1. REPRESENTATION OF COMPLEX NUMBER BY POINTS

Complex numbers can be represented by points in a plane. To do so, the usual ordered pair of cartesian coordinates (α, β) for each point in the plane can be thought of to represent the complex number $\alpha + i\,\beta$ (Fig. 2.1.1). In this representation the x axis is called the *real axis* and the y axis is called the *imaginary axis*. We shall denote the real axis by $A'A$ and the imaginary axis by $B'B$.

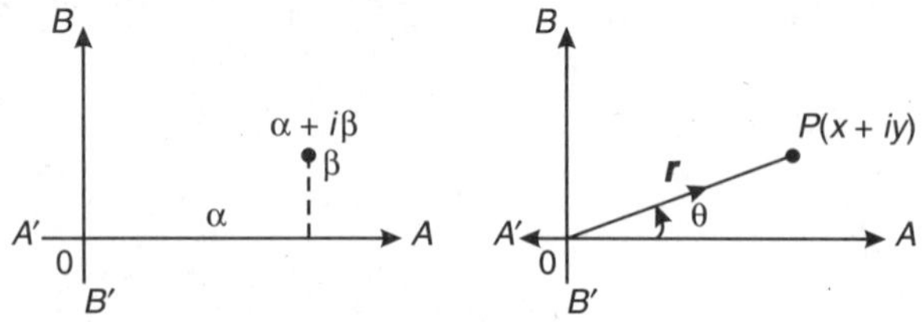

Fig. 2.1.1

Geometrical meaning of absolute value. The modulus of the complex number $x + iy$ represented by P equals OP because

$$OP = \sqrt{x^2 + y^2}\,.$$

It is conventional to denote this distance between O and P by r. So r is positive.

If $\qquad z_1 = x_1 + iy_1$ and $z_2 = x_2 + iy_2$, then

$$|z_1 - z_2| = |z_2 - z_1| = (x_1 - x_2) + i\,(y_1 - y_2)|$$

$$= \sqrt{(x_1 - x_2)^2 + (y_1 - y_2)^2}$$

$$= \text{distance between } z_1 \text{ and } z_2.$$

Argument. If a complex number $x + iy$ is represented by P, then the angle AOP made by OP with the positive direction of the real axis, is called the argument of the complex number $x + iy$.

The argument is taken to be positive if the angle is made in the anti-clockwise sense as shown in the figure; otherwise it is taken to be negative. It is evident that the argument of a complex number is not unique; for, if α is a value of it then

$$2n\pi + \alpha, \quad n = 0, \pm 1, \pm 2, \ldots,$$

are also its values. All these values are commonly denoted by the symbol Arg $(x + iy)$. However, one of these values is defined to be the *principal value* of the argument of the complex number. In this regard, there are two conventions, one to suppose that the principal value is that value that lies between 0 and 2π and the other, to suppose that the principal value is that value that lies between $-\pi$ and π. This value is denoted by arg $(x + iy)$ so that

$$\text{Arg}\,(x+iy) = \arg\,(x+iy) + 2n\pi, \quad n = 0, 1, 2, \ldots$$

The argument θ of $x + iy$ is to be calculated from

$$x = r\cos\theta \quad \text{and} \quad y = r\sin\theta$$

which imply that $\tan\theta = y/x$.

Remark. Geometrically, arg z is the angle measured in radians, which the line joining the origin O and the complex point z makes with the real in the anticlockwise sense. So arg z has an infinite number of values, namely,

$$\arg z = \theta + 2n\pi, \quad n = 0, \pm 1, \pm 2, \ldots$$

Polar form of a complex number. If r and θ are the absolute value and argument of $x + iy$, then (r, θ) are the polar coordinates of the point which represents $x + iy$ and

$$x + iy = r(\cos\theta + i\sin\theta) = re^{i\theta}.$$

As such, $r(\cos\theta + i\sin\theta)$ is said to be the *polar form* of $x + iy$.

Remark 1. It is important to note that if the absolute value and argument of a complex number are equal to the absolute value and argument of another, then they themselves are equal. In symbols, if $|z_1| = |z_2|$ and $\arg z_1 = \arg z_2$, then $z_1 = z_2$.

Remark 2. From figure 2.1.2 it can be seen that the points which represents a pair of conjugate numbers are such that one is the image of the other on the real axis. Also, if these numbers are z and $\bar{z}$, then

$$\arg\bar{z} = -\arg z, \quad |z| = |\bar{z}|,$$

and if the polar form of z is $r(\cos\theta + i\sin\theta)$, then the polar form of $\bar{z}$ is $r\{\cos(-\theta) + i\sin(-\theta)\}$.

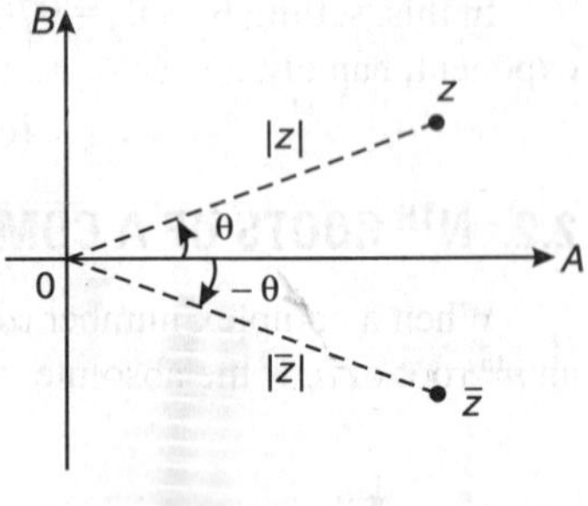

Fig. 2.1.2

Remark 3. The argument of zero (i.e., arg $(0 + 0i)$) is not defined.

Theorem 2.1. The argument of the product of two complex numbers is the sum of the arguments of the complex numbers.

Proof: Consider the complex numbers.

$z_1 = r_1(\cos\theta_1 + i\sin\theta_1), z_2 = r_2(\cos\theta_2 + i\sin\theta_2)$, where $\theta_1 = \arg z_1$, $\theta_2 = \arg z_2$. Their product is

$$z_1 z_2 = r_1 r_2 (\cos\theta_1 + i\sin\theta_1)(\cos\theta_2 + i\sin\theta_2)$$
$$= r_1 r_2 [\cos(\theta_1 + \theta_2) + i\sin(\theta_1 + \theta_2)].$$

$$\therefore \quad |z_1 z_2| = r_1 r_2 = |z_1||z_2|,$$
$$\arg(z_1 z_2) = \theta_1 + \theta_2 = \arg z_1 + \arg z_2$$

of which the first one is the result that the absolute value of the product of two complex numbers is the product of absolute values of the complex numbers which has already been proved algebraically in chapter 1. In the second result care should be taken to add 2π, if necessary, to the first angle, namely arg $(z_1 z_2)$.

Remark. The second result extends to

$$\arg(z_1 z_2 \ldots z_n) + 2n\pi = \arg z_1 + \arg z_2 + \ldots + \arg z_n$$

Theorem 2.2. If z_1 and z_2 are two complex numbers, then

$$\arg\left(\frac{z_1}{z_2}\right) = \arg z_1 - \arg z_2.$$

Proof:
$$\frac{z_1}{z_2} = \frac{r_1(\cos\theta_1 + i\sin\theta_1)(\cos\theta_2 - i\sin\theta_2)}{r_2(\cos\theta_2 + i\sin\theta_2)(\cos\theta_2 - i\sin\theta_2)}$$

$$= \frac{r_1}{r_2} \{\cos(\theta_1 - \theta_2) + i \sin(\theta_1 - \theta_2)\}.$$

$$\therefore \qquad \left|\frac{z_1}{z_2}\right| = \frac{r_1}{r_2} = \frac{|z_1|}{|z_2|}$$

as obtained in chapter 1 and

$$\arg\left(\frac{z_1}{z_2}\right) = \theta_1 - \theta_2 = \arg z_1 - \arg z_2.$$

Abbreviation. It is sometimes convenient to use the symbol cis θ to denote the complex number $\cos\theta + i\sin\theta$.

De Moivre's theorem. Extension of the result

$$(\cos\theta_1 + i\sin\theta_1)(\cos\theta_2 + \mathrm{i}\sin\theta_2) = \cos(\theta_1 + \theta_2) + i\sin(\theta_1 + \theta_2),$$

is
$$\text{cis}\,\theta_1\ \text{cis}\,\theta_2 \ldots \text{cis}\,\theta_n = \text{cis}\,(\theta_1 + \theta_2 + \ldots \theta_n).$$

In this, setting $\theta_1 = \theta_2 = \theta_3 = \ldots = \theta$, we obtain the De Moivere's theorem for a positive integral exponent, namely,

$$(\cos\theta + i\sin\theta)^n = \cos n\,\theta + i\sin n\,\theta.$$

2.2. NTH ROOTS OF A COMPLEX NUMBER

When a complex number a is given, the complex number z which is such that $z^n = a$, is called an nth root of a. If the absolute value and argument of a are r and α, then $a = r$ cis α. But

$$\left(r^{1/n}\ \text{cis}\ \frac{2k\pi + \alpha}{n}\right)^n = r\ \text{cis}\ \alpha$$

if k is an integer. So

$$r^{1/n}\ \text{cis}\ \frac{2k\pi + \alpha}{n}$$

is an nth root of a. Now assigning to k the n values

$$0, 1, 2, \ldots, n - 1,$$

we get n distinct n^{th} roots of a. It can be seen that the other integral values of k will just cause repetition of the roots. It is evident that the nth roots have the same absolute value but different arguments which are in an arithmetic progression. So, in the complex plane, the roots are the vertices of a regular n-sided polygon with its centre at the origin.

In particular, the nth roots of 1 are

$$\text{cis}\ \frac{2k\pi}{n}, \quad k = 0, 1, 2, \ldots, n - 1,$$

which can be denoted respectively by $1, \omega, \omega^2, \ldots \omega^{n-1}$. Then

$$1 + \omega + \omega^2 + \ldots + \omega^{n-1} = 0$$

since
$$1 + \omega + \omega^2 + \ldots + \omega^{n-1} = (1 - \omega^n)/(1 - \omega) \text{ and } \omega^n = 1.$$

2.3. GEOMETRICAL ADDITION

Now we show that, if the points P and Q denote the complex numbers

$$z_1 = x_1 + iy_1, \quad z_2 = z_2 + iy_2$$

(Fig. 2.3.1) and if the parallelogram $OPRQ$ is completed, then the point R denotes the sum

$$z_1 + z_2 = (x_1 + x_2) + i\,(y_1 + y_2).$$

For, if PL, QM, RN are perpendiculars from P, Q, R to the real axis and if PS is perpendicular to RN, then

$$ON = OL + LN = OL + PS = OL + OM = x_1 + x_2,$$
$$NR = NS + SR = NS + MQ = y_1 + y_2.$$

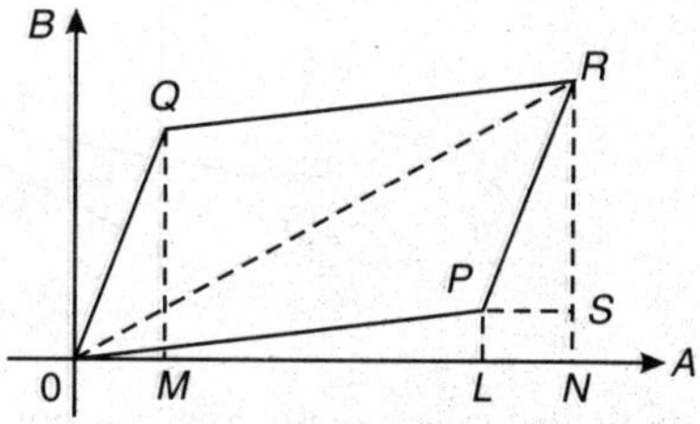

Fig. 2.3.1

So R denotes the complex number $(x_1 + x_2) + i\,(y_1 + y_2)$.

First triangle inequality. From the triangle OPR (Fig. 2.3.1).

$$OR \le OP + PR \quad \text{or} \quad OR \le OP + OQ$$

$$\therefore \qquad | z_1 + z_2 | \le | z_1 | + | z_2 |.$$

Geometrical subtraction. Given $P\,(z_1)$ and $Q\,(z_2)$, to obtain $z_1 - z_2$ geometrically, we should first find $Q'(-z_2)$ (Fig. 2.3.2) and then find R' by completing the parallelogram $OPR'Q'$. Then R' denotes $z_1 + (-z_2)$ or $z_1 - z_2$.

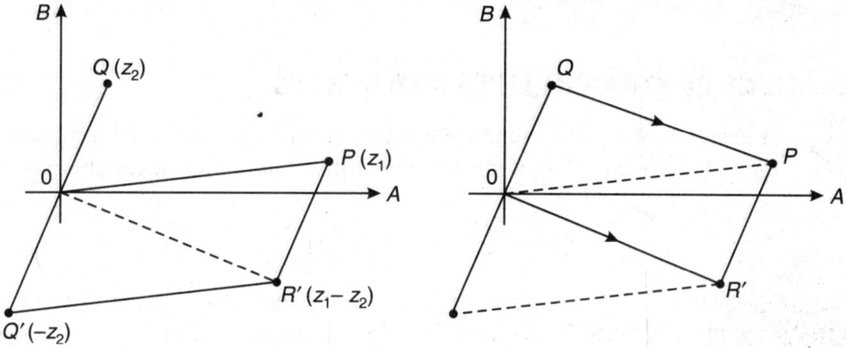

Fig. 2.3.2

Second triangle inequality. From the triangle $OR'Q'$,

$$OR' \ge | Q'R' - OQ' | \quad \text{or} \quad OR' \ge | OP - OQ' |$$

$$\therefore \qquad | z_1 - z_2 | \ge | \, | z_1 | - | z_2 | \, |.$$

Complex numbers as vectors. The parallelogram law, used in addition, suggests that addition of complex numbers corresponds to addition of vectors. That is, if the vectors $\overline{OP}$ and $\overline{OQ}$ stand for the complex number z_1 and z_2 (Fig. 2.3.1), then the vector

$$\overline{OP} + \overline{OQ} \quad \text{or} \quad \overline{OR}$$

stands for the complex number $z_1 + z_2$. This suggests a free usage of the word *vector* to mean a complex number so far addition and subtraction of complex numbers are concerned. So *points* and *vectors* can be used unmistakably to mean complex numbers.

2.4. ANGLE BETWEEN TWO RAYS

Let $P\,(z_1)$, $Q\,(z_2)$, $R\,(z_3)$, $S\,(z_4)$ be four points in the complex plane. Let the angle measured from the vector $\overline{PQ}$ to the vector $\overline{RS}$, as shown in figure 2.4.1 be denoted by the notation

$$\text{angle } (\overline{PQ}, \overline{RS})$$

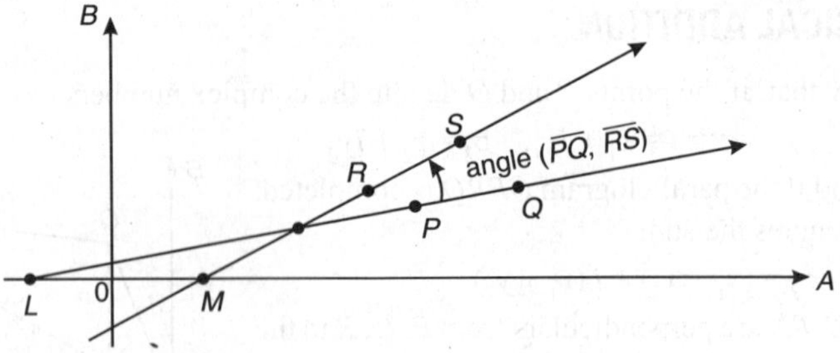

Fig. 2.4.1

with the understanding that it is positive if the sense of measurement is anticlockwise and, otherwise, negative. In figure 2.4.1,

$$\text{angle } (\overline{PQ}, \overline{RS}), \quad \text{angle } (\overline{RS}, \overline{PQ})$$

are respectively positive and negative. Further,

$$\begin{aligned} \text{angle } (\overline{PQ}, \overline{RS}) &= \text{angle } RMA - \text{angle } PLA \\ &= \arg (z_4 - z_3) - \arg (z_2 - z_1) \\ &= \arg \frac{z_4 - z_3}{z_2 - z_1} = \arg \frac{z_3 - z_4}{z_1 - z_2}. \end{aligned} \tag{2.4.1}$$

2.5. EQUATIONS OF STRAIGHT LINES AND CIRCLES

Suppose the equation of a curve in the complex plane with real and imaginary axes as x and y axes is $F(x, y) = 0$. Then the equation of the curve in terms of the complex variable z is

$$F\left[\frac{1}{2}(z+\bar{z}), \ \frac{1}{2i}(z-\bar{z})\right] = 0$$

because $z = x + iy$ and $\bar{z} = x - iy$ give that $x = (z + \bar{z})/2$ and $y = (z - \bar{z})/2i$. In some cases the equation in the complex form can be obtained directly as illustrated below.

(*i*) *Ray emerging out of a point.* Let us consider the ray which emerges out from $Q(z_1)$ and travels in the direction which makes angle α with the real axis as shown in figure 2.5.1. Suppose $P(z)$ is any

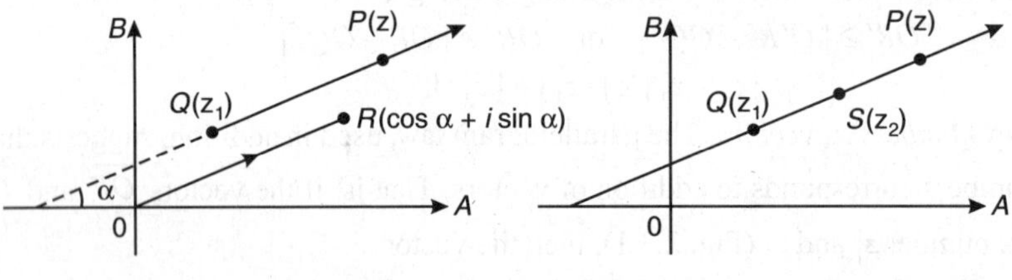

Fig. 2.5.1

point on the ray. Then $\overline{QP} = z - z_1$. So

$$\arg (z - z_1) = \alpha.$$

Thus $\arg (z - z_1) = \alpha$ is the equation of the ray. Also if R is a point such that $OR = 1$ and the unit vector $\overline{OR}$ and the vector $\overline{QP}$ have the same direction, then

$$\overline{OR} = \cos\alpha + i \sin\alpha,$$

$$\overline{QP} = QP\, \overline{OR} = QP(\cos\alpha + i\sin\alpha).$$

Thus, from $\overline{OP} = \overline{OQ} + \overline{QP}$, we get

$$z = z_1 + QP(\cos\alpha + i\sin\alpha),$$

where QP varies with the position of P on the ray. So the complex equation of the ray in the parametric form is

$$z = z_1 + t(\cos\alpha + i\sin\alpha), \quad 0 \le t < \infty, \tag{2.5.1}$$

and the equation of the straight line of which the ray forms a part is

$$z = z_1 + t(\cos\alpha + i\sin\alpha), \quad -\infty < t < \infty. \tag{2.5.2}$$

Remark. Arg $z = \alpha$ denotes the ray emerging out from the origin and making an angle α with the real axis.

(*ii*) *Straight line joining $Q(z_1)$ and $S(z_2)$.* If $P(z)$ is a point on this line, then

$$\overline{OP} = \overline{OQ} + \overline{QP} = \overline{OQ} + t\,\overline{QS} = z_1 + t(z_2 - z_1).$$

So the equatión of the line in the parametric form is

$$z = z_1 + t(z_2 - z_1), \quad -\infty < t < \infty. \tag{2.5.3}$$

Remark. If a and b are two complex constants, then the equation

$$z = a + tb, \quad -\infty < t < \infty$$

represents a straight line through the point a.

(*iii*) *Circle.* The equation of the circle whose centre is at the point z_1 and radius is ρ, is

$$|z - z_1| = \rho$$

because $|z - z_1|$ is the distance of $P(z)$, any point on the circle, from the point z_1 (Fig. 2.5.2). The algebraic form of this equation is $(z - z_1)(z - \bar{z}_1) = \rho^2$.

Remark 1. Equation of the unit circle whose centre is the origin is $|z| = 1$.

Remark 2. The set of points inside the circle whose centre is a and radius is ρ, is $|z - a| < \rho$.

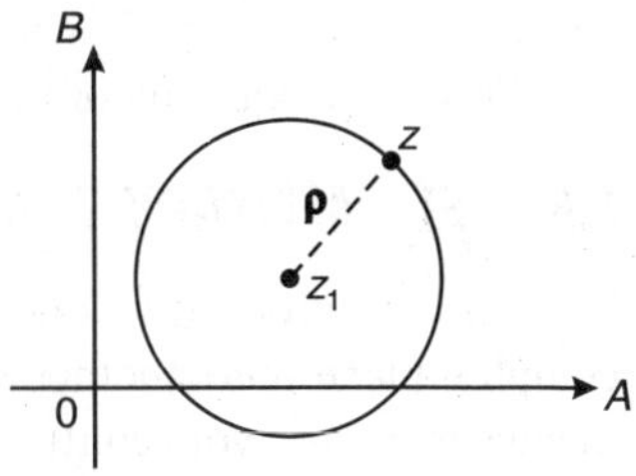

Fig. 2.5.2

The set of points on and inside this circle is $|z - a| \le \rho$.

The set of points outside this circle is $|z - a| > \rho$.

Equation of a general straight line. Let the cartesian equation of the line be $ax + by + c = 0$. Then its complex form is

$$a\frac{z + \bar{z}}{2} + b\frac{z - \bar{z}}{2i} + c = 0$$

i.e.,
$$(a - ib)z + (a + ib)\bar{z} + 2c = 0$$

or
$$\bar{P}z + P\bar{z} + \lambda = 0 \tag{2.5.4}$$

where λ is real and P is complex.

Corollary. $\bar{P}z + P\bar{z} = 0$ is a line through the origin.

Equation of a general circle. Let the cartesian equation of a circle be

$$x^2 + y^2 + 2gx + 2fy + c = 0.$$

Then its complex form is

$$z\bar{z} + g(z + \bar{z}) + \{f(z - \bar{z})/i\} + c = 0$$

i.e., $$z\bar{z} + \bar{a}z + a\bar{z} + c = 0, \quad (2.5.5)$$

where c is real and a is complex. The centre of the circle is at the point $-a$ and radius is $\sqrt{a\bar{a} - c}$.

Circle on the join of A (z_1) *and B* (z_2) *as diameter.* Let P (z) be any point on the circle (Fig. 2.5.3). Then the angle made by

BP with *AP* is

$$\arg \frac{z_2 - z}{z_1 - z} = \frac{\pi}{2}.$$

When P lies on the other side of AB,

$$\arg \frac{z_2 - z}{z_1 - z} = -\frac{\pi}{2}.$$

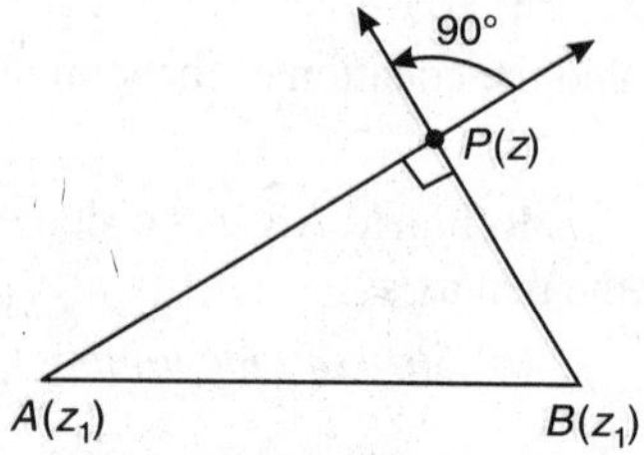

Fig. 2.5.3

Thus, in either case, $(z_2 - z)/(z_1 - z)$ is purely imaginary. That is, its real part is zero. Hence

$$\frac{z_2 - z}{z_1 - z} + \overline{\left(\frac{z_2 - z}{z_1 - z}\right)} = 0$$

i.e., $$\frac{z_2 - z}{z_1 - z} + \frac{\bar{z}_2 - \bar{z}}{\bar{z}_1 - \bar{z}} = 0$$

i.e., $$2z\bar{z} - z(\bar{z}_1 + \bar{z}_2) - \bar{z}(z_1 + z_2) + z_1\bar{z}_2 + \bar{z}_1 z_2 = 0.$$

This is the equation of the circle.

2.6. ELEMENTARY TRANSFORMATION

Now we proceed to study some elementary mappings which transform a set of points of the complex plane onto another set. We shall denote the points of the complex plane by z and their images by w. To avoid confusion we shall take two separate planes, one for z's and the other for w's.

General points in z and w planes. The general point in the z plane is (x, y) and the general point in the w plane is assumed to be (X, Y) or (u, v).

Linear transformation. Transformations of the form

$$w = az + b$$

are called linear transformations.

Translation. First, consider the mapping

$$w = z + b,$$

where b is a complex constant. Here we observe, from the geometrical addition of complex numbers, that to get the point w from the point z we have to travel from the point z through a distance $|b|$ in the direction specified by arg b (Fig. 2.6.1). Thus the transformation $w = z + b$ shifts the set of points bodily through a distance $|b|$ in the direction specified by arg b. This transformation is called a *translation*.

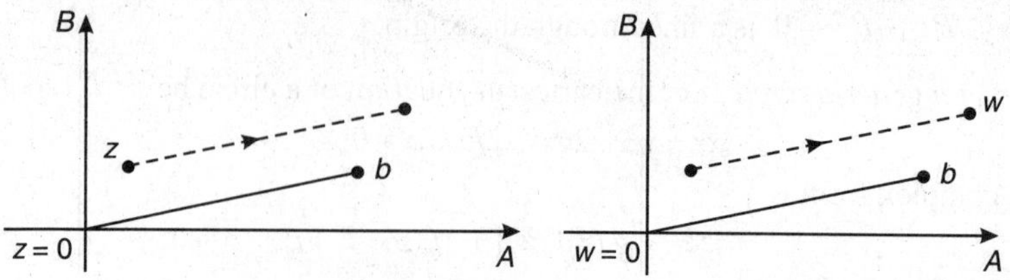

Fig. 2.6.1

Magnification and rotation. Consider the transformation

$$w = az$$

for the following different cases of a:

(*i*) a is real and positive,

(*ii*) a is a complex number whose absolute value = 1,

(*iii*) a is a complex number whose absolute value ≠ 1.

Case (*i*): Let $z = r \text{ cis } \theta$. Then, if a is real and positive,

$$w = (ar) \text{ cis } \theta.$$

$$|w| = a|z|, \quad \arg w = \arg z.$$

So the transformation is a magnification or contraction with reference to the origin according as $a > 1$ or $a < 1$. This transformation is called a *homothetic transformation*.

Case (*ii*): If a is a complex number whose absolute value is 1 and argument is α, then $a = \text{cis } \alpha$ and

$$w = r \text{ cis } (\theta + \alpha).$$

$$\therefore \quad |w| = |z|, \quad \arg w = (\arg z) + \alpha.$$

Thus the effect of the transformation is to rotate the point z about the origin through an angle α.

Case (*iii*): If a is a complex number whose absolute value is not equal to 1, then a is of the form $\rho \text{ cis } \alpha$. Therefore

$$w = (r\rho) \text{ cis } (\theta + \alpha).$$

So the effect of the transformation is a magnification or contraction with respect to the origin according as $\rho > 1$ or $\rho < 1$, followed by a rotation about the origin. That is, the effect is a homothetic transformation followed by a rotation.

Corollary. Transformation

$$w = az + b$$

is a rotation and magnification ($w = az$) about the origin and then a translation ($w = z + b$).

Transformation $w = 1/z$. Suppose P denotes

$$z = OP \text{ cis } \theta$$

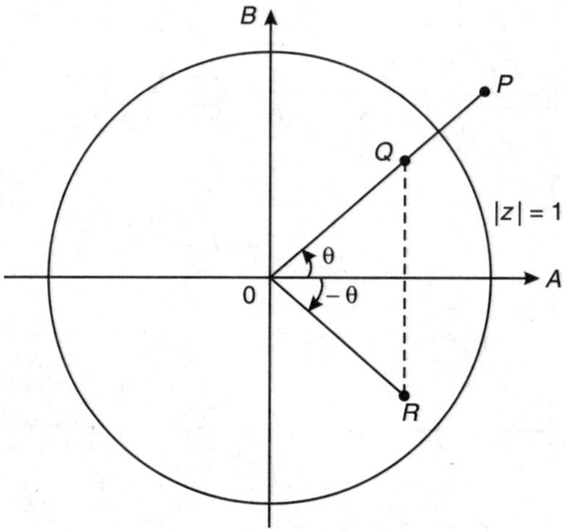

Fig. 2.6.2

(Fig. 2.6.2). Let Q be the inversion of P with respect to the unit circle whose centre is at the origin. Then $OQ \,.\, OP = 1$ and Q denotes $OQ \text{ cis } \theta$. If R is the reflection of Q in the real axis, then R denote the complex number

$$OR \text{ cis } (-\theta) = OQ \text{ cis } (-\theta) = \frac{1}{OP} \text{ cis } (-\theta)$$

$$= \frac{1}{OP} \frac{\text{cis } (-\theta) . \text{cis } \theta}{\text{cis } \theta} = \frac{1}{OP \text{ cis } \theta} = \frac{1}{z}.$$

So R is the image of P under the transformation $w = 1/z$. Hence the transformation $w = 1/z$ consists of the following two transformations:

(*i*) Inversion with respect to the unit circle,

(*ii*) Reflection in the real axis.

It is evident that the order in which these two transformations are to be effected, is immaterial.

Nomenclature. For the sake of convenience we call the transformation $w = 1/z$ simply *inversion.*

Effects of the transformation $w = 1/z$. Under the transformation a point outside the unit circle $|z| = 1$ goes into the circle radially and gets reflected on the real axis and thus becomes a point inside the circle. On the other hand a point inside the circle goes outside the circle radially and gets reflected on the x axis and thus becomes a point outside the circle. Further it is evident that, if R is the image of P, then P is the image of R.

2.7. INFINITY AND EXTENDED COMPLEX PLANE

The inversion $w = 1/z$ appears to be a one-to-one mapping of the z plane on to the w plane since it is composed of an inversion with respect to the unit circle $|z| = 1$ and reflection in the real axis. Under this transformation, the unit circle $|z| = 1$ goes onto the unit circle $|w| = 1$ (Fig. 2.7.1). But $w = 1/z$ gives that $|z||w| = 1$. Therefore

(*i*) all the points z inside the unit circle $|z| = 1$ except $z = 0$ go onto point w outside the unit circle $|w| = 1$ and

(*ii*) all the points z outside the unit circle $|z| = 1$ go onto points w inside the unit circle $|w| = 1$ except to $w = 0$.

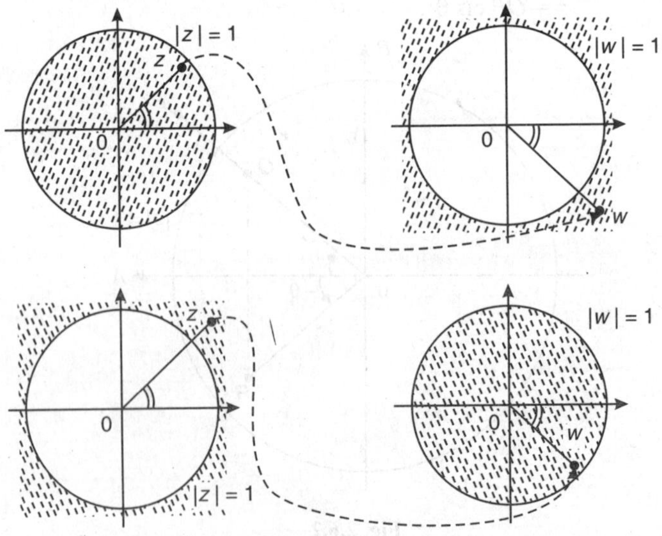

Fig. 2.7.1

That is, zero in the z plane has no image in the w plane and no point in the w plane has zero in the z plane as its preimage. As such, the transformation $w = 1/z$ is not a one-to-one mapping of the

z plane onto the w plane. This difficulty is relegated if we define an ideal point called *infinity* in the complex plane such that, under the transformation $w = 1/z$, zero goes to infinity and infinity goes to zero. If this complex infinity is denoted by ∞, then

$$\frac{1}{0} = \infty, \quad \frac{1}{\infty} = 0.$$

Finite complex numbers. All complex numbers except ∞ are called finite complex numbers.

If a is any finite complex number other than zero, then, from $1/0 = \infty$,

$$\frac{1}{0} \cdot \frac{a}{a} = \infty \quad \text{or} \quad \frac{a}{0} = \infty,$$

and, from $1/\infty = 0$,

$$\frac{1}{\infty} \cdot a = 0 \cdot a \quad \text{or} \quad \frac{a}{\infty} = 0 .$$

Intuitively and also from the stereographic projection given below, we can have the following definitions:

$$a \cdot \infty = \infty \qquad (a \neq 0)$$
$$a + \infty = \infty$$
$$\infty \cdot \infty = \infty.$$

However, there is no reasonable value we can assign to

$$\infty \pm \infty, \quad \infty/\infty, \quad 0/0, \quad 0 \cdot \infty.$$

Finite complex plane. The complex plane containing only the finite complex numbers, is called the *finite complex plane* or the *entire complex plane.*

Extended complex plane. The complex plane containing all the finite complex numbers and ∞, is called the *extended complex plane.*

2.8. STEREOGRAPHIC PROJECTION

There is a geometric model which provides a one-to-one correspondence between the points in the extended complex plane π and the points on a sphere called *Riemann sphere*. The sphere is a unit sphere and the complex plane is a diametral plane. To specify the model, make the following assumptions:

O : Centre of the sphere (Fig. 2.8.1)

Ox, Oy : x, y axes chosen in π (also real and imaginary axes)

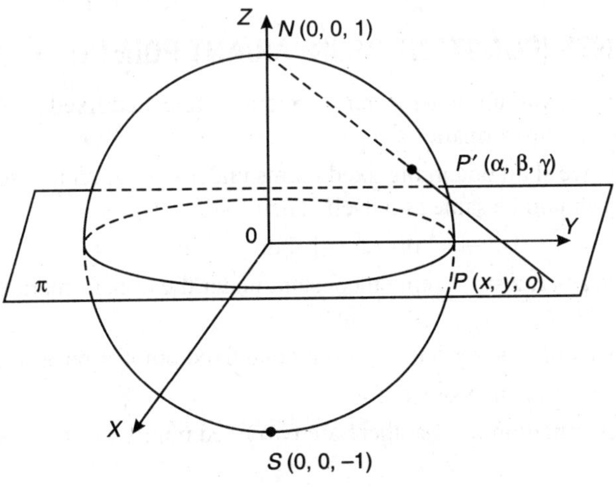

Fig. 2.8.1

OZ : Z axis (z is avoided since it means a complex number)
N : Point (0, 0, 1) on Z axis
S : Point (0, 0, – 1) on Z axis

Now the equation of the sphere is $x^2 + y^2 + z^2 = 1$. Consider a point P in π. Let the line joining P and N intersect the sphere at P'. Then a point in π is made correspond to a point on the sphere. It is evident that all the points in π have corresponding points on the sphere. However, there is no point in π which corresponds to N. But, if it is assumed that the complex infinity corresponds to N, then all the points in the extended complex plane have corresponding points on the sphere.

This correspondence is a central projection from the centre of projection (0, 0, 1).This projection is called a *stereographic projection.*

If P denotes $x + iy$, then P is $(x, y, 0)$. If P' is (α, β, γ), then $\alpha^2 + \beta^2 + \gamma^2 = 1$ and the parametric equations of the straight line NP are

$$\frac{x-0}{\alpha-0} = \frac{y-0}{\beta-0} = \frac{z-1}{\gamma-1} = t.$$

So any point on the line PN is $\{\alpha t, \beta t, 1 + (\gamma - 1)\, t\}$. The value of t corresponding to the point $P\,(x, y, 0)$ is obtained from

$$1 + (\gamma - 1)\, t = 0 \quad \text{as} \quad t = \frac{1}{1-\gamma}.$$

$$\therefore \qquad x = \frac{\alpha}{1-\gamma}, \quad y = \frac{\beta}{1-\gamma}$$

and
$$x + iy = \frac{\alpha + i\beta}{1-\gamma}. \tag{2.1.1}$$

Further, using $\alpha^2 + \beta^2 + \gamma^2 = 1$, it can be seen that

$$x^2 + y^2 = \frac{1+\gamma}{1-\gamma}$$

and hence

$$\gamma = \frac{x^2+y^2-1}{x^2+y^2+1}, \quad \alpha = \frac{2x}{1+x^2+y^2}, \quad \beta = \frac{2y}{1+x^2+y^2}. \tag{2.8.1}$$

2.9. FIXED POINTS (CONSTANT OR INVARIANT POINTS)

Points which are invariant under a transformation are called fixed points or constant points or invariant points of the transformation.

As an example, we shall obtain the fixed points under $w = 1/z$. Suppose z is a fixed point. Then z goes to $1/z$ which should be same as z itself. Therefore

$$1/z = z \quad \text{or} \quad z^2 = 1 \quad \text{or} \quad z = \pm 1.$$

Thus the points $z = 1$, $z = -1$ remain constant under the transformation and hence they are the fixed points.

Under the translation $w = z + b$ there is only one fixed point ∞ because $z = \infty$ goes to

$$w = \infty + b = \infty.$$

Under the transformation $w = az$, there are two fixed points $z = 0$, $z = \infty$ because the images of 0 and ∞ are

(*i*) $w = a \times 0 = 0$, (*ii*) $w = a \times \infty = \infty$.

2.10. EXAMPLES

2.1. Representation of Complex Numbers by Points

1. Find the curves represented by the following equations:

(*i*) $|z - z_1| = |z - z_2|$ (*ii*) $|z - 1| + |z + 1| = 3$

(*iii*) $|z - 2| - |z + 2| = 3$ (*iv*) $\textbf{Re}\, z + \textbf{Im}\, z = 1$

(*v*) $|z| = \textbf{Re}\, z + 1$ (*vi*) $\textbf{Re}\, z^2 = 1.$

[**Ans.** (*i*) Perpendicular bisector of the line joining z_1 and z_2 (*ii*) Ellipse with foci at $z = 1$ and at $z = -1$ (*iii*) Hyperbola with foci at 2, – 2 (*iv*) St. line $x + y = 1$ (*v*) Parabola $x^2 + y^2 = (x + 1)^2$ (*vi*) R. H. $x^2 - y^2 = 1$]

2. Identify the following sets in the complex plane.

(*i*) $|z| < 1$ (*ii*) $|z| > 1$

(*iii*) $|z - 2| < 3$ (*iv*) $|z - 2| > 3$

(*v*) $\textbf{Re}\, z > 0$ (*vi*) $\textbf{Im}\, z > 0$

(*vii*) $|z - 1| < |z + 1|$ (*viii*) $|z - 1| \geq |z + 1|$

[**Ans.** (*i*) Interior of the unit circle (*ii*) Exterior of the unit circle (*iii*) Interior of the circle, centre at 2 and radius 3 (*iv*) Exterior of the circle, centre at 2 and radius 3 (*v*) Right half-plane (*vi*) Upper half-plane (*vii*) Right half-plane (*viii*) Left half-plane including the imaginary axis]

3. Show that the centroid of the triangle whose vertices are z_1, z_2, z_3 is $\frac{1}{3}(z_1 + z_2 + z_3)$.

[*Hint* : If $z_1 = x_1 + iy_1$, etc., the vertices are (x_1, y_1), etc. The centroid is $\{\frac{1}{3}(x_1 + x_2 + x_3), \frac{1}{3}(y_1 + y_2 + y_3)\}$]

4. If $|z_1| = |z_2| = |z_3|$ and if $z_1 + z_2 + z_3 = 0$, show that the points z_1, z_2, z_3 are the vertices of an equilateral triangle.

[*Hint* : Origin is the circumcentre and also the centroid]

5. **Solved sum.** Show that the area of the triangle whose vertices are z_1, z_2, z_3 is

$$\frac{1}{4i}\sum \frac{(z_2 - z_3)|z_1|^2}{z_1}. \qquad \textbf{(1980)}$$

✍ If $z_1 = x_1 + iy_1, z_2 = x_2 + iy_2, z_3 = x_3 + iy_3$ then the area A of the triangle is given by

$$A = \frac{1}{2}\begin{vmatrix} x_1 & x_2 & x_3 \\ y_1 & y_2 & y_3 \\ 1 & 1 & 1 \end{vmatrix} = \frac{1}{2}\begin{vmatrix} \dfrac{z_1 + \bar{z}_1}{2} & \dfrac{z_2 + \bar{z}_2}{2} & \dfrac{z_3 + \bar{z}_3}{2} \\ \dfrac{z_1 - \bar{z}_1}{2i} & \dfrac{z_2 - \bar{z}_2}{2i} & \dfrac{z_3 - \bar{z}_3}{2i} \\ 1 & 1 & 1 \end{vmatrix}$$

$$= \frac{1}{8i}\begin{vmatrix} z_1 + \bar{z}_1 & z_2 + \bar{z}_2 & z_3 + \bar{z}_3 \\ z_1 - \bar{z}_1 & z_2 - \bar{z}_2 & z_3 - \bar{z}_3 \\ 1 & 1 & 1 \end{vmatrix}$$

$$= \frac{1}{8i}\begin{vmatrix} z_1 + \bar{z}_1 & z_2 + \bar{z}_2 & z_3 + \bar{z}_3 \\ 2z_1 & 2z_2 & 2z_3 \\ 1 & 1 & 1 \end{vmatrix}$$

$$= \frac{2}{8i}\begin{vmatrix} z_1+\bar{z}_1 & z_2+\bar{z}_2 & z_3+\bar{z}_3 \\ z_1 & z_2 & z_3 \\ 1 & 1 & 1 \end{vmatrix}$$

$$= \frac{2}{8i}\begin{vmatrix} \bar{z}_1 & \bar{z}_2 & \bar{z}_3 \\ z_1 & z_2 & z_3 \\ 1 & 1 & 1 \end{vmatrix} = \frac{1}{4i}\sum (z_2 - z_3)\,\bar{z}_1$$

$$= \frac{1}{4i}\sum \frac{(z_2 - z_3)\, z_1 \bar{z}_1}{z_1} = \frac{1}{4i}\sum \frac{(z_2 - z_3)\,|z_1|^2}{z_1}$$

6. Show that the equations

(*i*) $|(z-1)/(z+1)| = \lambda$ (*ii*) $\arg\{(z-1)/(z+1)\} = \mu$ represent orthogonal circles.

[*Hint* : (*i*) $|(x-1)+iy|/|(x+1)+iy| = \lambda$ reduces to $x^2+y^2+2gx+1=0$ (*ii*) $\arg(x-1+iy) - \arg(x+1+iy) = \mu$ reduces to

$$\tan^{-1}\frac{y}{x-1} - \tan^{-1}\frac{y}{x+1} = \tan^{-1}\frac{2y}{x^2+y^2-1} = \mu$$

i.e., to the form $x^2+y^2+fy-1=0$]

2.2. n[th] Roots of a Complex Number

7. Is ω is any of the nth roots of unity, show that

$$1+\omega+\omega^2+\omega^3+\ldots\,\omega^{n-1}=0,$$

[*Hint* : $1+\omega+\omega^2+\ldots\,\omega^{n-1} = (1-\omega^n)/(1-\omega)$]

2.4. Angle Between Two Rays

8. **Solved sum.** Show that, if

$$\arg\frac{z_2-z_3}{z_1-z_3} = \arg\frac{z_2-z_4}{z_1-z_4},$$

then the points z_3 and z_4 are on the same side of the line joining z_1 and z_2 and that z_1, z_2, z_3, z_4 are concyclic.

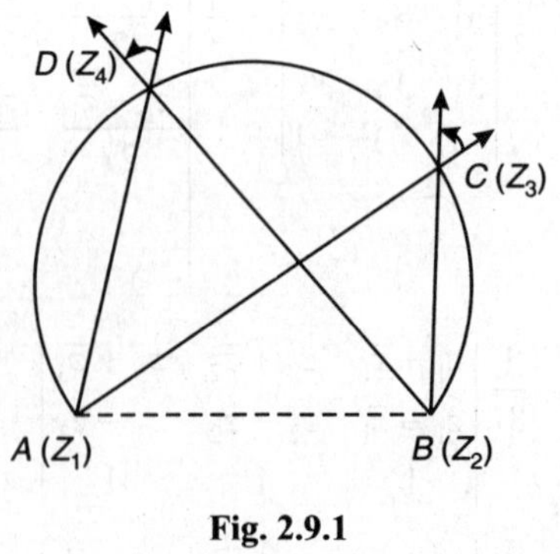

Fig. 2.9.1

If *A*, *B*, *C*, *D* represent z_1, z_2, z_3, z_4, then

$$\arg\frac{z_2-z_3}{z_1-z_3} = \text{angle made by } BC \text{ with } AC,$$

$$\arg \frac{z_2 - z_4}{z_1 - z_4} = \text{angle made by } BD \text{ with } AD.$$

It is evident from the figure that C and D lie on the same side of AB and $\underline{|ACB} = \underline{|ADB}$. Hence A, B, C, D are concyclic.

9. If $P_1(z_1), P_2(z_2), P_3(z_3)$ are the vertices of an equilateral triangle, show that

$$z_1^2 + z_2^2 + z_3^2 = z_1 z_2 + z_2 z_3 + z_3 z_1.$$

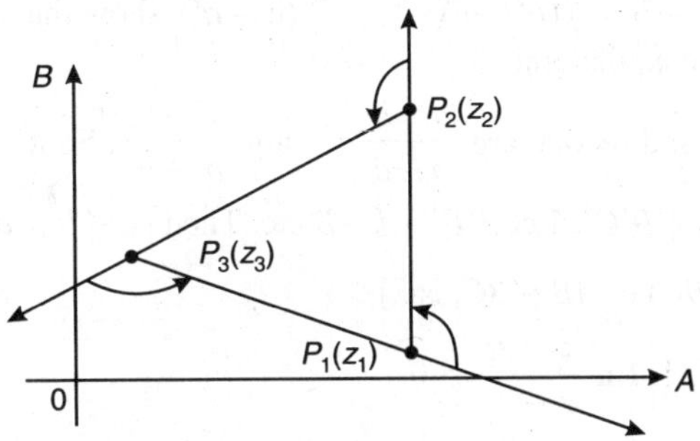

Fig. 2.9.2

Now, from figure 2.9.2, we see that

$$\text{angle } (\overline{P_1 P_2}, \overline{P_2 P_3}) = \text{angle } (\overline{P_2 P_3}, \overline{P_3 P_1}). \qquad (2.6.2)$$

$$\arg \frac{z_2 - z_3}{z_1 - z_2} = \arg \frac{z_3 - z_1}{z_2 - z_3}. \qquad (1)$$

Also
$$\left| \frac{z_2 - z_3}{z_1 - z_2} \right| = \left| \frac{z_3 - z_1}{z_2 - z_3} \right| \qquad (2)$$

because each of $|z_1 - z_2|, |z_2 - z_3|, |z_3 - z_1|$ is equal to the length of the side of the triangle. Thus from (1), (2)

$$\frac{z_2 - z_3}{z_1 - z_2} = \frac{z_3 - z_1}{z_2 - z_3}$$

or
$$(z_2 - z_3)^2 = (z_1 - z_2)(z_3 - z_1)$$

$\therefore$
$$z_1^2 + z_2^2 + z_3^2 = z_1 z_3 + z_2 z_3 + z_3 z_1.$$

10. **Solved sum.** Show that the triangles whose vertices are a, b, c and a', b', c' are similar if and only if

$$\begin{vmatrix} 1 & 1 & 1 \\ a & b & c \\ a' & b' & c' \end{vmatrix} = 0. \qquad \textbf{(1979)}$$

✍ Suppose the triangles are ABC and $A' B' C'$ and they are similar. Then the angle made by BC with AB equals the angle made by $B' C'$ with $A' B'$. Thus

$$\arg \frac{b - c}{a - b} = \arg \frac{b' - c'}{a' - b'}.$$

Also, from $BC/AB = B'C'/A'B'$,

$$\frac{|b-c|}{|a-b|} = \frac{|b'-c'|}{|a'-b'|} \quad \text{or} \quad \left|\frac{b-c}{a-b}\right| = \left|\frac{b'-c'}{a'-b'}\right|.$$

$$\therefore \qquad \frac{b-c}{a-b} = \frac{b'-c'}{a'-b'} \quad \text{or} \quad \begin{vmatrix} 1 & 1 & 1 \\ a & b & c \\ a' & b' & c' \end{vmatrix} = 0.$$

Retracing the steps we get the converse.

11. If $(a-b)(a'-b') = (b-c)(b'-c') = (c-a)(c'-a')$, show that the triangles whose vertices are a, b, c and a', b', c' are equilateral.

[*Hint* : From the first and second $\arg \dfrac{a-b}{b-c} = -\arg \dfrac{a'-b'}{b'-c'}$. So $\pi - \angle B = \pi - \angle B'$ or $\angle B = \angle B'$, etc; ΔABC is similar to $\Delta A'B'C'$; Let $A'B' = k\,AB$ etc. Then $|a-b||a'-b'| = |b-c||b'-c'|$ gives $AB\,(k\,AB) = BC\,(k\,BC)$ or $AB = BC$, etc.]

12. Find the locus of z if $\mathbf{Im}\ \dfrac{z-z_1}{z-z_2} = 0.$

[*Hint* : $(z-z_1)/(z-z_2)$ is real. So $\arg\{(z-z_1)/(z-z_2)\} = 0$ i.e., angle made by the line joining z, z_1 with the line joining $z, z_2 = 0$ or z, z_1, z_2 are collinear; Locus: st line through z_1, z_2])

13. Find the locus of z if $\mathbf{Re}\ \dfrac{z-z_1}{z-z_2} = 0.$

[*Hint* : $(z-z_1)/(z-z_2)$ is purely imaginary; $\arg\{(z-z_1)/(z-z_2)\} = \pm\frac{1}{2}\pi$; Locus: circle on the line joining z_1, z_2 as diameter]

14. Show that, if z_1, z_2, z_3 are the vertices of an isosceles triangle right-angled at z_1, then

$$2z_1^2 + z_2^2 + z_3^2 = 2z_1(z_2 + z_3).$$

[*Hint* : $\arg\{(z_1-z_3)/(z_1-z_2)\} = \pm\pi/2$. So $(z_1-z_3)/(z_1-z_2)$ is of the form r cis $(\pm\frac{1}{2}\pi)$. But $|z_1-z_3|/|z_1-z_2| = 1$. So $r = 1$ and $(z_1-z_3)/(z_1-z_2) = \text{cis}\,(\pm\frac{1}{2}\pi) = \pm i$. Square]

15. Show that, if the points z_1, z_2, z_3, z_4 are concyclic, then

$$\frac{z_1-z_3}{z_1-z_4}\cdot\frac{z_2-z_4}{z_2-z_3} \text{ is real,}$$

i.e., the cross ratio of z_1, z_4, z_2, z_3 (which will be defined later) is real.

[*Hint* : The angles subtended by the join of z_1 and z_2 are equal;

$$\arg\frac{z_2-z_4}{z_1-z_4} = \arg\frac{z_2-z_3}{z_1-z_3} \text{ or } \frac{z_1-z_3}{z_1-z_4}\cdot\frac{z_2-z_4}{z_2-z_3} \text{ is real]}$$

2.5. Equations of Straight Lines and Circles

16. When does $az + b\bar{z} - c = 0$ represent a straight line?

[*Hint* : Use (2.5.4); b is conjugate to a and c is real]

2.6. Elementary Transformations

17. Show that the mapping $w = 1/z$ is a combination of an inversion and a reflection. **(1977)**

18. **Solved sum.** Show that, under the transformation $w = 1/z$, the images of the following curves are curves as shown against them:

Curve	*Image*
(*i*) Circle not through O	Circle not through O
(*ii*) Circle through O	Straight line not through O
(*iii*) Straight line not through O	Circle through O
(*iv*) Straight line through O	Straight line through O

✍ Under the transformation $w = 1/z$, let $x + iy$ in the z plane go to $X + iY$ in the w plane. Then

$$X + iY = \frac{1}{x + iy} \quad \text{or} \quad (X + iY)(x + iy) = 1.$$

Equating real and imaginary parts separately

$$Xx - Yy = 1, \quad Yx + Xy = 0$$

and

$$x = \frac{X}{X^2 + Y^2}, \quad y = \frac{-Y}{X^2 + Y^2}.$$

(*i*) and (*ii*). On substitution, it is evident that the circle

$$x^2 + y^2 + 2gx + 2fy + c = 0$$

which does not pass through the origin, goes onto the circle

$$c(X^2 + Y^2) + 2gX - 2fY + 1 = 0$$

which also does not pass through the origin. But the circle

$$x^2 + y^2 + 2gx + 2fy = 0$$

which passes through the origin, goes onto the straight line

$$2gX - 2fY + 1 = 0$$

which does not pass through the origin.

(*iii*) and (*iv*). The straight line $ax + by + c = 0$ goes onto the circle

$$c(X^2 + Y^2) + aX - bY = 0$$

and the straight line $y = mx$ goes onto the line $Y = -mX$.

19. Show that the transformations (*i*) $w = i/z$ (*ii*) $w = 1/(z - 2)$ map circles and lines into circles and lines.

[*Hint* : (*i*) Transformation by i/z is the transformation by $1/z$ followed by a rotation about the origin through 90° in anticlockwise sense (*ii*) Transformation by $1/(z - 2)$ is the translation to the left by 2 units followed by the transformation by $1/z$].

20. Find the images of the following curves under the transformation $w = 1/z$:

(*i*) Family of circles $x^2 + y^2 = \lambda x$

(*ii*) Family of circles $x^2 + y^2 = \lambda y$

(*iii*) Family of parallel lines $y = x + \lambda$

(*iv*) Family of lines $y = \lambda x$ through the origin.

[**Ans.** (*i*) $X = 1/\lambda$ (*ii*) $Y = -1/\lambda$ (*iii*) $\lambda(X^2 + Y^2) + X + Y = 0$ (*iv*) $Y = -\lambda X$]

21. If $z_1, z_2, z_3, \ldots z_n$ are points on one side of a straight line through the origin, show that $1/z_1$, $1/z_2$, $1/z_3$, ..., $1/z_n$ lie on the other side of the line unless the line is the y axis.

[*Hint* : If the points lie above the line $y = (\tan\alpha)\, x$, then $\alpha < \arg z_k < \pi + \alpha$ and $-(\pi + \alpha) < \arg (1/z_k) < -\alpha$ and the points $1/z_k$ lie below $y = \{\tan(-\alpha)\}\, x$]

22. Show that the image of the region $y > c$ $(c > 0)$ under the transformation $w = 1/z$ is the interior of a circle. What is the image of the region $y > 0$?

[*Hint* : $-Y/(X^2+Y^2) > c$ or $c(X^2+Y^2)+Y<0$; Image of $y>0 : Y<0$]

23. Show that the image of the infinite rectangular strip $0<y<1/(2c)$ under the transformation $w=1/z$ is the exterior of the circle $X^2+Y^2+2cY=0$ in the lower half plane $Y<0$. **(1980)**

[*Hint* : $0<y<1/(2c)$ gives $0<\dfrac{-Y}{X^2+Y^2}<\dfrac{1}{2c}$ or $-2cY<X^2+Y^2$; Y is negative since c and X^2+Y^2 are positive.

24. Show that the image of the quadrant region $x>1, y>0$ under the transformation $w=1/z$ is the disc $|w-\frac{1}{2}|<\frac{1}{2}$ in the lower half plane $Y<0$.

[*Hint* : $X/(X^2+Y^2)>1$, $-Y/(X^2+Y^2)>0$ give $X^2+Y^2-X<0$, $Y<0$ or $(X-\frac{1}{2})^2+Y^2<(\frac{1}{2})^2$, $Y<0$]

25. Find the image of the region $x>1, y>1$ under the transformation $w=1/z$. **(1975, 77, 82)**

[**Ans.** Region bounded by the circles $X^2+Y^2-X=0$, $X^2+Y^2+Y=0$]

26. Show that the polar equation of the image of the hyperbola $x^2-y^2=1$ under the transformation $w=1/z$ is $\rho^2=\cos 2\phi$.

[*Hint* : $\dfrac{X^2}{X^2+Y^2}-\dfrac{Y^2}{X^2+Y^2}=1$, where $X=\rho\cos\phi$, $Y=\rho\sin\phi$]

27. **Solved sum.** Find the image of the circle $|z|=2$ by the transformation $w=z+(2+3i)$.

✍ The transformation is a translation which translates the figure in the x and y directions through the distances 2 and 3. Now $|z|=2$ is a circle whose centre is the origin (0, 0) and radius is 2. So its image is a circle whose centre is (2, 3) and radius is 2. Hence its equation is

$$(x-2)^2+(y-3)^2=2^2.$$

Or directly, from $w=z+(2+3i)$ and $|z|=2$,

$$|w-(2+3i)|=2$$

which is a circle whose centre is (2, 3) and radius is 2.

28. **Solved sum.** Determine the image of the rectangle with vertices

$$O(0,0), A(2,0), B(2,1), C(0,1)$$

under the transformation $w=(1+i)z+(1-2i)$.

✍ Let $w=X+iY$. Then we have

$$X+iY=(1+i)(x+iy)+(1-2i).$$

$$\therefore \quad X=x-y+1, \quad Y=x+y-2.$$

Equation of the image of the line OA, $y=0$: It is the eliminant of x, y from

$$\left.\begin{aligned} y&=0,\\ X&=x-y+1,\\ Y&=x+y-2.\end{aligned}\right\} \qquad \therefore \quad X-Y=3.$$

Equation of the image of the line CB, $y=1$: It is the eliminant of x, y from

$$\left.\begin{aligned} y&=1,\\ X&=x-y+1,\\ Y&=x+y-2.\end{aligned}\right\} \qquad \therefore \quad X-Y=1.$$

Equation of the image of the line OC, $x = 0$: It is the eliminant of x, y from

$$\left.\begin{aligned} x &= 0, \\ X &= x - y + 1, \\ Y &= x + y - 2. \end{aligned}\right\} \qquad \therefore \quad X + Y = -1.$$

Equation of the image of the line AB, $x = 2$: It is the eliminant of x, y from

$$\left.\begin{aligned} x &= 2, \\ X &= x - y + 1, \\ Y &= x + y - 2. \end{aligned}\right\} \qquad \therefore \quad X + Y = 3.$$

So the image is the rectangle whose sides are the lines

$$X - Y = 3,\ X - Y = 1,\ X + Y = -1,\ X + Y = 3.$$

29. **Solved sum.** Find the image of the circle $|z + 2i| = 2$ under the transformation $w = \dfrac{1}{z}$.

✍ **Method 1.** Since the mapping is $w = 1/z$, the equation of the image of the given circle $|z + 2i| = 2$ is

$$\left|\frac{1}{w} + 2i\right| = 2 \quad \text{or} \quad \left|\frac{1 + 2iw}{w}\right| = 2$$

i.e.,
$$\frac{|1 + 2iw|}{|w|} = 2 \quad \text{or} \quad |1 + 2iw| = 2|w|.$$

Taking the image of z as $w = X + iY$, we get the equation as

$$|1 + 2i(X + iY)| = 2|X + iY|$$

or
$$|(1 - 2Y) + i(2X)| = 2|X + iY|$$

i.e.,
$$(1 - 2Y)^2 + (2X)^2 = 4(X^2 + Y^2) \quad \text{or} \quad 1 - 4Y = 0.$$

$\therefore$
$$4Y = 1 \quad \text{or} \quad Y = 1/4.$$

Method 2. Taking $w = X + iY$,

$$x + iy = z = \frac{1}{w} = \frac{1}{X + iY} = \frac{X - iY}{X^2 + Y^2}.$$

$\therefore$
$$x = \frac{X}{X^2 + Y^2}, \quad y = \frac{-Y}{X^2 + Y^2}.$$

Thus $|z + 2i| = 2$ or $|x + iy + 2i| = 2$ becomes

$$\left|\frac{X}{X^2 + Y^2} + i\frac{-Y}{X^2 + Y^2} + 2i\right| = 2$$

or
$$\left(\frac{X}{X^2 + Y^2}\right)^2 + \left(\frac{-Y}{X^2 + Y^2} + 2\right)^2 = 4$$

or
$$\frac{X^2}{(X^2 + Y^2)^2} + \frac{Y^2}{(X^2 + Y^2)^2} - \frac{4Y}{X^2 + Y^2} + 4 = 4.$$

Cancelling 4's and multiplying both sides by $X^2 + Y^2$ we get

$$1 - 4Y = 0 \quad \text{or} \quad Y = 1/4.$$

30. Find the images of the following circles under the transformation $w = 1/z$.

(*i*) $|z-1|=1$ (*ii*) $|z+1|=1$

(*iii*) $|z-3i|=3$ (*iv*) $|z+2i|=2$

[**Ans.** (*i*) $X = 1/2$ (*ii*) $X = -1/2$ (*iii*) $Y = -1/6$ (*iv*) $Y = 1/4$]

31. Show that the image of the first quadrant region $x > 1, y > 1$ is mapped by $w = 1/z$ into the region bounded by the circles

$$X^2 + Y^2 - X = 0, \quad X^2 + Y^2 + Y = 0.$$

[*Hint* : $\dfrac{X}{X^2+Y^2} > 1, \ \dfrac{-Y}{X^2+Y^2} > 1$]

32. What are the invariant points in a mapping?

33. Which are the fixed points under the mapping $w = az + b$, where a and b are complex constants?

[**Ans.** ∞ if $b \neq 0$, and 0 and ∞ if $b = 0$]

34. Which are the invariant points of the mapping $w = z^2$?

[*Hint* : $z = z^2$ gives the fixed points as 0, 1]

2.8. Stereographic Projection

35. **Solved sum.** If z_1 and z_2 are the images in the complex plane of two diametrically opposite points on the Riemann sphere, show that $z_1 \bar{z}_2 = -1$. **(1974)**

Let the diametrically opposite points be (α, β, γ) and $(-\alpha, -\beta, -\gamma)$. Then, by (2.8.1),

$$z_1 = \frac{\alpha + i\beta}{1-\gamma}, \quad z_2 = \frac{-\alpha - i\beta}{1+\gamma}, \quad \bar{z}_2 = -\frac{\alpha - i\beta}{1+\gamma}$$

$$\therefore \quad z_1 \bar{z}_2 = -\frac{(\alpha+i\beta)(\alpha-i\beta)}{(1-\gamma)(1+\gamma)} = -\frac{\alpha^2+\beta^2}{1-\gamma^2} = -1$$

$(\because \alpha^2 + \beta^2 + \gamma^2 = 1)$

36. **Solved sum.** Show that stereographic projection maps circles on the Riemann sphere onto circles or straight lines in the complex plane. **(1975)**

Let the equations of a circle on the sphere be

$$lx + my + nZ = p, \quad x^2 + y^2 + Z^2 = 1.$$

If (α, β, γ) is point on this circle, then

$$l\alpha + m\beta + n\gamma = p, \quad \alpha^2 + \beta^2 + \gamma^2 = 1.$$

Suppose (α, β, γ) is mapped into $x + iy$, then by (2.8.2), and from $l\alpha + m\beta + n\gamma = p$,

$$\frac{2lx}{x^2+y^2+1} + \frac{2my}{x^2+y^2+1} + \frac{n(x^2+y^2-1)}{x^2+y^2+1} = p$$

i.e., $$(n-p)(x^2+y^2) + 2lx + 2my - (n+p) = 0$$

which is a circle or a straight line according as $n \neq p$ or $n = p$. When $n = p$, the plane $lx + my + nZ = p$ and the corresponding circle on the sphere passes through the centre of projection (0, 0, 1).

CHAPTER 3

SETS OF COMPLEX POINTS

3.1. CLOSED SETS

In this section we define a closed set and a compact set.

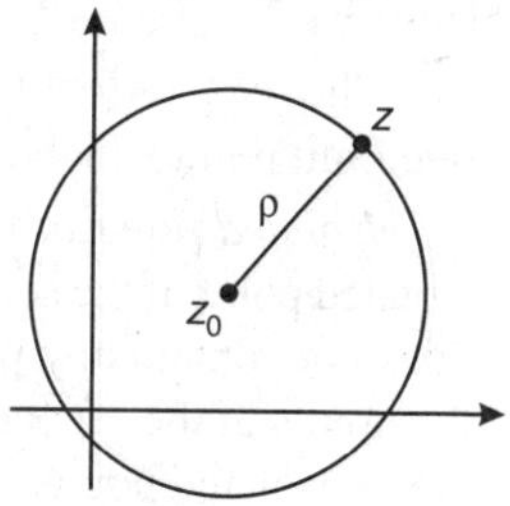

We know that the distance between two complex points z_1 and z_2 is

$$|z_1 - z_2|.$$

Thus the equation

$$|z - z_0| = \rho$$

represents the set of points on the circle whose centre is z_0 and radius is ρ. The equation

$$|z - z_0| \le \rho$$

represents the set of all points on the circle $|z - z_0| = \rho$ and the points interior to the circle.

Neighbourhood of a point. A set of all z with

$$|z - z_0| < \rho,$$

where ρ is a (positive) number, is called a neighbourhood of z_0 (Fig. 3.1.1 (a)). This neighbourhood is denoted by the symbol $N(z_0, \rho)$ or simply by $N(z_0)$. Changing ρ we get many neighbourhood of z_0.

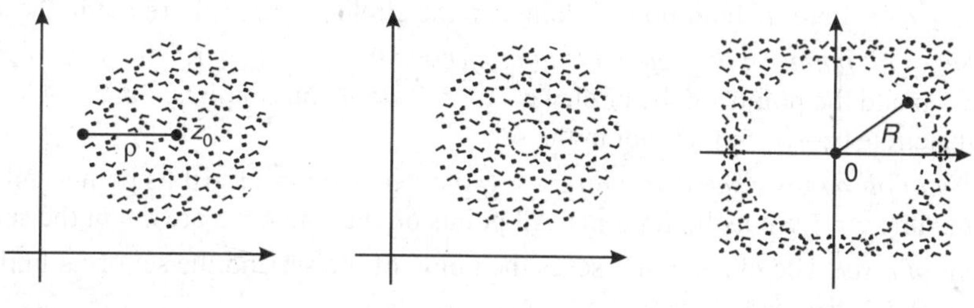

Fig. 3.1.1 (a) **Fig. 3.1.1 (b)** **Fig. 3.1.1 (c)**

Deleted neighbourhood of a point. The equation of the circle whose centre is the origin 0 and the radius is R is

$$|z - 0| = R \quad \text{or} \quad |z| = R.$$

The equation

$$|z| > R$$

represents the set of all points outside the circle $|z| = R$. A set of all z with

$$0 < |z - z_0| < \rho$$

where ρ is a positive number, is called a deleted neighbourhood of z_0 (Fig. 3.1.1 (b)). The difference between a deleted neighbourhood of z_0 and a neighbourhood of z_0 is that the former does not contain z_0 while the latter does.

Neighbourhood of ∞. The set of all z with

$$|z| > R,$$

where R is a positive number, is called a neighbourhood of ∞ (Fig. 3.1.1.(c)). This set is the range of the neighbourhood of the origin $|z| < 1/R$ under the transformation $w = 1/z$. (The physical nature of this neighbourhood can be visualized geometrically by going through the stereographic projection).

Limit point of a set. A point z_0 is said to be a limit point of a set if every neighbourhood of z_0 contains an infinite number of points of the set.

Examples. Every point of the set of all points on a circle is a limit point of the set because, if z_0 is a point of the set, every neighbourhood of z_0 will contain an infinite number of points of the set lying on the arc of the circle which lies in the neighbourhood (Fig. 3.1.2).

The set $\{\cos(\pi/n) + i \sin(\pi/n)$, where $n = 1, 2, 3, ...\}$ has only one limit point which is 1.

Isolated point of a set. The point z_0 of a set is said to be an isolated point of the set if there exists a neighbourhood of z_0 which does not contain any point of the set other than z_0. For example, the points of the set $\{\cos(\pi/n) + i \sin(\pi/n)$, where $n = 1, 2, 3, ...\}$ are all isolated points.

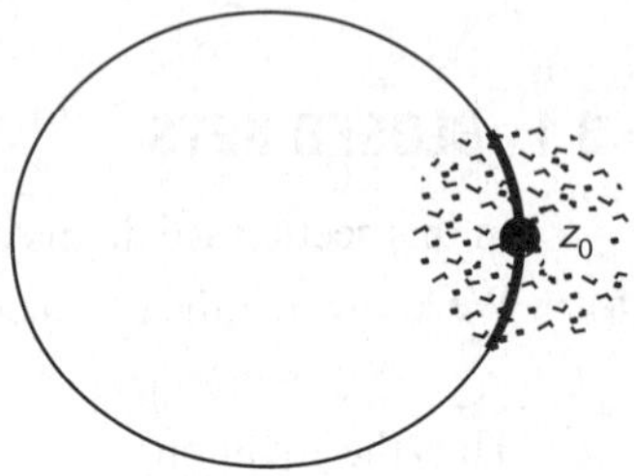

Fig. 3.1.2

Remark. The limit points of a set many or may not be in the set as given in the following examples:

(*i*) *Set of all points on the circle* $|z - a| = 1$. All its points are its limit points and no other point is a limit point.

(*ii*) *Set of all points in the circular disc* $|z - a| \le 1$. All its points are its limit points, and no other point is a limit point.

(*iii*) *Set of all points in the circular disc* $|z - a| < 1$. All its points and the points on the circle $|z - a| = 1$ are its limit points. Points on the circle $|z - a| = 1$ are not in the set.

(*iv*) *Set of all points in the deleted neighbourhood* $0 < |z - a| < 1$. All its points and the point $z = a$ and the points on the circle $|z - a| = 1$ are its limit points. Point $z = a$ and points on the circle $|z - a| = 1$ are not in the set.

(*v*) *Set of all points z satisfying* Im $z > 1$ (*points above* $y = 1$). All its points and all points on the line $y = 1$ are its limit points. All points on the line $y = 1$ are not in the set.

Closure of a set. The closure of a set is the union of the set and the set of its limit points. The closure of a set S is denoted by $\overline{S}$.

Closed set. A set is said to be closed if it contains all its limit points. i.e., when $S = \overline{S}$. For example,

(*i*) $|z - a| = 1$ and $|z - a| \le 1$ are closed sets and

(*ii*) $|z - a| < 1$, $0 < |z - a| < 1$, and $0 < |z - a| \le 1$ are not closed sets.

Bounded set. A set S of points z is said to be bounded if the set of real numbers $|z|$ is bounded. In other words, a set S is bounded if there exists a positive number M such that for all z in S, $|z| \le M$. For example, the sets $|z| = 1$, $|z| < 1$ and $|z| \le 1$ are bounded sets while $|z| > 1$ and $|z| \ge 1$ are unbounded sets. All the sets

$$\text{Im } z = 1, \quad \text{Im } z < 1, \quad \text{Im } z > 1$$

are bounded because $|\text{Re } z|$ and hence $|z|$ is not bounded.

Compact set. A set which is both closed and bounded is said to be compact.

3.2. OPEN SETS

Interior point of a set. A point z_0 of a set is said to be an interior point of the set if there exists a neighbourhood of z_0 which is *wholly* contained in the set.

Remark. An interior point of a set is a limit point of the set. But a limit point of a set may or may not be an interior point of the set. For example, the set of points on a circle has no interior point but all its points are its limit points.

Open set. A set is said to be open if all its points are interior points.

Examples for open sets: (*i*) Neighbourhoods.

(*ii*) Annular rings like $1 < |z| < 2$.

(*iii*) Union of the circular discs $|z-1| < 1$ and $|z-3| < 1$.

Remark 1. The set $0 < |z| \leq 1$ is neither closed nor open.

Remark 2. The union of two or more open sets, is an open set.

Boundary points of an open set. Points which do not belong to an open set but are its limit points, are called boundary points of the set.

Boundary of an open set. The set of all boundary points of an open set is called the boundary of the set. For example,

Open set	*Boundary points of the set*
$\|z\| < 1$	Points on the circle $\|z\| = 1$
$0 < \|z\| < 1$	Zero and points on the circle $\|z\| = 1$
$1 < \|z\| < 2$	Points on the circles $\|z\| = 1$ and $\|z\| = 2$.

Exterior points of an open set. Points which are neither interior points nor boundary points of an open set are called exterior points of the set. That is, an exterior point of an open set is a point which has a neighbourhood such that no point of the neighbourhood belongs to the set.

Arcwise-connected set. A set is called arcwise-connected if every pair of its points can be connected by a polynomial line, all of whose points belong to the set.

The union of $|z-1| < 1$ and $|z+1| < 1$ is not arcwise-connected while the union of $|z-1| \leq 1$ and $|z+1| \leq 1$ is arcwise-connected.

Domain. An arcwise-connected open set is called a domain.

Region. The union of a domain and some or all of its boundary points, is referred to as a region.

Set	Open?	Arcwise connected?	Domain or region
$\|z\| \leq R$	No	Yes	Region
$0 < \|z\| \leq R$	No	Yes	Region
$\|z\| < R$	Yes	Yes	Domain
$\|z\| > R$	Yes	Yes	Domain
$\text{Re}\, z > 0$	Yes	Yes	Domain
$\text{Re}\, z \neq 0, (x \neq 0)$	Yes	No	None

Note. The set, $\text{Re}\, z \neq 0$ is, the set of all complex points except those on the y axis.

Closed region. The set obtained by adding to a region all its boundary points is called a closed region. For example,

Region S	*Boundary points of S*	*Closed region* $\overline{S}$
$\|z\|<1$	$\|z\|=1$	$\|z\|\le 1$
$\|z\|>1$	$\|z\|=1$	$\|z\|\ge 1$
$0<\|z\|<1$	$0,\ \|z\|=1$	$\|z\|\le 1$
$1<\|z\|<2$	$\|z\|=1,\ \|z\|=2$	$1\le\|z\|\le 2$
$\text{Re } z>0$	$\text{Re } z=0$	$\text{Re } z\ge 0$

If S is a region, then, its closure $\overline{S}$ is a closed region.

Set S	**Interior points of S**	**Limit points of S**	**Boundary points**
$\|z-a\|<1$	$\|z-a\|<1$	$\|z-a\|\le 1$	$\|z-a\|=1$
$\|z-a\|\le 1$	$\|z-a\|<1$	$\|z-a\|\le 1$	$\|z-a\|=1$

3.3. THEOREMS ON BOUNDED INFINITE SETS

Theorem 3.1. (Bolzano-Weierstrass theorem). Every bounded infinite set has at least one limit point.

Proof: Let S be the given set and assume that S has no limit point. Take a square A_0 with its sides parallel to the real and imaginary axes such that the square contains the given set S (Fig. 3.3.1). Such a square exists because S is bounded. Let the projection of this square on the real axis be the closed interval $[h, k]$. Then the side of the square is $k - h$. Divide the square A_0 into four subsquares. Then S is divided, in general, into four subsets and at least one of them should contain an infinite number of points of S; otherwise, it becomes that S is a finite set. Let this particular subset be contained in the subsquare A_1. Suppose the projection of A_1 on the real axis be $[h_1, k_1]$. Then the side of A_1 is

$$k_1 - h_1 = \frac{k-h}{2}.$$

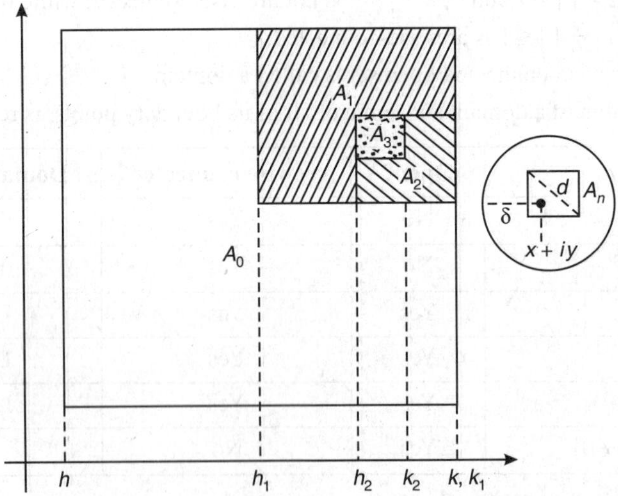

Fig. 3.3.1

Again divide A_1 into four squares. Then, again at least one of them contains infinite number of points of S. Let this square be A_2. Continuing this process repeatedly, we get that the square, say A_n, contains an infinite number of points of S and the side of A_n is

$$k_n - h_n = \frac{k - h}{2^n}.$$

It is evident that the real sequence $\{k_n - h_n\}$ converges to zero.

i.e.,
$$\lim_{n \to \infty} (h_n - k_n) = 0.$$

This implies that

$$\lim_{n \to \infty} h_n = \lim_{n \to \infty} k_n = x, \text{ say.}$$

Similarly let the projection of A_n on the imaginary axis be the interval $[l_n, m_n]$.Then

$$l_n - m_n = \frac{k - h}{2^n}.$$

Let
$$\lim_{n \to \infty} l_n = \lim_{n \to \infty} m_n = y.$$

Now $x + iy$ is a point which belongs to all the squares $A_0, A_1, A_2, A_3, \ldots$ each of which contains an infinite number of points of S. Further, if δ is any given positive number, then the square A_n will be contained in the neighbourhood $N(x + iy, \delta)$ as soon as n is large enough so that the diagonal d of A_n is less than δ, i.e., as soon as n is large enough so that

$$\frac{\sqrt{2}\,(k - h)}{2^n} < \delta.$$

So the neighbourhood $N(x + iy, \delta)$ contains an infinite number of points of S. But δ is arbitrary. Hence every neighbourhood of $x + iy$ contains infinite number of points of S. So, by definition, $x + iy$ is a limit point of S. This completes the proof.

Covering of a set. Suppose S is a given set. A collection of sets whose union contains S is said to cover the set S. If the sets are open, then collection is called an open covering.

Theorem 3.2. (Heine-Borel theorem). Let S be a compact set. Suppose there are neighbourhoods $N(z)$ such that each one of them is associated with a particular z of S. Then a finite number of these neighbourhoods cover S.

Proof: Suppose the theorem is false, i.e., suppose the set S cannot be covered by a finite number of neighbourhoods in question. But S is bounded. So, as in the proof of Bolzano-Weierstrass theorem (Theorem 3.1), consider a square A_0 which contains the set S and divide the square into four equal squares. This division divides S, in general, into four subsets. According to supposition made, at least one of these subsets, say S_1, cannot be covered by a finite number of $N(z)$. Let S_1 be in the subsquare A_1. Continuing this division repeatedly we get, as in Bolzano-Weierstrass theorem, that there exists a complex number a common to all the squares $A_0, A_1, A_2, A_3, \ldots$, and that it is a limit point of S. Since S is *compact*, a belongs to S. Let the neighbourhood associated with a, be $N(a, \delta)$. Then it will contain the square A_n as soon as n is large enough so that the diagonal of A_n is less than δ. That is, the neighbourhood $N(a, \delta)$ contains A_n and hence covers A_n.

But, by supposition made, each of $A_0, A_1, A_2, \ldots, A_n$ cannot be covered by a finite number of neighbourhoods in question. But we have shown that A_n is covered by a single given neighbourhood $N(a, \delta)$. So the supposition made is wrong. This implies that S can be covered by a finite number of given neighbourhoods $N(z)$.

3.4. EXAMPLES

1. Solved sum. State whether the following sets are open, closed and arcwise connected and whether they are regions:

$$|z| \leq \text{R}, \quad 0 < |z| \leq \text{R}, \quad |z| < R, \quad |z| > R, \quad \text{Re}\, z > 0, \quad \text{Re}\, z \neq 0.$$

Set	Open	Closed	Arcwise connected	Domain or Region
$\|z\| \leq R$	no	yes	yes	no
$0 < \|z\| \leq R$	no	no	yes	no
$\|z\| < R$	yes	no	yes	yes
$\|z\| > R$	yes	no	yes	yes
$\text{Re}\, z > 0$	yes	no	yes	yes
$\text{Re}\, z \neq 0$	yes	no	no	no

CHAPTER 4

ANALYTIC FUNCTIONS

4.1. COMPLEX FUNCTIONS

Suppose S is a set of complex numbers. If there is a rule f which associates each complex number of S with a complex number in the complex plane, then f is said to be a function defined in S. If z, the complex number in S, is associated by f with the complex number w in the complex plane, then w is said to be the value of f at z. w is symbolically denoted by $f(z)$ so that

$$w = f(z)$$

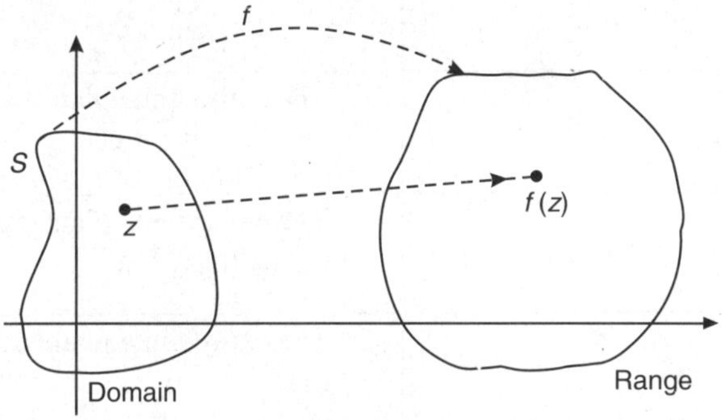

Fig. 4.1.1

(Fig. 4.1.1). The set S is the *domain of definition of f* and the set of values of f is the *range of S under f.*

Single-valued function. If f is defined in a set S and if for each z in S there is only one value $f(z)$ of f, then f is said to be single-valued; otherwise, it is said to be multiple-valued or multi-valued.

For example, $f(z) = z^2$ is a single valued function defined in any set since z^2 has only one value for every z. In particular, if

$$z = 1 + i, \quad f(z) = (1 + i)^2 = 2i.$$

As an example for a multiple-valued function, consider

$$f(z) = z^{1/2}.$$

It has two values for each z because, if $z = 4\ (\cos \frac{1}{4}\pi + i \sin \frac{1}{4}\pi)$,

$$f(z) = 2 \text{ cis } (\pi/8) \text{ and } 2 \text{ cis } (\pi + (\pi/8)).$$

We will see later that Log z and a^z are multiple-valued functions.

Inverse functions. If f is a function defined in S and if S' is the set of values of f, the function f^{-1} which associates each element of S' with the corresponding elements of S, is called the inverse function of f. If f and f^{-1} are single-valued functions, then there is a one-to-one correspondence between S and S'.

Polynomial functions. Functions of the form

$$P_n(z) = a_0 + a_1 z + a_2 z^2 + ... + a_n z^n, \quad a_n \neq 0,$$

where n is a positive integer and $a_0, a_1, a_2, ... a_n$ are complex numbers, is called a polynomial in z of degree n.

Rational function. The quotient of two polynomials is called a rational function. That is, the function

$$R(z) = \frac{P(z)}{Q(z)}, \quad Q(z) \neq 0$$

where $P(z)$ and $Q(z)$ are polynomials in z, is a rational function. In this we assume that $P(z)$ and $Q(z)$ have no common factors.

Bounded function. Suppose $f(z)$ is a function defined in a set S. If there exists a finite number M such that $|f(z)| < M$ for all z in S, then $f(z)$ is said to be bounded in S.

Examples:

Function $f(z)$	Domain of definition of $f(z)$
$\frac{2}{z^2+1}$	The entire finite plane except $z \neq \pm i$.
$\arg \frac{3}{z}$ or $-\arg z$, since $\arg 3 = 0$	The entire finite plane except $z = 0$ since $\arg 0$ is not defined.
$\frac{z}{z+\bar{z}}$ or $\frac{z}{2x}$	The entire finite plane except the points on the line $x = 0$
$\frac{1}{1-\|z\|^2}$	The entire finite plane except the circle $x^2 + y^2 = 1$.

4.2. LIMIT OF A FUNCTION

Let f be a function defined in a set S and let z_0 be a limit point of S (Fig. 4.2.1). Then A is said to be the limit of $f(z)$ at z_0 if for any $\in > 0$ there exists a $\delta > 0$ such that

$$|f(z) - A| < \in$$

for all z in S other than z_0 with $|z - z_0| < \delta$. δ is a function of $\in$, say $\delta(\in)$.

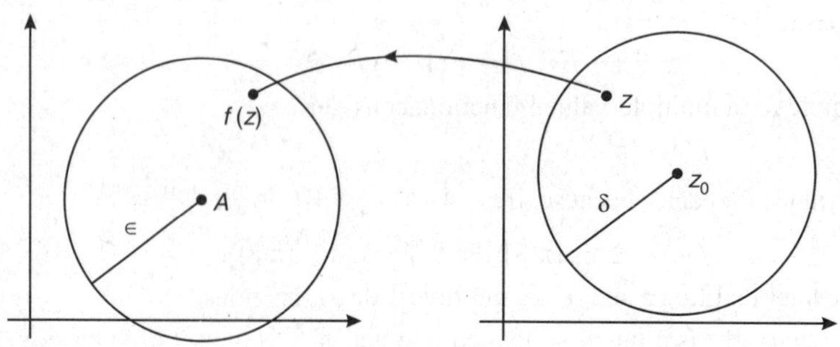

Fig. 4.2.1

Remark. The above limit point z_0 may be an interior point or a boundary point.

Notation. When $f(z)$ has a limit A at z_0, it is said that $f(z)$ tends to A as z to z_0 and this fact is symbolically written as

$$\lim_{z \to z_0} f(z) = A.$$

Uniqueness of a Limit

Theorem 4.1. The limit of a function is unique.

Proof: Suppose a function $f(z)$ has two limits as

$$\lim_{z \to z_0} f(z) = A_1, \quad \lim_{z \to z_0} f(z) = A_2.$$

Then, for any small positive $\in$, there exist two positive numbers δ_1 and δ_2, depending on $\in$, such that

$$|f(z) - A_1| < \in \text{ if } |z - z_0| < \delta_1,$$
$$|f(z) - A_2| < \in \text{ if } |z - z_0| < \delta_2.$$

Let δ be the smaller of δ_1, δ_2. Then, if

$$|z - z_0| < \delta,$$

we have

$$|f(z) - A_1| < \in,$$
$$|f(z) - A_2| < \in,$$

and, using the result $|z_1 + (-z_2)| \leq |z_1| + |z_2|$, we get

$$|[f(z) - A_1] - [f(z) - A_2]| \leq |f(z) - A_1| + |f(z) - A_2|$$
$$= \in + \in$$

or $\qquad |A_1 - A_2| \leq 2 \in.$

But ε is arbitrary and so we can choose any small value for ε. This requires that

$$A_1 - A_2 = 0 \text{ or } A_1 = A_2.$$

That is, the limit of $f(z)$ as $z \to z_0$ is unique.

It is important to note that the limit, when it exists, is unique whatever be the mode of approach of z to z_0 through the points in S. If the value is different for different approach of z to z_0, then the limit does not exist. The following illustration exhibits non-uniqueness and hence non-existence of the limit.

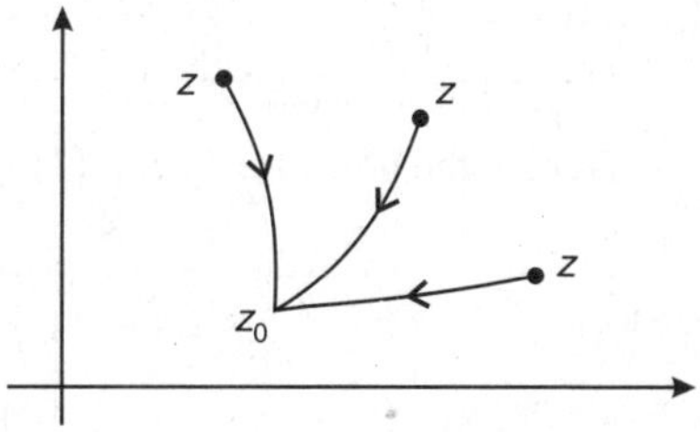

Illustration. Consider the function

$$f(z) = \frac{\textbf{Re } z}{|z|}, \qquad \text{when } z \neq 0$$
$$= 0, \qquad \text{when } z = 0.$$

The limit of $f(z)$ as $z = x + iy$ approaches zero along the line $y = mx$, is

$$\lim_{x \to 0} \frac{x}{\sqrt{x^2 + y^2}} = \lim_{x \to 0} \frac{x}{\sqrt{x^2 + m^2 x^2}} = \frac{1}{\sqrt{1 + m^2}}.$$

This quantity depends on m which varies with the line and so it is not unique. Therefore the limit of $f(z)$ as z tends to zero does not exist.

Theorem 4.2. Suppose $f(z) = u(x, y) + iv(x, y)$ is a function defined in a set S and $z_0 = x_0 + iy_0$ is a limit point of S. Then $f(z)$ has a limit $h + ik$ at $z_0 = x_0 + iy_0$ if and only if $u(x, y) \to h$, $v(x, y) \to k$ as $x \to x_0$ and $y \to y_0$.

Necessity part:

Hypothesis: $f(z)$ tends to $h + ik$ as z tends to $x_0 + iy_0$.

To prove: $u(x, y) \to h$, $v(x, y) \to k$ as $x \to x_0$ and $y \to y_0$.

Proof: Since $f(z)$ tends to $h + ik$ as z tends to $x_0 + iy_0$, for any $\in > 0$, there exists a δ such that

$$|f(z) - (h + ik)| < \in \text{ for all } z \text{ in } S \text{ with } |z - z_0| < \delta$$

But $|u(x, y) - h| = |\textbf{Re}\{f(z) - (h + ik)\}| \le |f(z) - (h + ik)|$

$\therefore$ $|u(x, y) - h| < \in$ for all x, y with $\sqrt{(x - x_0)^2 + (y - y_0)^2} < \delta$.

This implies that the real function $u(x, y) \to h$ as $x \to x_0, y \to y_0$

Similarly $v(x, y) \to k$ as $x \to x_0, y \to y_0$.

Sufficiency part :

Hypothesis : $u(x, y) \to h, v(x, y) \to k$ as $x \to x_0, y \to y_0$

To prove : $f(z) \to h + ik$ as $z \to x_0 + iy_0$.

Proof : Since the real functions $u(x, y)$, $v(x, y)$ tend to h, k as $x \to x_0, y \to y_0$, for any $\in > 0$, there exist positive numbers δ_1 and δ_2 such that

$$|u(x, y) - h| < \tfrac{1}{2} \in \qquad \text{for all } x, y \text{ with } \sqrt{(x - x_0)^2 + (y - y_0)^2} < \delta_1$$

$$|v(x, y) - k) < \tfrac{1}{2} \in \qquad \text{for all } x, y \text{ with } \sqrt{(x - x_0)^2 + (y - y_0)^2} < \delta^2$$

$$\begin{aligned} \therefore \quad |f(z) - (h + ik)| &= |u(x, y) + iv(x, y) - (h + ik)| \\ &= |u(x, y) - h + i\{v(x, y) - k\}| \\ &\le |u(x, y) - h| + |v(x, y) - k| \\ &< \tfrac{1}{2}\in + \tfrac{1}{2}\in = \in \end{aligned}$$

for all z in S with $|z - z_0| < \min(\delta_1, \delta_2)$. So $f(z)$ tends to $h + ik$ as z tends to $x_0 + iy_0$.

Theorem 4.3. If $\lim_{z \to \infty} f(z) = A$, then $\lim_{z \to 0} f\left(\frac{1}{z}\right) = A$.

Proof : From the hypothesis, we get that, for any small positive number ε, there exists an $M(\varepsilon)$ such that,

if $|z| > M(\in)$, then $|f(z) - A| < \in$

i.e., if $\left|\frac{1}{z}\right| < \frac{1}{M(\in)}$, then $|f(z) - A| < \in$. (1)

Denote the point $1/z$ by t in (1). Then z is the point $1/t$, and (1) becomes

if $|t - 0| < \delta(\in)$, then $\left|f\left(\frac{1}{t}\right) - A\right| < \in$,

where $\delta(\in) = \frac{1}{M(\in)}$. This implies that

$$\lim_{t \to 0} f\left(\frac{1}{t}\right) = A.$$

Here t is a dummy symbol. Changing it to z, we have

$$\lim_{z \to 0} f\left(\frac{1}{z}\right) = A.$$

Theorem 4.4. If $\lim_{z \to 0} f(z) = A$, then $\lim_{z \to \infty} f\left(\frac{1}{z}\right) = A$.

Proof : From the hypothesis, we get that, for any small positive number ε, there exists a $\delta(\varepsilon)$ such that,

if $|z - 0| < \delta(\in)$, then $|f(z) - A| < \in$

i.e., if $\left|\dfrac{1}{z}\right| > \dfrac{1}{\delta(\epsilon)}$, then $|f(z) - A| < \epsilon$. (1)

Let us denote the point $1/z$ by t. Then z is the point $1/t$ and (1) becomes that

$$\text{if } |t| > \frac{1}{\delta(\epsilon)}, \text{ then } \left|f\left(\frac{1}{t}\right) - A\right| < \epsilon.$$

$$\therefore \qquad \lim_{t \to \infty} f\left(\frac{1}{t}\right) = A.$$

Here t is a dummy symbol. Changing it to z, we have

$$\lim_{z \to \infty} f\left(\frac{1}{z}\right) = A.$$

Remark. Further, from the algebraic operations on limits of real functions, if $f(z) \to A$ as $z \to z_0$ and $g(z) \to B$ as $z \to z_0$, then

$$\begin{aligned} f(z) \pm g(z) &\to A \pm B \\ f(z)\, g(z) &\to AB \\ f(z) / g(z) &\to A/B \end{aligned}$$

as $z \to z_0$, wherein the last result requires that $B \neq 0$.

Finite limit at ∞. Let $f(z)$ be a function defined in a set S and let ∞ be a limit point of S. Then A is said to be the limit of $f(z)$ at ∞ if for any $\epsilon > 0$ there exists an $R(\epsilon) > 0$ such that

$$|f(z) - A| < \epsilon \qquad \text{for all } z \text{ in } S \text{ (other than } \infty) \text{ with } |z| > R.$$

Example. Let $f(z) = 1/z$ be a function defined in the finite plane. Now, for every $\epsilon > 0$,

$$\left|\frac{1}{z} - 0\right| = \left|\frac{1}{z}\right| = \frac{1}{|z|} < \epsilon \text{ for all } z \text{ with } |z| > \frac{1}{\epsilon}.$$

Hence $1/z \to 0$ as $z \to \infty$.

Infinite limit at a finite point. Let $f(z)$ be a function defined in a set S and let z_0 be a limit point of S. Then ∞ is said to be the limit of $f(z)$ at z_0 if for any $M > 0$ there exists a $\delta(M) > 0$ such that

$$|f(z)| > M \qquad \text{for all } z \text{ in } S \text{ other than } z_0 \text{ with } |z - z_0| < \delta.$$

Example. Let $f(z) = 1/z$ be a function defined in the finite plane except at zero. Now, for every $M > 0$,

$$\left|\frac{1}{z}\right| = \frac{1}{|z|} > M \qquad \text{for all } z \text{ other than } 0 \text{ with } |z - 0| < 1/M.$$

Hence $1/z \to \infty$ as $z \to 0$.

Infinite limit at ∞. Let $f(z)$ be a function defined in a set S and let ∞ be a limit point of S. Then ∞ is said to be the limit of $f(z)$ at ∞ if for any $M > 0$ there exists an $R(M) > 0$ such that

$$|f(z)| > M \qquad \text{for all } z \text{ in } S \text{ other than } \infty \text{ with } |z| > R.$$

Example. Let $f(z) = z^2$ be a function defined in the finite plane. Now, for any $M > 0$,

$$|z^2| = |z|^2 > M \qquad \text{for all } z \text{ other than } \infty \text{ with } |z| > \sqrt{M}$$

Hence $z^2 \to \infty$ as $z \to \infty$.

ILLUSTRATIONS

(i) Finite limit at a finite point

Example. Show that, if $f(z) = 2iz$ defined on the domain $S : |z| < 1$, then

$$\lim_{z \to 1} f(z) = 2i.$$

Intuitively, substituting 1 for z in $2iz$, we get $2i$ which is the required limit. However, to obtain the limit, using the definition, consider any positive ε. Take a z such that

$$|2iz - 2i| < \varepsilon \text{ or } |2i||z-1| < \varepsilon$$

or
$$|z-1| < \varepsilon/2.$$

Conversely, for any z of S in $|z-1| < \varepsilon/2$,

$$|f(z) - 2i| < \varepsilon.$$

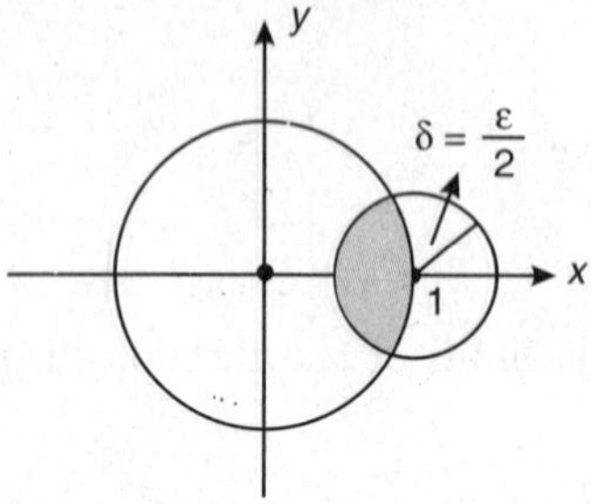

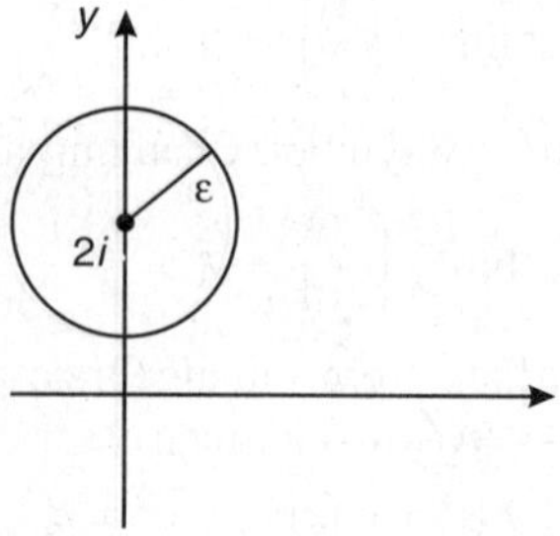

That is, choosing δ as ε/2, we get that, for any z of S which is in $|z-1| < \delta$,

$$|f(z) - 2i| < \varepsilon.$$

Thus, by definition of limit, we get that $\lim_{z \to 1} f(z) = 2i$.

Note. $z = 1$ is a limit point of S. Actually it is a boundary point of S and it is not contained in S.

(ii) Finite limit at ∞

Example. Show that, if $f(z) = \dfrac{1}{z^2}$, then $\lim_{z \to \infty} f(z) = 0$.

Consider any small positive ε and a z such that

$$\left|\frac{1}{z^2} - 0\right| < \varepsilon \text{ or } \frac{1}{|z^2|} < \varepsilon \text{ or } |z| > \frac{1}{\sqrt{\varepsilon}}.$$

Conversely, for any z in the neighbourhood of ∞, $|z| > \dfrac{1}{\sqrt{\varepsilon}}$,

we have
$$\left|\frac{1}{z^2} - 0\right| < \varepsilon.$$

That is, choosing M as the large number $\dfrac{1}{\sqrt{\varepsilon}}$, we have, for any z in the neighbourhood $|z| > M$ of ∞,

$$\left|\frac{1}{z^2} - 0\right| < \varepsilon.$$

Thus, by definition of limit, we get that $\lim_{z \to \infty} \dfrac{1}{z^2} = 0$.

(iii) Infinite limit at a finite point

Example. Show that, if $f(z) = \dfrac{1}{z^2}$, then

$$\lim_{z \to 0} f(z) = \infty.$$

Consider any large positive number M and a z such that

$$\left|\frac{1}{z^2}\right| > M \text{ or } \frac{1}{|z^2|} > M \text{ or } |z| < \frac{1}{\sqrt{M}}.$$

Conversely, for any z in the neighbourhood of 0, $|z - 0| < \frac{1}{\sqrt{M}}$,

We have $$\left|\frac{1}{z^2}\right| > M.$$

That is, choosing δ as $\frac{1}{\sqrt{M}}$, we have, for any z in the neighbourhood $|z - 0| < \delta$ of zero

$$\left|\frac{1}{z^2}\right| > M.$$

Thus, by definition of limit, we get that $\lim_{z \to 0} \frac{1}{z^2} = \infty$.

(iv) Infinite limit at ∞

Example. Show that, if $f(z) = z^2$, then $\lim_{z \to \infty} f(z) = \infty$.

Consider any large positive M and a z such that $|z|^2 > M$ or $|z| > \sqrt{M}$.

Conversely, for any z in the neighbourhood of ∞, $|z| > \sqrt{M}$,

we have $$|z^2| > M.$$

Thus, by definition of limit, we get that $\lim_{z \to \infty} z^2 = \infty$.

Illustrations

$$\lim_{z \to \infty} \frac{z^2}{(z+1)^2} = \lim_{z \to 0} \frac{1/z^2}{(1/z + 1)^2} \quad \left\{\begin{array}{l}\text{Replacing} \\ z \text{ by } 1/z\end{array}\right.$$

$$= \lim_{z \to 0} \frac{1}{z^2}\left(\frac{z}{1+z}\right)^2$$

$$= \lim_{z \to 0} \frac{1}{(1+z)^2} = \frac{1}{(1+0)^2} = 1.$$

$$\lim_{z \to \infty} \frac{z^2+1}{z-1} = \lim_{z \to \infty}\left(z + \frac{z+1}{z-1}\right) \quad \left\{\begin{array}{l}\text{dividing } z^2+1 \\ \text{by } z-1\end{array}\right.$$

$$= \lim_{z \to \infty} z + \lim_{z \to \infty} \frac{z+1}{z-1}$$

$$= \lim_{z\to\infty} z + \lim_{z\to 0} \frac{1/z + 1}{1/z - 1}$$

$$= \lim_{z\to\infty} z + \lim_{z\to 0} \frac{1+z}{1-z}$$

$$= \infty + \frac{1+0}{1-0} = \infty.$$

4.3. CONTINUITY OF A FUNCTION

Suppose f is a function defined in S and z_0 is a limit point of S contained in S. If the limit of $f(z)$ at z_0 exists and if it is finite and is equal to $f(z_0)$, i.e., if

$$\lim_{z\to z_0} f(z) = f(z_0) \neq \infty.$$

then $f(z)$ is said to be continuous at z_0.

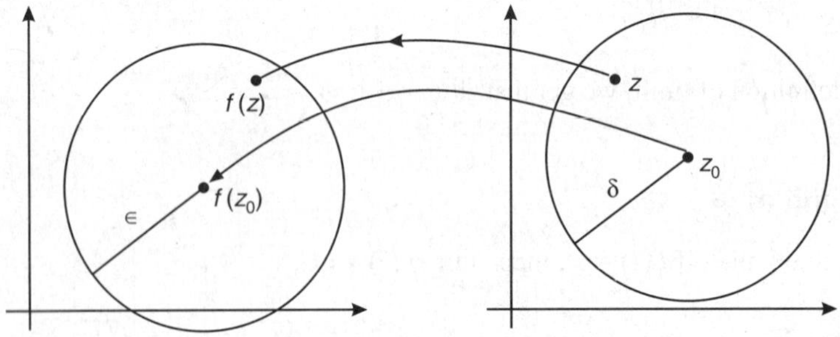

Fig. 4.3.1

Alternative definition. Suppose $f(z)$ is a function defined in S and z_0 is a limit point of S contained in S (Fig. 4.3.1). Then $f(z)$ is said to be continuous at z_0, if for any $\in > 0$ there exists a $\delta > 0$ such that

$$|f(z) - f(z_0)| < \in \qquad \text{for all } z \text{ in } S \text{ with } |z - z_0| < \delta.$$

In this, δ depends on $\in$ and z_0.

Remark. When $f(z_0) = \infty$, the function $f(z)$ is not continuous at $z = z_0$.

$f(z)$ is said to be continuous at ∞, if, for every $\in > 0$, there exists an R such that

$$|f(z) - f(\infty)| < \in \qquad \text{for all } z \text{ in } S \text{ with } |z| > R.$$

In this, R depends on $\in$.

Continuity in a set. If a function is continuous at all points of a set S, then the function is said to be continuous in S. In particular, S may be a region or a closed region or a curve.

Continuous real functions. Polynomial functions in x and y, e^x, $\sin x$, $\cos y$, etc., are some simple continuous real functions.

Example for discontinuity. The function

$$f(z) = \frac{\textbf{Re } z}{|z|}, \qquad \text{when } z \neq 0$$

$$= 0, \qquad \text{when } z = 0,$$

is not continuous at $z = 0$ because the limit of $f(z)$ as $z \to 0$ does not exist (vide the illustration in the previous section).

The following theorem is an immediate consequence of the theorem 4.1 and the definition of continuity.

Theorem 4.5. Suppose $f(z) = u(x, y) + iv(x, y)$ is a function defined in a set S and $z_0 = x_0 + iy_0$ is a limit point of S contained in S. Then $f(z)$ is continuous at z_0 if and only if

$$\lim_{\substack{x \to x_0 \\ y \to y_0}} u(x, y) = u(x_0, y_0) \neq \infty, \quad \lim_{\substack{x \to x_0 \\ y \to y_0}} v(x, y) = v(x_0, y_0) \neq \infty.$$

Remark 1. From the algebraic operations on limits of real functions and the definition of continuity of a complex function, we get that, if $f(z)$ and $g(z)$ are continuous at z_0, then $f(z) \pm g(z)$, $f(z)\, g(z)$ and $f(z) / g(z)$ are continuous at z_0 wherein the last result requires that $g(z_0) \neq 0$.

Remark 2. A function $f(z)$ which is continuous in a *closed* region D, is bounded in D because $u(x, y)$ and $v(x, y)$ are continuous in D and consequently they are bounded in D.

A function which is continuous in a region which is not closed may or may not be bounded in it. For example, $1/z$ is continuous in $0 < |z| \leq 1$ but not bounded in it.

Theorem 4.6. Continuity of a function of a function. Suppose $g(z)$ is a function defined in a set S_0 and is continuous at z_0 in S_0. Let S_1 be the range of S_0 under the mapping g. If $f(z)$ is a function defined in the set S_1 and is continuous at $g(z_0)$, then the function $(fg)(z)$ is continuous at z_0 (Fig. 4.3.2.).

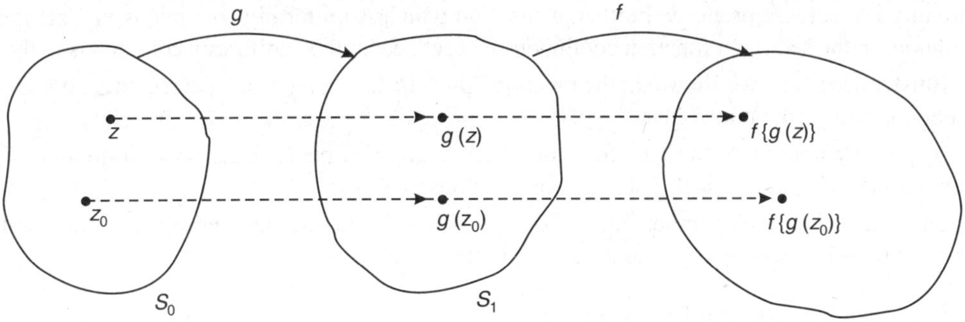

Fig. 4.3.2

Proof: Since $f(z)$ is continuous at $g(z_0)$ in S_1, given an $\in > 0$, we can find a $\delta > 0$, depending on $\in$, such that

$$|f\{g(z)\} - f\{g(z_0)\}| < \in, \qquad \text{for all } g(z) \text{ in } S_1 \text{ with } |g(z) - g(z_0)| < \delta,$$

Since $g(z)$ is continuous at z_0 in S_0, for $\delta > 0$, we can find a δ' depending on δ which in its turn depends on $\in$ such that

$$|g(z) - g(z_0)| < \delta \qquad \text{for all } z \text{ in } S_0 \text{ with } |z - z_0| < \delta'.$$

From the above two inequalities it follows that

$$|(fg)(z) - (fg)(z_0)| < \in \qquad \text{for all } z \text{ in } S_0 \text{ with } |z - z_0| < \delta'.$$

This implies that the composite function $(fg)(z)$ is continuous at z_0 in S_0.

Theorem 4.7. If $f(z)$ is continuous at $z = z_0$ and if for any $M > 0$ there exists a d such that $|f(z)| > M$ for all z in the disc $|z - z_0| < d$, then $1 / f(z)$ is continuous at $z = z_0$.

Proof: Since $f(z)$ is continuous at $z = z_0$, for every $\in > 0$ there exists a $\delta > 0$, such that

$$|f(z) - f(z_0)| < \in \qquad \text{for all } z \text{ in } |z - z_0| < \delta.$$

But it is given that for any $M > 0$ there exists a d such that

$$|f(z)| > M \qquad \text{for all } z \text{ in } |z - z_0| < d.$$

$$\therefore \quad \left| \frac{1}{f(z)} - \frac{1}{f(z_0)} \right| = \frac{|f(z_0) - f(z)|}{|f(z)||f(z_0))|}$$

$$< \frac{\in}{M|f(z_0)|} \quad \text{for all } z \text{ in } |z - z_0| < \delta',$$

where $\delta' = \min(\delta, d)$. Thus for any $\in' = \in / \{M|f(z_0)|\}$, there exists a $\delta' > 0$ such that

$$\left|\frac{1}{f(z)} - \frac{1}{f(z_0)}\right| < \frac{\in}{M|f(z_0)|} \qquad \text{for all } z \text{ in } |z - z_0| < \delta'.$$

$\therefore$ $1/f(z)$ is continuous at $z = z_0$.

4.4. UNIFORM CONTINUITY

A function $f(z)$ defined in a set S is said to be *uniformly continuous* in S if for any $\in > 0$ there exists a $\delta > 0$, depending on $\in$ alone, such that

$$|f(z_1) - f(z_2)| < \in \qquad \text{for all pairs of } z_1, z_2 \text{ of } S \text{ with } |z_1 - z_2| < \delta.$$

In other words, $f(z)$ is said to be uniformly continuous, if for any $\in > 0$ there exists a $\delta > 0$ such that for every pair of z_1, z_2 in S with the distance between them less than δ,

$$|f(z_1) - f(z_2)| < \in.$$

Continuity and uniform continuity. Continuity of a function is a pointwise property while uniform continuity is a setwise property. Further, a function which is uniformly continuous in a set is also continuous in the set. But a function continuous in a set need not be uniformly continuous in the set.

Illustration: Now we show that the function $f(z) = 1/z$ is continuous in the region $S: 0 < |z| < 1$ but not uniformly continuous in it.

If z_0 is any point in S, then the functions 1 and z are continuous at z_0. So their quotient $1/z$ is continuous at z_0 because $z_0 \neq 0$. Thus $1/z$ is continuous in S.

Suppose λ is a real number greater than 2. Then the points denoted by the real numbers $\alpha_1 = 1/\lambda$, $\alpha_2 = 2/\lambda$ belong to S. The distance between them is

$$|\alpha_1 - \alpha_2| = \left|\frac{1}{\lambda} - \frac{2}{\lambda}\right| = \frac{1}{\lambda}.$$

But $$|f(\alpha_1) - f(\alpha_2)| = \left|\frac{1}{\alpha_1} - \frac{1}{\alpha_2}\right| = \left|\lambda - \frac{\lambda}{2}\right| = \frac{\lambda}{2}.$$

Therefore, when the distance between α_1 and α_2 decreases infinitesimally, λ increases infinitely and consequently $|f(\alpha_1) - f(\alpha_2)|$ increases infinitely. So for any $\in > 0$ we cannot find a $\delta > 0$ such that

$$|f(z_1) - f(z_2)| < \in \qquad \text{for every pair } z_1, z_2 \text{ of } S \text{ with } |z_1 - z_2| < \delta$$

because $|f(z_1) - f(z_2)|$ increases infinitely when the distance between $z_1 = \alpha_1$ and $z_2 = \alpha_2$ decreases. Therefore $f(z) = 1/z$ is not uniformly continuous in S.

Theorem 4.8. In a compact set every continuous function is uniformly continuous (i.e., In a bounded closed region a continuous function is uniformly continuous).

Proof: Let S be the given set and a be any point in S. Then, since $f(z)$ is continuous in S, for any $\in > 0$, we can find a δ such that

$$|f(z) - f(a)| < \tfrac{1}{2} \in \qquad \text{for all } z \text{ in } S \text{ with } |z - a| < \delta \} \qquad (4.4.1)$$

i.e., for all z in the neighbourhood $N(a, \delta)$. To cover S, we shall consider all the half-neighbourhoods $N(a_k, \frac{1}{2}\delta_k)$, $a_k \in S$. Since S is compact, by Heine-Borel theorem, a finite number of these half-neighbourhoods can cover S. Let these half-neighbourhoods be

$$N(a_1, \tfrac{1}{2}\delta_1), N(a_2, \tfrac{1}{2}\delta_2), \ldots, N(a_n, \tfrac{1}{2}\delta_n).$$

Suppose d is the smallest of $\frac{1}{2}\delta_1, \frac{1}{2}\delta_2, \ldots, \frac{1}{2}\delta_n$. Consider any two points z_1, z_2 with

$$|z_1 - z_2| < d.$$

Let z_1 lie in the half-neighbourhood $N(a_k, \frac{1}{2}\delta_k)$ (Fig. 4.4.1). Then both z_1 and z_2 lie in the full-neighbourhood $N(a_k, \delta_k)$.The latter case is due to the fact that

$$|z_2 - a_k| = |(z_2 - z_1) + (z_1 - a_k)| \le |z_2 - z_1| + |z_1 - a_k|$$
$$< d + \tfrac{1}{2}\delta_k \le \tfrac{1}{2}\delta_k + \tfrac{1}{2}\delta_k = \delta_k.$$

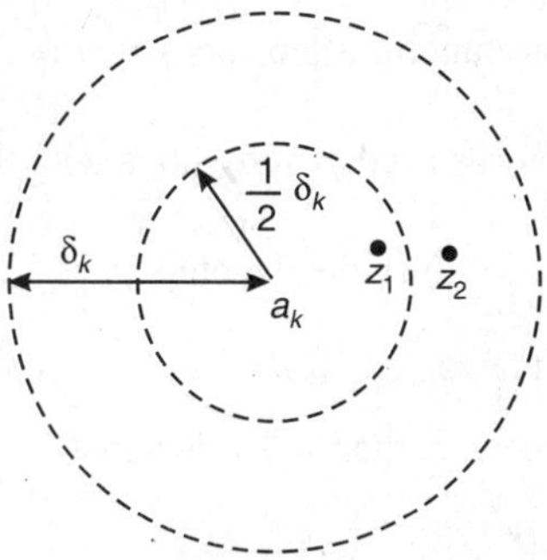

Fig. 4.4.1

Since z_1 and z_2 lie in $|z - a_k| < \delta$, by (4.4.1), i.e., due to continuity of $f(z)$ at a_k,

$$|f(z_1) - f(a_k)| < \tfrac{1}{2}\in, \; |f(z_2) - f(a_k)| < \tfrac{1}{2}\in.$$

$$\therefore \quad |f(z_1) - f(z_2)| = |f(z_1) - f(a_k) + f(a_k) - f(z_2)|$$
$$\le |f(z_1) - f(a_k)| + |f(a_k) - f(z_2)|$$
$$< \tfrac{1}{2}\in + \tfrac{1}{2}\in = \in.$$

Thus, given any $\in > 0$, it is possible to find a $d > 0$ such that

$$|f(z_1) - f(z_2)| < \in \qquad \text{for all } z_1, z_2 \text{ in } S \text{ with } |z_1 - z_2| < d.$$

So, by definition, $f(z)$ is uniformly continuous in S.

4.5. DIFFERENTIABILITY AND ANALYTICITY OF A FUNCTION

Differentiability is a property of a function stronger than continuity. That is, a function which is differentiable, is continuous while a function which is continuous, need not be differentiable.

Differentiability at a point. Suppose $f(z)$ is a single-valued function defined in a region and z_0 is a point in that region. Then the function is said to be differentiable at z_0

(*i*) if $f(z_0) \neq \infty$,

(*ii*) if $\lim\limits_{z \to z_0} \dfrac{f(z) - f(z_0)}{z - z_0}$ i.e., $\lim\limits_{a \to 0} \dfrac{f(z_0 + a) - f(z_0)}{a}$ exists and

(*iii*) if this limit is not equal to ∞.

The above limit is called the derivative of $f(z)$ at z_0 and the operation of obtaining the derivative is called differentiation.

Differentiability in a region. If a function is differentiable at all points in a region, then the function is said to be differentiable in that region.

Analyticity at a point. If a function is differentiable at all points in some neighbourhood of a point, then the function is said to be analytic at that point.

The function $|z|^2$ is differentiable at $z = 0$ but it is not differentiable at all other points in any neighbourhood of $z = 0$. So it is not analytic at $z = 0$ (see sum 27).

Analytic in a region. A function which is analytic at all points of a region is said to be analytic in that region.

Entire Function. A function which is analytic everyhwere in the finite plane is called an entire function. An entire function is analytic everywhere except at $z = \infty$.

Examples of entire functions

(*i*) The polynomial in z $a_0 z^n + a_1 z^{n-1} + a_2 z^{n-2} + ... + a_n$.

(*ii*) z^2.

(*iii*) The exponential function e^z.

(*iv*) The circular and hyperbolic functions $\sin z$, $\cos z$, $\sinh z$, $\cosh z$ (since defined using e^z).

(*v*) $(1 + z) \sin z$

Synonyms for analyticity. The words used synonymous with the word analytic are holomorphic and regular.

Differentiation. The derivative of $f(z)$ at z is denoted by

$$\frac{d}{dz} f(z) \quad \text{or} \quad f'(z).$$

If $f(z)$ and $g(z)$ are differentiable at z, then it follows that

(*i*) $\dfrac{d}{dz}[f(z) + g(z)] = f'(z) + g'(z)$

(*ii*) $\dfrac{d}{dz}[cf(z)] = cf'(z)$, ($c$ is a constant)

(*iii*) $\dfrac{d}{dz}[f(z)\, g(z)] = f'(z)\, g(z) + f(z)\, g'(z)$

(*iv*) $\dfrac{d}{dz} \dfrac{f(z)}{g(z)} = \dfrac{g(z)\, f'(z) - f(z)\, g'(z)}{[g(z)]^2}$ if $g(z) \neq 0$

(*v*) $\dfrac{d}{dz} f\{g(z)\} = f'\{g(z)\}\, g'(z)$, if $g(z)$ is analytic in a region D_1 and $f(z)$ is analytic in D_2 which is the range of D_1 under the mapping $g(z)$.

In particular, if n is a positive integer,

$$\frac{dz^n}{dz} = nz^{n-1}.$$

Differentiability of an inverse function.

Theorem 4.9. Suppose $f(z)$ is a function differentiable in a region D and the mapping

$$w = f(z)$$

is one-to-one and the inverse mapping is $z = \phi(w)$ (Fig. 4.5.1).

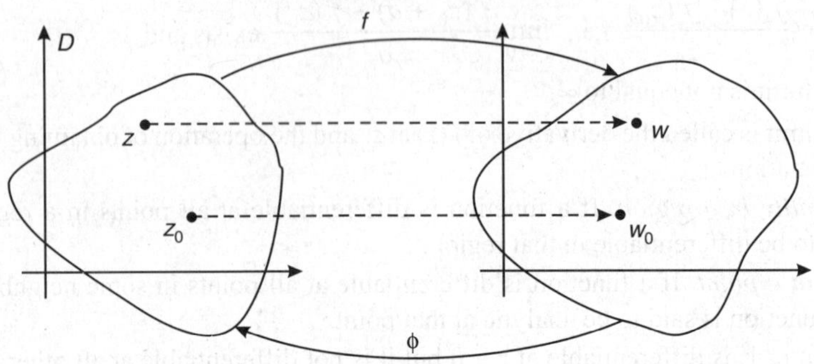

Fig. 4.5.1

If z_0 is a point in D such that $f'(z_0) \neq 0$, then

(*i*) the inverse function $\phi(w)$ is differentiable at w_0, where $w_0 = f(z_0)$ and

(*ii*) $\phi'(w_0) = \dfrac{1}{f'(z_0)}$.

Proof: To show that $\phi(w)$ is differentiable at w_0, we have to show that the limit

$$\lim_{w \to w_0} \frac{\phi(w) - \phi(w_0)}{w - w_0}$$

exists. Now $f(z)$ is differentiable at z_0. So $f(z)$ is continuous at z_0. Therefore, as $z \to z_0$, $f(z) \to f(z_0)$, that is, $w \to w_0$. This implies that, as $w \to w_0$, $z \to z_0$. Hence

$$\lim_{w \to w_0} \frac{\phi(w) - \phi(w_0)}{w - w_0} = \lim_{z \to z_0} \frac{z - z_0}{f(z) - f(z_0)}$$

$$= \lim_{z \to z_0} \frac{1}{\dfrac{f(z) - f(z_0)}{z - z_0}} = \frac{1}{\lim\limits_{z \to z_0} \dfrac{f(z) - f(z_0)}{z - z_0}}$$

$= \dfrac{1}{f'(z_0)}$ since $f(z)$ is differentiable at z_0 and $f'(z_0) \neq 0$. So the inverse function $\phi(w)$ is differentiable at w_0.

Remark 1. If $f(z)$ is analytic in D throughout which $f'(z)$ is not zero and if D' is the range of D under the mapping $w = f(z)$, then $\phi(w)$ is analytic in D' and $\phi'(w)$ is not zero anywhere in D'.

Remark 2. The behaviour of a function $f(z)$ at ∞ is same as the behaviour of the function $f(1/z)$ at 0. In particular, if $f(1/z)$ is analytic at 0, then $f(z)$ is analytic at ∞.

Example 1. The function $1/(1 + z)$ is analytic at ∞ because the function $1/(1 + 1/z)$ or $z/(z + 1)$ is analytic at zero.

Example 2. The function $z + 1/z$ is analytic at every point except 0 and ∞ because at 0 and ∞ the values of the function are ∞.

Singular point or singularity. A singular point or singularity of a function is a point at which the function ceases to be analytic or the function is not defined.

Entire function. An entire function is one which is analytic in the entire finite complex plane, that is, analytic at every point of the extended complex plane except at ∞.

Example. The derivative of a polynomial in z exists at every point in the finite plane. So a polynomial is analytic in the finite plane and hence it is an entire function.

We will see latter that e^z, $\sin z$, $\cos z$, ze^z and $(1 + z) \sin z$ are entire functions.

4.6. NECESSARY CONDITIONS FOR DIFFERENTIABILITY

The following theorem provides a necessary condition for differentiability.

Theorem 4.10. A necessary condition for a function to be differentiable at a point is the continuity of the function at the point.

Proof: Suppose $f(z)$ is a function differentiable at z_0. Then

$$\lim_{z \to z_0} \{f(z) - f(z_0)\} = \lim_{z \to z_0} \left\{\frac{f(z) - f(z_0)}{z - z_0}(z - z_0)\right\}$$

i.e.,
$$\left\{\lim_{z \to z_0} f(z)\right\} - f(z_0) = \left(\lim_{z \to z_0} \frac{f(z) - f(z_0)}{z - z_0}\right) \times \left(\lim_{z - z_0} (z - z_0)\right)$$

$$= f'(z_0)(0) = 0.$$

$$\therefore \quad \lim_{z \to z_0} f(z) = f(z_0).$$

Therefore limit exists and is not equal to ∞.

So, by definition, $f(z)$ is continuous at z_0.

Remark 1. Since continuity is a necessary condition for differentiability, if a function is not continuous at a point, then the function is not differentiable at that point.

Remark 2. *Insufficiency of continuity for differentiability.* Continuity is not sufficient for differentiability. The following illustration exhibits this fact.

Illustration. Consider the function $f(z) = \bar{z} = x - iy$. It is continuous at $z = 0$ because, for a given $\in > 0$.

$$|\bar{z} - 0| < \in \qquad \text{for all } z \text{ with } |z - 0| < \in$$

which establishes continuity of $\bar{z}$ at 0. But

$$\lim_{z \to 0} \frac{f(z) - f(0)}{z - 0} = \lim_{z \to 0} \frac{\bar{z} - \bar{0}}{z - 0} = \lim_{z \to 0} \frac{\bar{z}}{z} = \lim_{y \to 0} \frac{x - iy}{x + iy}$$

$$= \lim_{x \to 0} \frac{x - imx}{x + imx} \begin{cases} \text{when } z \text{ approaches } 0 \\ \text{along the line } y = mx \end{cases}$$

$$= \frac{1 - im}{1 + im}$$

which varies with m. So it is not unique. Hence the above limit does not exist. Therefore $\bar{z}$ is not differentiable at zero.

Stronger necessary conditions for differentiability. If $f(z)$ is continuous at $z_0 = x_0 + iy_0$, then $u(x, y)$ and $v(x, y)$ are continuous at (x_0, y_0) which implies from real analysis that the first partials

$$u_x, u_y, v_x, v_y$$

exist at (x_0, y_0). If $f(z)$ is differentiable at z_0, then, in addition to the existence of u_x, u_y, v_x, v_y, there exist two relationships among them, called *Cauchy-Riemann equations*, which can be seen from the following theorem.

Theorem 4.11. Suppose $f(z) = u(x, y) + iv(x, y)$ is a single-valued function defined in a neighbourhood of $z_0 = x_0 + iy_0$. Then the necessary condition for the differentiability of $f(z)$ at z_0 is (*i*) the existence of the partial derivatives u_x, u_y, v_x, v_y at (x_0, y_0), which satisfy the relations

$$u_x = v_y, \; u_y = -v_x.$$

(*ii*) u_x, u_y, v_x, v_y are continuous. Then $f(z)$ is differentiable at z.

Proof: Now $f(z)$ is differentiable at z_0. Hence the limit

$$\lim_{z \to z_0} \frac{f(z) - f(z_0)}{z - z_0} = f'(z_0)$$

exists and is unique. In particular, let z approach z_0 travelling along a line parallel to the real axis (Fig. 4.6.1). Then $z = z_0 + h$, where h is real and tends to zero. Therefore

$$f'(z_0) = \lim_{h \to 0} \frac{f(z_0 + h) - f(z_0)}{(z_0 + h) - z_0}$$

$$= \lim_{h \to 0} \frac{f(x_0 + h + iy_0) - f(x_0 + iy_0)}{h}$$

$$= \lim_{h \to 0} \frac{u(x_0 + h, y_0) + iv(x_0 + h, y_0) - u(x_0, y_0) - iv(x_0, y_0)}{h}$$

$$= \left\{ \lim_{h \to 0} \frac{u(x_0 + h, y_0) - u(x_0, y_0)}{h} \right\} + i \left\{ \lim_{h \to 0} \frac{v(x_0 + h, y_0) - v(x_0, y_0)}{h} \right\}$$

$$= u_x(x_0, y_0) + iv_x(x_0, y_0) = u_x + iv_x \text{ at } (x_0, y_0). \quad (4.6.1)$$

Thus the existence of the limit (4.6.1) implies the existence of u_x and v_x.

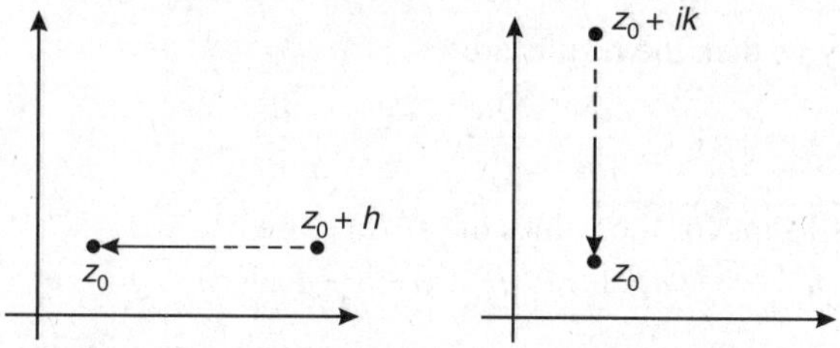

Fig. 4.6.1

If z approaches z_0 travelling along a line parallel to the y axis, then $z = z_0 + ik$, where k is real and tends to zero. Hence

$$f'(z_0) = \lim_{k \to 0} \frac{f(z_0 + ik) - f(z_0)}{(z_0 + ik) - z_0}$$

$$= \lim_{k \to 0} \frac{f\{x_0 + i(y_0 + k)\} - f(x_0 + iy_0)}{ik}$$

$$= \lim_{k \to 0} \frac{u(x_0, y_0 + k) + iv(x_0, y_0 + k) - u(x_0, y_0) - iv(x_0, y_0)}{ik}$$

$$= \frac{1}{i} \left\{ \lim_{k \to 0} \frac{u(x_0, y_0 + k) - u(x_0, y_0)}{k} \right\} + \lim_{k \to 0} \frac{i[v(x_0, y_0 + k) - v(x_0, y_0)]}{k} \Bigg\}$$

$$= \frac{1}{i}(u_y + iv_y) = -iu_y + v_y \text{ at } (x_0, y_0). \quad (4.6.2)$$

since $\frac{1}{i} = \frac{i}{i^2} = -i.$. Now the existence of the limit $\lim_{z \to z_0} \frac{f(z) - f(z_0)}{z - z_0}$ implies the existence of u_y and v_y.

Since the limit is unique, (4.6.1) and (4.6.2) should be equal.

Equating real parts and imaginary parts in (4.6.1) and (4.6.2), at (x_0, y_0),

$$u_x = v_y,\ u_y = -v_x. \quad (4.6.3)$$

The results $u_x = v_y$, $u_y = -v_x$ are called Cauchy-Riemann equations.

Remark 1. Since the C-R equations are necessary conditions for differentiability, if they are not satisfied at a point, then the function is not differentiable at that point.

Remark 2. Derivative of *f* (*z*) in different forms. From (4.6.1) in the above theorem, we get the derivative of $f(z)$ in partial derivatives w.r.t x as

$$f'(z) = u_x + iv_x.$$

From (4.6.2) we get the derivative of $f(z)$ in partial derivatives w.r.t y as

$$f'(z) = \frac{u_y + iv_y}{i}.$$

Further, rewriting these two results, we have

$$(i)\ f'(z) = u_x + iv_x = \frac{\partial}{\partial x}(u + iv) = \frac{\partial}{\partial x} f(x + iy) = \frac{\partial f}{\partial x}.$$

$$(ii)\ f'(z) = \frac{u_y + iv_y}{i} = \frac{1}{i}\frac{\partial}{\partial y}(u + iv) = \frac{1}{i}\frac{\partial}{\partial y} f(x + iy) = \frac{1}{i}\frac{\partial f}{\partial y}.$$

If $f(z)$ is denoted by w, then the results are

$$\frac{dw}{dz} = \frac{\partial w}{\partial x};\quad \frac{dw}{dz} = \frac{1}{i}\frac{\partial w}{\partial y}.$$

Remark 3. Various forms for the values of $|f'(z)|^2$ are

$$u_x^{\,2} + v_x^{\,2},\ u_y^{\,2} + v_y^{\,2},\ u_x^{\,2} + u_y^{\,2},\ v_x^{\,2} + v_y^{\,2},\ u_x v_y - u_y v_x. \tag{4.6.6}$$

4.7. SUFFICIENT CONDITIONS FOR DIFFERENTIABILITY

In this section we obtain a set of sufficient conditions for differentiability of a function.

Theorem 4.12. Suppose that $f(z) = u(x, y) + iv(x, y)$ is a single-valued function defined in a neighbourhood of $z_0 = x_0 + iy_0$. If the partial derivatives u_x, u_y, v_x, v_y exist and are continuous at (x_0, y_0) and if

$$u_x = v_y,\ u_y = -v_x$$

at (x_0, y_0), then $f(z)$ is differentiable at z_0.

Hypothesis : At (x_0, y_0), u_x, u_y, v_x, v_y exist and are continuous and satisfy the C-R equations.

To prove : $f(z)$ is differentiable at $z_0 = x_0 + iy_0$.

Proof : Let $z_0 + a$ be a point in the neighbourhood of z_0. Suppose $a = h + ik$, where h and k are real. Then

$$\begin{aligned} f(z_0 + a) - f(z_0) &= f\{x_0 + h + i(y_0 + k)\} - f(x_0 + iy_0) \\ &= u(x_0 + h, y_0 + k) + iv(x_0 + h, y_0 + k) - u(x_0, y_0) - iv(x_0, y_0) \\ &= u(x_0 + h, y_0 + k) - u(x_0, y_0) + i\{v(x_0 + h, y_0 + k) - v(x_0, y_0)\} \end{aligned} \tag{4.7.1}$$

Denoting $u(x_0 + h, y_0 + k) - u(x_0, y_0)$ by $\lambda + \mu$, where

$$\lambda = u(x_0 + h, y_0 + k) - u(x_0, y_0 + k)$$

$$\mu = u(x_0, y_0 + k) - u(x_0, y_0),$$

$$\lambda = hu_x(x_0 + \theta_1 h, y_0 + k),\ 0 < \theta_1 < 1,$$

by mean value theorem and since u_x exists

$$= h\{u_x(x_0, y_0) + \lambda_1\},\ \text{where } \lambda_1 \to 0 \text{ as } h \to 0, k \to 0$$

since u_x is continuous

$$\mu = k\{u_y(x_0, y_0) + \lambda_2],\ \text{where } \lambda_2 \to 0 \text{ as } k \to 0.$$

$$\therefore\quad u(x_0 + h, y_0 + k) - u(x_0, y_0) = hu_x + ku_y + h\lambda_1 = k\lambda_2.$$

Similarly,

$$i\{v(x_0 + h, y_0 + k) - v(x_0, y_0)\} = i\{hv_x + kv_y + h\lambda_3 + k\lambda_4\}.$$

$$\begin{aligned} \therefore\quad f(z_0 + a) - f(z_0) &= hu_x + ku_y + h\lambda_1 + k\lambda_2 + i(hv_x + kv_y + h\lambda_3 + k\lambda_4) \\ &= h(u_x + iv_x) + k(u_y + iv_y) + h\lambda_1 + k\lambda_2 + ih\lambda_3 + ik\lambda_4 \\ &= h(u_x + iv_x) + k(-v_x + iu_x) + h\lambda_1 + k\lambda_2 + ih\lambda_3 + ik\lambda_4 \end{aligned}$$

by C-R equations

$$\begin{aligned} &= h(u_x + iv_x) + ik(iv_x + u_x) + h\lambda_1 + k\lambda_2 + ih\lambda_3 + ik\lambda_4 \\ &= (h + ik)(u_x + iv_x) + h\lambda_1 + k\lambda_2 + ih\lambda_3 + ik\lambda_4 \end{aligned}$$

$$\therefore \qquad \frac{f(z_0+a)-f(z_0)}{a} = u_x + iv_x + \frac{h\lambda_1 + k\lambda_2 + ih\lambda_3 + ik\lambda_4}{h+ik}$$

Taking limit as $a = h + ik \to 0$, all of

$$\frac{h\lambda_1}{h+ik}, \frac{k\lambda_2}{h+ik}, \frac{ih\lambda_3}{h+ik}, \frac{ik\lambda_4}{h+ik}$$

tend to **zero** because, as $h \to 0$, $k \to 0$, all of $\lambda_1, \lambda_2, \lambda_3, \lambda_4$ tend to zero since

$$\left|\frac{h\lambda_1}{h+ik}\right| = \frac{|h|}{\sqrt{h^2+k^2}} |\lambda_1| \le |\lambda_1|, \text{ etc.}$$

Thus the limit of $\{f(z_0+a)-f(z_0)\}/a$, as a tends to zero is $u_x + iv_x$. So the limit exists and hence $f(z)$ is differentiable at z_0.

Remark 1. Existence of u_x, u_y, v_x, v_y with $u_x = v_y$, $u_y = -v_x$ are necessary conditions. But the sufficient conditions are that, besides these, u_x, u_y, v_x, v_y are continuous.

Remark 2. We will prove later that derivatives of analytic functions are also analytic.

4.8. C-R EQUATIONS IN POLAR COORDINATES

Theorem 4.13. The C-R equations in polar coordinates are

$$\frac{\partial u}{\partial r} = \frac{1}{r}\frac{\partial v}{\partial \theta}, \quad \frac{\partial u}{\partial \theta} = -r\frac{\partial v}{\partial r} \qquad (4.8.1)$$

Proof : Treating u, v as functions of x, y and x, y as functions of r, θ, we get $x = r\cos\theta$, $y = r\sin\theta$ and

$$\frac{\partial u}{\partial r} = \frac{\partial u}{\partial x}\frac{\partial x}{\partial r} + \frac{\partial u}{\partial y}\frac{\partial y}{\partial r} = u_x \cos\theta + u_y \sin\theta,$$

$$\frac{1}{r}\frac{\partial v}{\partial \theta} = \frac{1}{r}\left\{\frac{\partial v}{\partial x}\frac{\partial x}{\partial \theta} + \frac{\partial v}{\partial y}\frac{\partial y}{\partial \theta}\right\} = \frac{1}{r}\{v_x(-r\sin\theta) + v_y(r\cos\theta)\}$$

$$= u_y \sin\theta + u_x \cos\theta$$

from which the first of (4.8.1) follows. Further

$$\frac{\partial u}{\partial \theta} = \frac{\partial u}{\partial x}\frac{\partial x}{\partial \theta} + \frac{\partial u}{\partial y}\frac{\partial y}{\partial \theta} = u_x(-r\sin\theta) + u_y(\text{r}\cos\theta)$$

$$-r\frac{\partial v}{\partial r} = -r\left\{\frac{\partial v}{\partial x}\frac{\partial x}{\partial r} + \frac{\partial v}{\partial y}\frac{\partial y}{\partial r}\right\} = -r(v_x\cos\theta + v_y\sin\theta)$$

$$= -r(-u_y\cos\theta + u_x\sin\theta)$$

from which the second of (4.8.1) follows.

Corollary. $f'(z) = \dfrac{r}{z}\left(\dfrac{\partial u}{\partial r} + i\dfrac{\partial v}{\partial r}\right)$, for

$$r\left(\frac{\partial u}{\partial r} + i\frac{\partial v}{\partial r}\right) = r\{(u_x\cos\theta + u_y\sin\theta) + i(v_x\cos\theta + v_y\sin\theta)\}$$

$$= r\cos\theta(u_x + iv_x) + r\sin\theta(u_y + iv_y)$$

$$= x(u_x + iv_x) + y(u_y + iv_y)$$

$$= xf'(z) + iyf'(z) \text{ by (4.6.1), (4.6.2)}$$

$$= (x+iy)f'(z) = zf'(z)$$

$$\therefore \qquad f'(z) = \frac{r}{z}\left(\frac{\partial u}{\partial r} + i\frac{\partial v}{\partial r}\right). \qquad (4.8.2)$$

Derivative of f (z) in polar coordinates. From C-R equations in polar coordinates we get the derivative as

$$f'(z) = \frac{1}{e^{i\theta}}\,(u_r + iv_r),$$

$$f'(z) = \frac{1}{ir}\,(u_\theta + iv_\theta).$$

Using any one of the above two forms, we get $f'(z)$.

4.9. COMPLEX FUNCTION AS A FUNCTION OF *z* and $\bar{z}$.

Suppose $f(z)$ is a complex function. It is a function of two real variables x and y. Let the variables x and y be changed by the substitution

$$x = \tfrac{1}{2}\,(z + \bar{z}),\; y = -\tfrac{1}{2}\,i\,(z - \bar{z})$$

obtained from $z = x + iy$ and $\bar{z} = x - iy$. Then z and $\bar{z}$ can be treated formally as independent variables. Thus f is a function of x and y and x and y are functions of z and $\bar{z}$.

Theorem 4.14. For an analytic function $f(z)$ the C-R equations can be put in the condensed form $\dfrac{\partial f}{\partial \bar{z}} = 0.$

Proof: $$\frac{\partial x}{\partial z} = \frac{1}{2}, \frac{\partial x}{\partial \bar{z}} = \frac{1}{2} \text{ and } \frac{\partial y}{\partial z} = -\frac{i}{2}, \frac{\partial y}{\partial \bar{z}} = \frac{i}{2}. \text{ So}$$

$$\frac{\partial f}{\partial \bar{z}} = \frac{\partial u}{\partial \bar{z}} + i\frac{\partial v}{\partial \bar{z}} = \left(\frac{\partial u}{\partial x}\frac{\partial x}{\partial \bar{z}} + \frac{\partial u}{\partial y}\frac{\partial y}{\partial \bar{z}}\right) + i\left(\frac{\partial v}{\partial x}\frac{\partial x}{\partial \bar{z}} + \frac{\partial v}{\partial y}\frac{\partial y}{\partial \bar{z}}\right)$$

$$= \left(u_x\left(\frac{1}{2}\right) + u_y\left(\frac{1}{2}i\right)\right\} + i\{v_x\left(\frac{1}{2}\right) + v_y\left(\frac{1}{2}i\right)\Bigg)$$

$$= \frac{1}{2}\,\{\,(u_x - v_y) + i\,(u_y + v_x)\}.$$

Thus $\dfrac{\partial f}{\partial \bar{z}} = 0$ implies $u_x - v_y = 0$, $u_y + v_x = 0$ which are the C-R equations.

Remark 1. If the C-R equations are not true, then $\partial f/\partial \bar{z} \neq 0$. Thus, if $\partial f/\partial z \neq 0$, then $f(z)$ is not analytic. For example, if $f(z) = \bar{z}$, then $\partial f/\partial \bar{z} = 1 \neq 0$ and hence $\bar{z}$ is nowhere analytic.

Remark 2. If $f(z)$ is analytic, then $\partial f/\partial \bar{z} = 0$ and so $f(z)$ is independent of $\bar{z}$ and hence $f(z)$ is a function of z alone (i.e., $f(z)$ is purely a function of z, not involving any of **Re** z, **Im** z, $|z|$, arg z, $\bar{z}$). This formal reasoning leads one to think that analytic functions are true functions of the complex variable $z = x + iy$ and not arbitrary functions of the real variables x and y. For example, $x^2 - y^2 + 2ixy$ which is z^2 is analytic but $x^2 + y^2$ which cannot be expressed as a function of the variable z, is not analytic. As $\sin(x - iy) = \sin \bar{z}$ is not a regular function of z, it is not analytic.

Remark 3. Similar to the result that if $f(z)$ is analytic, then $\partial f/\partial \bar{z} = 0$, we have the result that if $f(z)$ is analytic, then $\partial \bar{f}/\partial z = 0$ (i.e., $\dfrac{\partial \overline{f(z)}}{\partial z} = 0$).

Theorem 4.12. If $f(z) = u + iv$ is analytic, then the two families of curves

$$u(x, y) = c_1, \tag{1}$$

$$v(x, y) = c_2, \tag{2}$$

where c_1 and c_2 are parameters, are orthogonal.

Proof: In $u(x, y) = c_1$, u is a function of x and y, where y is a function of x. So, differentiating u w.r.t x,

$$\frac{\partial u}{\partial x} + \frac{\partial u}{\partial y}\frac{dy}{dx} = 0 \quad \text{or} \quad \frac{dy}{dx} = -\frac{u_x}{u_y}.$$

So the slope of a curve of (1) at (x, y) is $-\dfrac{u_x}{u_y}$.

Similarly, the slope of a curve of (2) at (x, y) is $-\dfrac{v_x}{v_y}$.

$$\therefore \quad \text{Product of the slopes} = \frac{u_x}{u_y}\frac{v_x}{v_y}$$

$$= \frac{u_x}{(-v_x)}\frac{v_x}{(u_x)} \quad \text{since} \begin{cases} u_y = -v_x \\ v_y = u_x \end{cases}$$

$$= -1.$$

Thus the curves are orthogonal and consequently the families are orthogonal.

Properties of an analytic function

1. If $f(z) = u + iv$ is analytic, then the two families of curves
$$u(x, y) = c_1, \quad v(x, y) = c_2,$$
where c_1 and c_2 are parameters, are orthogonal. (See Theorem 3.12 given below).
2. If $f(z)$ is analytic, then $f(z)$ is a regular function of z and is independent of $\bar{z}$ so that
$$\frac{\partial f(z)}{\partial \bar{z}} = 0.$$
3. If $f(z) = u + iv$ is analytic, then $u(x, y)$ and $v(x, y)$ are harmonic functions. (Harmonic functions are defined in section 6.12).
4. If $f(z)$ is analytic in a domain R and if $f'(z) \neq 0$ in R, then the mapping
$$w = f(z)$$
is conformal. (Mapping is dealt with in Chapter 7).

Example. Treating $f(z)$ as a function of x, y and x, y as functions of $z, \bar{z}$, we now formally show that

(i) $\left(\dfrac{\partial^2}{\partial x^2} + \dfrac{\partial^2}{\partial y^2}\right)|f(z)|^2 = 4|f'(z)|^2$ **(1977, 80)**

(ii) $\left(\dfrac{\partial^2}{\partial x^2} + \dfrac{\partial^2}{\partial y^2}\right)\log|f'(z)| = 0.$

(i)

$$\frac{\partial(f)}{\partial z} = \frac{\partial(f)}{\partial x}\frac{\partial x}{\partial z} + \frac{\partial(f)}{\partial y}\frac{\partial y}{\partial z} = \frac{\partial(f)}{\partial x}\frac{1}{2} + \frac{\partial(f)}{\partial y}\left(-\frac{i}{2}\right) = \frac{1}{2}\left(\frac{\partial}{\partial x} - i\frac{\partial}{\partial y}\right)(f)$$

$$\frac{\partial(f)}{\partial \bar{z}} = \frac{\partial(f)}{\partial x}\frac{\partial x}{\partial \bar{z}} + \frac{\partial(f)}{\partial y}\frac{\partial y}{\partial \bar{z}} = \frac{\partial(f)}{\partial x}\frac{1}{2} + \frac{\partial(f)}{\partial y}\left(\frac{i}{2}\right) = \frac{1}{2}\left(\frac{\partial}{\partial x} + i\frac{\partial}{\partial y}\right)(f)$$

$$\therefore \qquad \nabla^2 = \frac{\partial^2}{\partial x^2} + \frac{\partial^2}{\partial y^2} = \left(\frac{\partial}{\partial x} - i\frac{\partial}{\partial y}\right)\left(\frac{\partial}{\partial x} + i\frac{\partial}{\partial y}\right) = 4\frac{\partial}{\partial z}\cdot\frac{\partial}{\partial \bar{z}} = 4\frac{\partial^2}{\partial z\,\partial \bar{z}}$$

$$\nabla^2 |f(z)|^2 = \left(\frac{\partial^2}{\partial x^2} + \frac{\partial^2}{\partial y^2}\right)|f(z)|^2$$

$$= \left(\frac{\partial^2}{\partial x^2} + \frac{\partial^2}{\partial y^2}\right)[f(z)\cdot \bar{f}(\bar{z})]$$

$$= 4\frac{\partial}{\partial z}\frac{\partial}{\partial \bar{z}}[f(z)\cdot \overline{f(z)} = 4\frac{\partial}{\partial z}[f(z)]\frac{\partial}{\partial \bar{z}}[\overline{f(z)}]$$

$$= 4f'(z)\overline{\frac{\partial}{\partial z}f(z)} = 4f'(z)\overline{f'(z)} = 4|f'(z)|^2.$$

(*ii*) Logarithmic function log z is defined in chapter 6.

Now $\qquad \nabla^2 \log|f'(z)| = \nabla^2 \dfrac{1}{2}\log|f'(z)|^2 = \dfrac{1}{2}\nabla^2 \log[f'(z)f'(z)]$

$$= \frac{1}{2}\nabla^2[\log f'(z) + \log f'(z)]$$

$$= \frac{1}{2}\,.\,4\frac{\partial}{\partial z}\frac{\partial}{\partial \bar{z}}[\log f'(z) + \log \overline{f'(z)}]$$

$$= 2\frac{\partial}{\partial z}[0 + \frac{\partial}{\partial \bar{z}}\log \overline{f'(z)}]$$

$$= 2\frac{\partial}{\partial z}\overline{\left[\frac{\partial}{\partial z}\log f'(z)\right]} = 0.$$

since derivatives of analytic functions are analytic.

4.10. EXAMPLES

4.2. Limit of a function

1. Using definition, show that $\lim\limits_{z\to\infty} 1/z = 0.$

[*Hint* : $\left|\dfrac{1}{z} - 0\right| = \dfrac{1}{|z|} < \in$ for all z with $|z| > \dfrac{1}{\in}$. So $1/z \to 0$ as $z \to \infty$]

2. Show that (*i*) $\lim\limits_{z\to\infty} \dfrac{z^2}{(z-1)^2} = 1$

(*ii*) $\lim\limits_{z\to 1} \dfrac{1}{(z-1)^3} = \infty$

[*Hint* : (*ii*) Let $z - 1 = t$. Limit $= \lim\limits_{t\to 0}\dfrac{1}{t^3} = \infty$]

4.3 Continuity of a Function

3. Show that the function $1/z$ is continuous in $0 < |z| < 1$. **(1981)**

4.4 Uniform Continuity

4. Show that the function $1/z$ is not uniformly continuous in $0 < |z| < 1$. **(1981)**

5. **Solved example.** Show that the function $f(z) = z^2$ is uniformly continuous in the unit disc $S: |z| < 1$.

Take any two points z_1 and z_2 in S. Then

$$\begin{aligned} |f(z_1) - f(z_2)| &= |z_1^2 - z_2^2| = |(z_1 - z_2)(z_1 + z_2)| \\ &= |z_1 - z_2||z_1 + z_2| \\ &\le |z_1 - z_2|\{|z_1| + |z_2|\} \\ &< |z_1 - z_2|(1 + 1) = 2|z_1 - z_2|. \end{aligned}$$

Therefore, given any $\in > 0$,

$$|f(z_1) - f(z_2)| < \in \text{ if } 2|z_1 - z_2| < \in \text{ i.e., if } |z_1 - z_2| < \tfrac{1}{2}\in.$$

So, given any $\in > 0$, we are able to find a $\delta = \frac{1}{2}\in$ such that $|f(z_1) - f(z_2)| < \in$. For all z_1, z_2 in S with $|z_1 - z_2| < \delta$.

Hence $f(z)$ is uniformly continuous on S.

6. Prove that $f(z) = 3z - 2$ is uniformly continuous in $|z| \le 10$.

[*Hint* : $|(3z_1 - 2) - (3z_2 - 2)| = 3|z_1 - z_2|$. For any $\in$, $\delta = \frac{1}{3}\in$.]

7. Prove that (*i*) $f(z) = 1/z^2$ is not uniformly continuous in the region $z < 1$ (*ii*) and it is uniformly continuous in the closed region $\frac{1}{2} \le z \le 1$.

[*Hint* : (*i*) Vide illustration in Section 4.4. Take $\alpha_1 = 1/\sqrt{\lambda}$, $\alpha_2 = 2\sqrt{\lambda}$. (*ii*) $f(z)$ is continuous in the compact set $\frac{1}{2} \le z \le 1$. So, by theorem 4.4, $f(z)$ is uniformly continuous in $\frac{1}{2} \le z \le 1$.]

4.5 Differentiability; 4.6 Conditions for Differentiability

8. Show that, at the origin, the following functions are not continuous and not differentiable

(*i*) $f(z) = \left.\begin{array}{ll} \dfrac{\mathbf{Re}\, z}{|z|}, & z \ne 0 \\ = 0, & z = 0 \end{array}\right\}$ (*ii*) $f(z) = \begin{array}{ll} \dfrac{z}{|z|}, & z \ne 0 \\ = 0, & z = 0 \end{array}$

[*Hint* : (*i*) When z approaches 0 along $y = mx$,

$$\lim_{z \to 0} f(z) = \lim_{x, y \to 0} \frac{x}{\sqrt{x^2 + y^2}} = \frac{x}{\sqrt{x^2 + m^2x^2}} = \frac{1}{\sqrt{1 + m^2}}$$

and is not unique. So, at the origin, $f(z)$ is not continuous and hence not differentiable.]

9. Show that the function $f(z) = \bar{z}$ is nowhere differentiable. **(1977, 78)**

[*Hint* : $u_x = 1$, $v_y = -1$ at all (x, y). So C-R equations are not satisfied anywhere.]

10. Show that the function **Re** z and $z - \bar{z}$ are nowhere differentiable.

11. Show that the function $f(z) = xy + iy$ is continuous everywhere but is not analytic anywhere. **(1978)**

[*Hint* : Since x and y are continuous, the product xy and y are continuous in the finite plane. So $xy + iy$ is continuous in the finite plane. But nowhere the C-R equations are satisfied.]

12. **Solved sum.** $f(z)$ is a function analytic in a region D. Prove that $f(z)$ is constant in D, if in D, either

(*i*) its real part is constant or

(*ii*) its imaginary part is constant or

(*iii*) $|f(z)|$ is constant or

(*iv*) $\arg f(z)$ is constant. **(1981)**

(*i*) If $u(x, y)$ is constant, $u_x = 0$, $u_y = 0$. So $v_y = 0$, $v_x = 0$. Hence v is constant and thus $u + iv$ is constant.

(*ii*) Similarly, when v is constant, u is constant.

(*iii*) If $|f(z)|$ is constant, $u^2 + v^2$ is constant. Differentiating partially with respect to x and y.

$$u\,u_x + v\,v_x = 0,\; u\,u_y + v\,v_y = 0.$$

Eliminating u and v,

$$u_x v_y - u_y v_x = 0. \qquad (1)$$

i.e., $\quad u_x(u_x) - u_y(-u_y) = 0$ by C-R equations.

$\therefore \quad u_x^{\,2} + u_y^{\,2} = 0.$

This implies that $u_x = 0$, $u_y = 0$ and hence u is constant. Further, by C-R equations, $v_y = 0$, $v_x = 0$. Hence v is constant. Thus $u + iv$ is constant.

(*iv*) $\tan^{-1}(v/u)$ is constant or v/u is constant, say c. Then $cu = v$. So $cu_x = v_x$, and $cu_y = v_y$. Eliminating c, $u_x v_y - u_y v_x = 0$ which is (1). Thus $u + iv$ is constant.

13. If $f(z)$ and $\overline{f(z)}$ are analytic in a region, show that $f(z)$ is constant in that region.

[*Hint* : Since $u + iv$ and $u + i(-v)$ are analytic,

$$\left.\begin{aligned} u_x &= v_y \\ u_y &= -v_x \end{aligned}\right\} \qquad \left.\begin{aligned} u_x &= -v_y \\ u_y &= v_x \end{aligned}\right\}$$

These equations imply $u_x = u_y = v_x = v_y = 0$.]

14. Solved sum. If $f(z) = x^3 y\,(y - ix)/(x^6 + y^2)$, $(z \neq 0)$, $f(0) = 0$, show that $\{f(z) - f(0)\}/z - 0$ as $z \to 0$ along any radius vector $y = mx$ but not as $z \to 0$ in any manner and hence $f(z)$ is not differentiable at $z = 0$.

As $z \to 0$ along $y = mx$,

$$\lim_{z\to 0}\frac{f(z)-f(0)}{z} = \lim_{x\to 0}\frac{x^3\cdot mx\,(mx - ix)}{(x^6 + m^2x^2)(x + i\,mx)}$$

$$= \lim_{x\to 0}\frac{x^5\,m\,(m-i)}{x^3\,(x^4 + m^2)(1 + im)} = \lim_{x\to 0}\frac{x^2 m\,(m-i)}{(x^4 + m^2)(1 + im)} = 0.$$

As $z \to 0$ along $y = mx^3$,

$$\lim_{z\to 0}\frac{f(z)-f(0)}{z} = \lim_{x\to 0}\frac{x^3 mx^3\,(mx^3 - i\,x)}{(x^6 + m^2x^6)(x + imx^3)}$$

$$= \lim_{x\to 0}\frac{m\,(mx^2 - i)}{(1 + m^2)(1 + imx^2)} = \frac{-mi}{1 + m^2}$$

which is not unique. So $f(z)$ is not differentiable at $z = 0$.

15. If $f(z) = xy^2(x + iy)/(x^2 + y^2)$, $(z \neq 0)$, $f(0) = 0$, then $f(z)$ is differentiable at the origin. **(1981)**

[*Hint* : $\left|\dfrac{f(z) - f(0)}{z - 0}\right| = \left|\dfrac{xy^2}{x^2 + y^2}\right| = \dfrac{y^2}{x^2 + y^2}\;|x| < |x| \to 0$ as $z \to 0$. So limit of $\{f(z) - f(0)\}/z$ as $z \to 0$ is zero.]

16. Solved sum. Show that, at the origin, the function

$$f(z) = \frac{x^3(1+i) - y^3(1-i)}{x^2 + y^2}, (z \neq 0), f(0) = 0$$

is continuous and the C-R equations are satisfied and yet $f'(0)$ does not exist. **(1979, 81)**

Now $$u(x, y) = \frac{x^3 - y^3}{x^2 + y^2}, \{u(x, y) \neq 0, u(0, 0) = 0\}$$

$$v(x, y) = \frac{x^3 + y^3}{x^2 + y^2}, \{v(x, y) \neq 0, u(0, 0) = 0\}.$$

Also $$\left|\frac{x^3}{x^2 + y^2}\right| = \frac{x^2}{x^2 + y^2}|x| < |x|.$$

But $|x| \to 0$ as $(x, y) \to (0, 0)$. So limit of $x^3/(x^2 + y^2)$ as $(x, y) \to (0, 0)$ exists and it is zero. Hence $x^3/(x^2 + y^2)$ is continuous at (0, 0). Similarly $y^3/(x^2 + y^2)$ is continuous at (0, 0). Hence $u(x, y)$ and $v(x, y)$ are continuous at (0, 0) and $f(z)$ is continuous at $z = 0$. Further, at (0, 0),

$$u_x = \lim_{h \to 0} \frac{u(0 + h, 0) - u(0, 0)}{h} = \lim_{h \to 0} \frac{h^3}{h^3} = 1$$

$$u_y = \lim_{k \to 0} \frac{u(0, 0 + k) - u(0, 0)}{k} = \lim_{k \to 0} \frac{-k^3}{k^3} - 1$$

$$v_x = \lim_{h \to 0} \frac{v(0 + h, 0) - v(0, 0)}{h} = \lim_{h \to 0} \frac{h^3}{h^3} = 1$$

$$v_y = \lim_{k \to 0} \frac{v(0, 0 + k) - v(0, 0)}{k} = \lim_{k \to 0} \frac{k^3}{k^3} = 1.$$

So the C-R equations are satisfied at the origin. But

$$\lim_{z \to 0} \frac{f(z) - f(0)}{z - 0} = \lim_{z \to 0} \frac{x^3(1+i) - y^3(1-i)}{(x^2 + y^2)(x + iy)}$$

$$= \lim_{x \to 0} \frac{x^3(1+i) - m^3x^3(1-i)}{(x^2 + m^2x^2)(x + imx)} \text{ if } z \to 0 \text{ along } y = mx$$

$$= \frac{(1+i) - m^3(1-i)}{(1+m^2)(1+im)}$$

which is not unique. So the limit does not exist and hence $f(z)$ is not differentiable at $z = 0$.

17. Show that for the following functions the C-R equations are satisfied at $z = 0$ but the functions are not differentiable there:

(*i*) $f(z) = \sqrt{|xy|}$

(*ii*) $f(z) = e^{-1/z^4}, (z \neq 0), f(0) = 0$ **(1974, 77)**

(*iii*) $f(z) = z^5/|z|^4, (z \neq 0), f(0) = 0.$

(The exponential function e^z is defined in chapter 6.)

[*Hint* : (*i*) At (0, 0), $u_x = u_y = v_x = v_y = 0$. But

$$\lim_{z \to 0} \frac{f(z) - f(0)}{z - 0} = \lim_{z \to 0} \frac{\sqrt{|xy|}}{x + iy} = \lim_{x \to 0} \frac{\sqrt{|mx^2|}}{x + imx} = \frac{\sqrt{|m|}}{1 + im}$$

when $z \to 0$ along $y = mx$. The limit is not unique.

(*ii*) Since $f(0) = 0, u(0, 0) = 0, v(0, 0) = 0.$

Further, from

$$f(x + 0i) = u(x, 0) + iv(x, 0) = e^{-1/x^4}$$

$$f(0 + iy) = u(0, y) + iv(0, y) = e^{-1/(iy)^4} = e^{-1/y^4},$$

we get $u(x,0) = e^{-1/x^4}$; $v(x, 0) = 0$

$u(0, y) = e^{-1/y^4}$; $v(0, y) = 0.$

So, at the origin,

$$u_x = \lim_{x \to 0} \frac{u(x, 0) - u(0, 0)}{x} = \lim_{x \to 0} \frac{1}{xe^{1/x^4}} = 0,$$

$$u_y = \lim_{y \to 0} \frac{u(0, y) - u(0, 0)}{y} = \lim_{y \to 0} \frac{1}{ye^{1/y^4}} = 0.$$

Similarly, at the origin, $v_x = 0$ and $v_y = 0$. Thus the C-R equations are satisfied at the origin. But

$$\lim_{z \to 0} \frac{f(z) - f(0)}{z} = \lim_{z \to 0} \frac{1}{ze^{1/z^4}}.$$

In particular, as z tends to zero travelling along the ray $\arg z = \pi/4$, that is, along $y = x$, $z = r \exp(i\pi/4)$, where $r \to 0$.

But $1/z^4 = z^{-4} = [r \exp(i\pi/4)]^{-4} = r^{-4} \exp(-\pi i) = -r^{-4}$. So

$\exp(-1/z^4) = \exp(r^{-4})$.

Hence $e^{1/z^4} \to 1$ as $z \to 0$ and $1/(ze^{1/z^4}) \to \infty$ as $z \to 0$. Therefore $f(z)$ is not differentiable at $z = 0$.

18. Show that, at the origin, the following functions are differentiable but not analytic there:

(*i*) $f(z) = x^2 + y^2 = |z|^2$ **(1975, 76)**

(*ii*) $f(z) = x(x + iy)$.

[*Hint* : (*i*) At (x, y), $u_x = 2x$, $u_y = 2y$, $v_x = 0$, $v_y = 0$. So at (0, 0), the C-R equations are satisfied. Further u_x, u_y, v_x, v_y are continuous. So $f(z)$ is differentiable at $z = 0$. But the C-R equations are not true at points other than $z = 0$. So $f(z)$ is not differentiable at points other than $z = 0$. Hence there is no neighbourhood of $z = 0$ in which $f(z)$ is differentiable. Therefore $f(z)$ is not analytic at $z = 0$.]

4.7. Sufficient Conditions for Differentiability

19. Solved sum. If $w = e^z$, find $\frac{dw}{dz}$.

✍ Usinsg $z = x + iy$, we have

$$e^z = e^{x + iy} = e^x e^{iy} = e^x (\cos y + i \sin y)$$

$\therefore$	$u(x, y) = e^x \cos y$	$v(x, y) = e^x \sin y$
	$u_x = e^x \cos y$	$v_x = e^x \sin y$
	$u_y = e^x(-\sin y)$	$v_y = e^x \cos y$

$\therefore$ $u_x = v_y$, $u_y = -v_x$.

So e^z satisfies the C-R equations. Further e^x, $\cos y$, $\sin y$ are continuous functions and so u_x, u_y, v_x, v_y are continuous everywhere. Thus the sufficient conditions for the existence of derivative of e^z are satisfied everywhere.

$$\therefore \quad f'(z) = u_x + iv_x = e^x \cos y + i\, e^x \sin y$$
$$= e^x (\cos y + i \sin y) = e^x e^{iy}$$
$$= e^{x+iy} = e^z \quad \text{itself.}$$

20. Solved sum. If $f(z) = \sin z$, find $f'(z)$.

✎ Here we have to use the following results:

1. $\dfrac{d}{dx} \sinh x = \cosh x, \quad \dfrac{d}{dx} \cosh x = \sinh x$

2. $\sin iy = i \sinh y, \quad \cos iy = \cosh y.$

Taking $z = x + iy$,

$$\sin z = \sin (x + iy) = \sin x \cos iy + \cos x \sin iy$$
$$= \sin x \cosh y + \cos x\, (i \sinh y)$$

$\therefore$

$u(x, y) = \sin x \cosh y$	$v(x, y) = \cos x \sinh y$
$u_x = \cos x \cosh y$	$v_x = -\sin x \sinh y$
$u_y = \sin x \sinh y$	$v_y = \cos x \cosh y$

$$\therefore \quad u_x = v_y, \qquad u_y = -v_x.$$

So $\sin z$ satisfies the C-R equations. Further $\sin x$, $\cos x$, $\sinh y$, $\cosh y$ are continuous functions and so u_x, u_y, v_x, v_y are continuous everywhere. Thus the sufficiency conditions for the existence of derivative of $\sin z$ are satisfied everywhere.

$$\therefore \quad f'(z) = u_x + iv_x = \cos x \cosh y - i \sin x \sinh y$$
$$= \cos x \cos iy - \sin x\, (i \sinh y)$$
$$= \cos x \cos iy - \sin x \sin iy$$
$$= \cos (x + iy) = \cos z.$$

21. Solved sum. Show that $f(z) = z^n$ is analytic and find its derivative.

Using $z = re^{i\theta}$, we have

$$f(z) = z^n = [r (\cos \theta + i \sin \theta]^n = r^n (\cos n\theta + i \sin n\theta).$$

$\therefore$

$u(r, \theta) = r^n \cos n\theta$	$v(r, \theta) = r^n \sin n\theta$
$u_r = nr^{n-1} \cos n\theta$	$v_r = nr^{n-1} \sin n\theta$
$u_\theta = -nr^n \sin n\theta$	$v_\theta = nr^n \cos n\theta$

$$\therefore \quad u_r = \frac{1}{r} v_\theta, \qquad u_\theta = -rv_r.$$

So z^n satisfies the C-R equations. Further r^n, $\sin n\theta$, $\cos n\theta$ are continuous and so $u_r, u_\theta, v_r, V_\theta$ also. Thus z^n satisfies all the sufficient conditions for the existence of the derivative.

$$\therefore \quad f'(z) = \frac{u_r + iv_r}{e^{i\theta}} = \frac{nr^{n-1} \cos n\theta + inr^{n-1} \sin n\theta}{e^{i\theta}}$$
$$= \frac{nr^n \cos n\theta + inr^n \sin n\theta}{re^{i\theta}} = \frac{n[r(\cos \theta + i \sin \theta)]^n}{re^{i\theta}}$$
$$= \frac{nz^n}{z} = nz^{n-1}$$

22. Solved sum. Show that $f(z) = 1/z$ is analytic everywhere except at $z = 0$ and find $f'(z)$.

Using $z = re^{i\theta}$, we get

$$f(z) = \frac{1}{z} = \frac{1}{re^{i\theta}} = \frac{e^{-i\theta}}{r} = \frac{\cos\theta}{r} - i\frac{\sin\theta}{r}$$

At $z = 0$, $r = 0$ and so $f(z)$ is not defined at $z = 0$. Now the C-R equations are

$$u_r = \frac{1}{r}v_\theta, \quad u_\theta = -rv_r$$

which are satisfied because

$$u = \frac{\cos\theta}{r}, \qquad v = -\frac{\sin\theta}{r}$$

$$u_r = -\frac{\cos\theta}{r^2} \qquad v_r = \frac{\sin\theta}{r^2}$$

$$u_\theta = -\frac{\sin\theta}{r} \qquad v_\theta = -\frac{\cos\theta}{r}.$$

$$\therefore \quad f'(z) = \frac{u_r + iv_r}{e^{i\theta}} = \frac{-\cos\theta + i\sin\theta}{r^2 e^{i\theta}}$$

$$= -\frac{e^{-i\theta}}{r^2 e^{i\theta}} = -\frac{1}{(re^{i\theta})^2} = -\frac{1}{z^2}.$$

23. Solved sum. Show that $f(z) = \log z$ is analytic everywhere except at the origin and find its derivative.

Using $z = re^{i\theta}$, we get

$$f(z) = \log z = \log re^{i\theta} = \log r + \log e^{i\theta} = \log r + i\theta.$$

But, at the origin, $r = 0$. Thus, at the origin,

$$f(z) = \log 0 + i\theta = -\infty + i\theta.$$

So $f(z)$ is not defined at the origin and hence is not differentiable there.

At points other than the origin, we have

$$u(r, \theta) = \log r \qquad v(r, \theta) = \theta$$

$$u_r = \frac{1}{r} \qquad v_r = 0$$

$$u_\theta = 0 \qquad v_\theta = 1$$

So $\log z$ satisfies the C-R equations. Further $1/r$ is not continuous at $z = 0$. So $u_r, u_\theta, v_r, v_\theta$ are continuous everywhere except at $z = 0$. Thus $\log z$ satisfies all the sufficient conditions for the existence of the derivative except at the origin. The derivative is

$$f'(z) = \frac{u_r + iv_r}{e^{i\theta}} = \frac{(1/r) + i(0)}{e^{i\theta}} = \frac{1}{re^{i\theta}} = \frac{1}{z}.$$

24. Show that the function $f(z) = e^x (\cos y + i \sin y)$ is analytic in the finite plane. Find its derivative. **(1978)**

[*Hint* : $u(x, y) = e^x \cos y$, $v(x, y) = e^x \sin y$. $\therefore$ $u_x = e^x \cos y = v_y$, $u_y = -e^x \sin y = -v_x$. The C-R equations are satisfied and u_x, u_y, v_x, v_y are continuous. $\therefore$ $f'(z)$ exists and equals $u_x + iv_x$ or $f(z)$.]

25. Show that $f(z) = e^{-x} (\cos y - i \sin y)$ is differentiable everywhere in the finite plane and $f'(z) = -f(z)$.

26. Show that $f(z) = e^{-y} (\cos x + i \sin x)$ is differentiable everywhere in the finite plane and $f'(z) = i f(z)$.

27. Show that $e^x (\cos y - i \sin y)$ and $e^y (\cos x + i \sin x)$ are nowhere differentiable.

28. Show that $u + iv = \dfrac{x - iy}{x - iy + a}$, where $a \neq 0$, is not an analytic function of $z = x + iy$ whereas $u - iv$ is an analytic function.

[**Hint :** $u + iv = \dfrac{\bar{z}}{\bar{z} + a}$. Therefore,

$$\frac{\partial (u + iv)}{\partial \bar{z}} = \frac{\partial}{\partial \bar{z}} \frac{\bar{z}}{\bar{z} + a} = \frac{(\bar{z} + a) - \bar{z}}{(\bar{z} - a)^2} = \frac{a}{(\bar{z} + a)^2} \neq 0..$$

So $u + iv$ is not analytic and $u - iv = \dfrac{z}{z + \bar{a}}$ is analytic except at $z \neq -\bar{a}$ because $u - iv$ satisfies the sufficiency conditions for analyticity]

29. **Solved sum.** ϕ and Ψ are functions of x and y satisfying Laplace's equation, namely $\phi_{xx} + \phi_{yy} = 0$, $\Psi_{xx} + \Psi_{yy} = 0$. If $u = \phi_y - \Psi_x$, $v = \phi_x + \Psi_y$ show that $u + iv$ is analytic. **(1980)**

$$u_x = \phi_{xy} - \Psi_{xx}. \text{ But } v_y = \phi_{yx} + \Psi_{yy} = \phi_{yx} - \Psi_{xx} \ (\because \Psi_{xx} + \Psi_{yy} = 0)$$
$$= \phi_{xy} - \Psi_{xx}.$$
$$u_y = \phi_{yy} - \Psi_{yx}. \text{ But } v_x = \phi_{xx} + \Psi_{xy} = -\phi_{yy} + \Psi_{yx} \ (\because \phi_{xx} + \phi_{yy} = 0)$$
$$= -(\phi_{yy} - \Psi_{yx}).$$

$\therefore$ C-R equations are satisfied and u_x, u_y, v_x, v_y are continuous. Hence $u + iv$ is analytic.

4.8. C–R–Equations in Polar Coordinates

30. **Solved sum.** Show that the single-valued continuous function

$$f(z) = z^{1/2} = r^{1/2} (\cos \tfrac{1}{2}\theta + i \sin \tfrac{1}{2}\theta),\ r > 0,\ 0 < \theta < 2\pi$$

is analytic. Find $f'(z)$.

$f(z)$ is continuous. Now

$$u = r^{1/2} \cos \tfrac{1}{2}\theta, \quad v = r^{1/2} \sin \tfrac{1}{2}\theta$$

$$\therefore \quad \frac{\partial u}{\partial r} = \frac{1}{2} r^{-1/2} \cos \tfrac{1}{2}\theta = \frac{1}{r} \frac{\partial v}{\partial \theta}$$

$$\therefore \quad \frac{\partial u}{\partial \theta} = -\frac{1}{2} r^{1/2} \cos \tfrac{1}{2}\theta = -r \frac{\partial v}{\partial r}.$$

So the C.R. equations are satisfied. Further the first partials are continuous. So $f'(z)$ exists and, by (4.8.2),

$$f'(z) = \frac{r}{z}\left(\frac{\partial u}{\partial r} + i \frac{\partial v}{\partial r}\right) = \frac{r}{z}\left[\frac{1}{2} \frac{1}{r^{1/2}} \left\{\cos \frac{\theta}{2} + i \sin \frac{\theta}{2}\right\}\right]$$

$$= \frac{1}{2}\frac{1}{z}\left[r^{1/2}\left(\cos\frac{\theta}{2} + i\sin\frac{\theta}{2}\right)\right] = \frac{1}{2}\frac{\sqrt{z}}{z} = \frac{1}{2\sqrt{z}}.$$

31. If $f(z) = \log r + i\theta$ is single-valued and continuous in a region D, show that $f'(z) = 1/z$.
[*Hint* : In D, the C-R equations are satisfied and the first partials of u and v are continuous.

$$f'(z) = \frac{r}{z}\left(\frac{\partial u}{\partial r} + i\frac{\partial v}{\partial r}\right) = \frac{r}{z}\left(\frac{1}{r}\right) = \frac{1}{z}.]$$

32. Solved sum. Suppose $f(z) = u(x, y) + iv(x, y)$ is analytic in a region D. Show that, in D, the level curves $u(x, y) = \lambda$ are orthogonal to the level curves $v(x, y) = \mu$, where λ and μ are parameters.

$\nabla u(x, y)$ is a vector normal to the curve $u(x, y) = \lambda$ at (x, y) and $\nabla v(x, y)$ is a vector normal to the curve $v(x, y) = \mu$ at (x, y). But the scalar product of these vectors is

$$\{\nabla u(x, y)\} . \{\nabla v(x, y)\} = \left(i\frac{\partial u}{\partial x} + j\frac{\partial u}{\partial y}\right)\cdot\left(i\frac{\partial v}{\partial x} + j\frac{\partial v}{\partial y}\right)$$

$= u_x v_x + u_y v_y = v_y v_x - v_x v_y = 0$ by C-R equations.

So the normals to the curves at (x, y) are perpendicular. This implies that the tangents to the curve at (x, y) are perpendicular. Hence the curves are orthogonal.

CHAPTER 5

SEQUENCES AND SERIES

5.1. SEQUENCES

As we have sequences in real numbers, we have sequences in complex numbers also. These sequences can be classified into two groups as

(*i*) Sequences of complex numbers

(*ii*) Sequences of complex functions.

Sequence of complex numbers. The set of complex numbers

$$\{z_n, n = 1, 2, 3, ...\},$$

where z_n is a function of n, is a sequence of complex numbers. The numbers $z_1, z_2, z_3, ...$ are called the *terms* of the sequence.

Sequence of complex functions. The set of complex functions

$$\{f_n(z), n = 1, 2, 3, ...\},$$

where $f_n(z)$ is defined in a given set and depends on n, is a sequence of functions. For each z in the set we have a number sequence.

5.2. NUMBER SEQUENCE

First we define convergence of a number sequence.

Convergence. The sequence $\{z_n\}$ is said to converge to z_0 if, for any $\in > 0$, there exists a positive integer n_0 depending on $\in$ such that

$$|z_n - z_0| < \in \text{ for all } n > n_0$$

(Fig. 5.1.1). In other words, the sequence converges to z_0 if every neighbourhood of z_0 contains all but a finite number of terms of the sequence. The fact of convergence of the sequence $\{z_n\}$ to z_0, is symbolically denoted by

$$z_n \to z_0 \text{ as } n \to \infty \quad \text{or} \quad \lim_{n\to\infty} (z_n) = z_0.$$

Limit of the sequence. In the above case z_0 is said to be the *limit of the sequence* $\{z_n\}$.

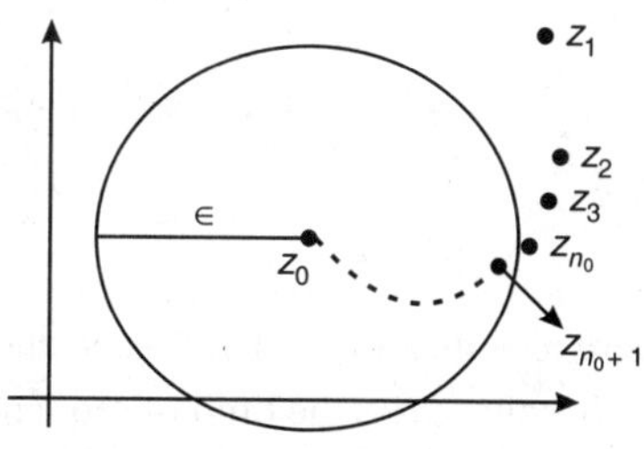

Fig. 5.1.1

Example. (*i*) $\dfrac{2^n}{n!} + \dfrac{n}{2^n} i \to 0$ as $n \to \infty$

because $2^n/n! \to 0$ as $n \to \infty$ and $n/2^n \to 0$ as $n \to \infty$.

(*ii*) If $0 < q < 1$, then $nq^n + n^{1/n}i \to i$ because $nq^n \to 0$ as $n \to \infty$ and $n^{1/n} \to 1$ as $n \to \infty$.

(*iii*) $i/n \to 0$ as $n \to \infty$.

Example for nonconvergence. The sequence

$$\{(-1)^n + \frac{1}{n}\}$$

does not converge because the terms cluster around 1 and – 1 alternately. Thus the sequence does not approach 1 or – 1. But 1 and – 1 are limit points of the sequence.

Sequence as a set. A convergent sequence of complex numbers is a point set having only one limit point.

Theorem 5.1. If $z_n = x_n + iy_n$, then the sequence $\{z_n\}$ converges to $z_0 = h + ik$ if and only if the real sequence $(x_n\}$ and $\{y_n\}$ converge to h and k respectively.

Necessity part:

Hypothesis: The complex sequence $\{z_n\}$ converges to z_0.

To prove: The real sequences $\{x_n\}$ and $\{y_n\}$ converge to h and k.

Proof: Since the sequence $\{z_n\}$ converges to z_0, for any $\in > 0$, there exists a positive integer n_0 depending on $\in$ such that

$$|z_n - z_0| < \in \text{ for all } n > n_0.$$

But $|\text{Re}(z_n - z_0)| \le |z_n - z_0|$ or $|x_n - h| \le |z_n - z_0|$

$\therefore$ $|x_n - h| < \in$ for all $n > n_0$.

This implies that the real sequence $\{x_n\}$ converges to h. Similarly the real sequence $\{y_n\}$ converges to k.

Sufficiency part:

Hypothesis: The real sequences $\{x_n\}$, $\{y_n\}$ converge to h, k.

To prove: The complex sequence $\{z_n\}$ converges to $h + ik$.

Proof: Since the real sequences $\{x_n\}$, $\{y_n\}$ converge to h, k, for any $\in < 0$, there exist positive integers n_1, n_2 such that

$$|x_n - h| < \frac{1}{2}\in \quad \text{for all } n > n_1$$

$$|y_n - k| < \frac{1}{2}\in \quad \text{for all } n > n_2$$

$$\therefore \quad |z_n - z_0| = |(x_n - h) + i(y_n - k)| \le |x_n - h| + |y_n - k|$$

$$< \frac{1}{2}\in + \frac{1}{2}\in = \in \text{ for all } n > n_0$$

where $n_0 = \max(n_1, n_2)$. So the sequence z_n converges to z_0.

In some cases the limit of sequence cannot be explicitly determined. In such cases the following test which is due to Cauchy can be used. The proof of this involves Cauchy's convergence criterion for real sequences.

Theorem 5.2. (Cauchy's criterion for convergence). The sequence $\{z_n\}$ is convergent if and only if, for any $\in > 0$, there exists a positive integer n_0 depending on $\in$ such that

$$|z_{n+p} - z_n| < \in \text{ for all } n > n_0 \text{ and } p > 0.$$

Proof of necessity part: Let $z_n = x_n + iy_n$. In the necessity part the hypothesis is that the sequence $\{z_n\}$ is convergent i.e., the real sequences $\{x_n\}$ and $\{y_n\}$ converge. Hence, by Cauchy's convergence criterion for *real* sequences, for any $\in > 0$, there exist positive integers $n_1(\in)$ and $n_2(\in)$ such that

$$|x_{n+p} - x_n| < \frac{1}{2}\in \quad \text{for all } n > n_1 \text{ and } p > 0,$$

$$|y_{n+p}-y_n| < \frac{1}{2} \in \quad \text{for all } n > n_2 \text{ and } p > 0.$$

But
$$\begin{aligned}|z_{n+p}-z_n| &= |(x_{n+p}-x_n)+i(y_{n+p}-y_n)| \\ &\le |x_{n+p}-x_n|+|y_{n+p}-y_n| \\ &< \frac{1}{2}\in+\frac{1}{2}\in=\in \quad \text{for all } n > \max(n_1, n_2) \text{ and } p > 0.\end{aligned}$$

Proof of sufficiency part: In the sufficiency part the hypothesis is that for any $\in > 0$ it is possible to find an $n_0(\in)$ such that

$$|z_{n+p}-z_n| < \in \text{ for all } n > n_0 \text{ and } p > 0.$$

But
$$|x_{n+p}-x_n| = |\mathbf{Re}\,(z_{n+p}-z_n)| \le |z_{n+p}-z_n| < \in \quad \text{for all } n > n_0 \text{ and } p > 0.$$

Thus the real sequence $\{x_n\}$ is convergent. Similarly $\{y_n\}$ is also convergent. Hence, by theorem 5.1, the sequence $\{z_n\}$ is convergent.

5.3. CONVERGENCE OF FUNCTIONAL SEQUENCE

Suppose $\{f_n(z)\}$ is a sequence of functions $f_n(z)$, $n = 1, 2, 3, ...$ defined in a set S. If the sequence converges for each z in S, then the sequence is said to be *pointwise convergent* or simply *convergent* in S.

Limit function of a functional sequence. If a functional sequence is convergent in a set S, then a single-valued function $f(z)$ can be defined in S such that its value at each point z of S, is the limit of $f_n(z)$ as $n \to \infty$. This function is called the *limit function* of the sequence.

Criterion for convergence. From the definition of convergence of a number series the criterion for convergence of a functional sequence follows.

The sequences $\{f_n(z)\}$ of functions $f_n(z)$ defined in a set S, is convergent in S if, given any $\in > 0$, for each z in S there exists an n_0 depending on $\in$ and z such that

$$|f_n(z)-f(z)| < \in \qquad \text{for all } n > n_0.$$

Cauchy's criterion for convergence. Cauchy's criterion for convergence of the sequence $f_n(z)$ in S is that the sequence is convergent in S if, given any $\in > 0$, for each z in S there exists an n_0 depending on $\in$ and z such that

$$|f_{n+p}(z)-f_n(z)| \qquad \text{for all } n > n_0 \text{ and } p > 0.$$

5.4. UNIFORM CONVERGENCE OF A FUNCTIONAL SEQUENCE

A functional sequence which is pointwise convergent in a set S is said to be uniformly convergent in S if, given any $\in > 0$, simultaneously for all z's in S, there exists an n_0 depending on $\in$ alone such that

$$|f_n(z)-f(z)| < \in \qquad \text{for all } n > n_0.$$

Remark. A convergent functional sequence may or may not be uniformly convergent (vide example under theorem 5.3).

Theorem 5.3. (Cauchy's criterion for uniform convergence). The sequence $\{f_n(z)\}$ is uniformly convergent in S if and only if, given any $\in < 0$, *simultaneously* for all z's in S, these exists an n_0 (depending on $\in$ alone) such that

$$|f_{n+p}(z)-f_n(z)| < \in \quad \text{for all } n > n_0 \text{ and } p > 0.$$

Proof of necessity part. Now $\{f_n(z)\}$ is convergent. Let the limit function be $f(z)$. Then, by definition of uniform convergence, given any $\in > 0$, simultaneously for all z's in S, there exists an n_0 such that

$$|f_n(z)-f(z)| < \frac{1}{2}\in, \text{ for all } n > n_0$$

and consequently,

$$|f_{n+p}(z)-f(z)| < \frac{1}{2}\in, \text{ for all } n > n_0 \text{ and } p > 0$$

So, given an $\in > 0$, for all z's in S,

$$\begin{aligned} |f_{n+p}(z)-f_n(z)| &= |\{f_{n+p}(z)-f(z)\} + \{f(z)-f_n(z)\}| \\ &\le |f_{n+p}(z)-f(z)| + |f(z)-f_n(z)| \\ &< \frac{1}{2}\in + \frac{1}{2}\in = \in \text{ for all } n > n_0 \text{ and } p > 0. \end{aligned}$$

Proof of sufficiency part. Now, given any $\in > 0$, simultaneously for all z's in S there exists an n_0 such that

$$|f_{n+p}(z)-f_n(z)| < \in \quad \text{for all } n > n_0 \text{ and } p > 0.$$

Therefore for each fixed z the sequence $\{f_n(z)\}$ is convergent i.e., the series is pointwise convergent in S. Let the limit function be $f(z)$, i.e. let $f_n(z) \to f(z)$ as $n \to \infty$.

Now the given result is that, given any $\in > 0$ and for all z't in S, there exists simultaneously an n_0 such that

$$|f_{n+p}(z)-f_n(z)| < \in \text{ for all } n > n_0 \text{ and } p > 0.$$

Keep n fixed and let $p \to \infty$. Then the result becomes that, given any $\in > 0$ and, simultaneously for all z's in S, there exist an n_s such that

$$|f(z)-f_n(z)| < \in \text{ for all } n > n_0.$$

So the series converges uniformly in S.

Remark. Uniform convergence of a sequence is a stronger property than pointwise convergence. That is, every convergent sequence in a set need not be uniformly convergent there. This fact is illustrated through the following example.

Example. In the open unit disc D, $|z| < 1$, the sequence

$$\left\{\frac{1-z^{n+1}}{1-z}\right\}$$

converges but not uniformly. For,

$$\begin{aligned} |f_{n+p}(z)-f_n(z)| &= \left|\frac{1-z^{n+p+1}}{1-z} - \frac{1-z^{n+1}}{1-z}\right| = \left|\frac{z^{n+1}-z^{n+p+1}}{1-z}\right| \\ &= \left|\frac{z^{n+1}(1-z^p)}{1-z}\right| = \frac{|z^{n+1}||1-z^p|}{|1-z|} \\ &\ge \frac{|z^{n+1}|(1-|z|^p)}{|1-z|} \quad (\because |1-z^p| \ge 1-|z^p|) \end{aligned}$$

Choosing n for p and the real number $(n-1)/n$ for z,

$$\frac{|z|^{n+1}(1-|z|^p)}{|1-z|} = n\left(1-\frac{1}{n}\right)\left[1-\left(1-\frac{1}{n}\right)^n\right] \to \infty \text{ as } n \to \infty$$

because $\left(1-\frac{1}{n}\right)^n \to \frac{1}{e}$.

So, as n increases, the value of $|f_{n+p}(z) - f_p(z)|$ increases indefinitely for the above choice of p and z in D. Hence the sequence cannot satisfy the Cauchy criterion for uniform convergence. Therefore the sequence is not uniformly convergent in D.

However, for any fixed z in D,

$$\frac{1-z^{n+1}}{1-z} = \frac{1}{1-z} - \frac{z^{n+1}}{1-z} \to \frac{1}{1-z} \text{ as } n \to \infty$$

because $|z| < 1$ and $|z|^{n+1} \to 0$ as $n \to \infty$. Hence the sequence is pointwise convergent in D.

5.5. COMPLEX SERIES

Now we study the following infinite series:

(*i*) Number series: $z_1 + z_2 + z_3 + ...$

(*ii*) Functional series: $f_1(z) + f_2(z) + f_3(z) + ...$

where the *n*th terms are functions of *n* and z_1, z_2, z_3 ... are complex constants and $f_1(z), f_2(z), f_3(z), ...$ are complex functions defined in a set.

The functional series which is of the form

$$a_0 + a_1(z-a) + a_2(z-a)^2 + ...$$

is called a power series in $z - a$.

Partial sum of a series. The sum of the first n terms of a series is called the *n*th partial sum of the series. It is denoted by s_n so that

$$s_n = z_1 + z_2 + z_3 + ... + z_n$$

if the series is $z_1 + z_2 + z_3 + ...$

Convergence of a series. The series

$$z_1 + z_2 + z_3 + z_4 + ...$$

is said to be convergent if the sequence of its partial sums, namely $\{s_n\}$, is convergent; otherwise, the series is said to be divergent.

Sum of a convergent series. The limit of the sequence of $\{s_n\}$ is called the sum of the series.

Example. The partial sum for the geometric series

$$1 + z + z^2 + z^3 + ...$$

is $s_n = (1 - z^{n+1})/(1 - z)$ and the limit of the sequence $\{s_n\}$ is $1/(1 - z)$ if $|z| < 1$, since $|z|^{n+1} \to 0$ as $n \to \infty$. So the geometric series is convergent if $|z| < 1$, and its sum is $1/(1 - z)$.

Theorem 5.4. A necessary condition for the series

$$z_1 + z_2 + z_3 + z_4 + ...$$

to be convergent is that, as $n \to \infty$, $z_n \to 0$.

Proof: Since the series is convergent, the sequence $\{s_n\}$ is convergent. So, by Cauchy's criterion (Theorem 5.2), for any $\in > 0$, there exists an n_0 such that

$$|s_{n+p} - s_n| < \in \text{ for all } n > n_0 \text{ and } p > 0.$$

Choose p as 1. Then this criterion becomes

$$|s_{n+1} - s_n| < \in \quad \text{for all} \quad n > n_0$$

i.e., $$|z_{n+1}| < \in \quad \text{for all} \quad n > n_0$$

i.e., $$|z_n - 0| < \in \quad \text{for all} \quad n > n_0 + 1$$

Since $\in$ is arbitrary $z_n \to 0$ as $n \to \infty$.

Absolutely convergent series. A series

$$z_1 + z_2 + z_3 + z_4 + ...$$

is said to be absolutely convergent if the positive-termed series

$$|z_1|+|z_2|+|z_3|+|z_4|+\dots$$

is convergent.

Theorem 5.5. An absolutely convergent series is convergent.

Proof: Let $z_1 + z_2 + z_3 + \dots$ be an absolutely convergent series. Then by Cauchy's criterion (Theorem 5.2), for any $\in > 0$ there exists an n_0 such that

$$||z_{n+1}|+|z_{n+2}|+|z_{n+3}|+\dots+|z_{n+p}|| < \in \qquad \text{for all } n > n_0 \text{ and } p > 0.$$

But $$|z_{n+1}+z_{n+2}+\dots+z_{n+p}| \le |z_{n+1}|+\dots+|z_{n+p}|.$$

$\therefore$ $$|z_{n+1}+z_{n+2}+\dots+z_{n+p}| < \in \qquad \text{for all } n > n_0 \text{ and } p > 0.$$

So, by Cauchy's criterion, the given series is convergent.

Remark. A convergent series may or may not be absolutely convergent. For example, the series

$$1-\frac{1}{2}+\frac{1}{3}-\frac{1}{4}+\dots$$

is convergent but not absolutely convergent. As such, absolute convergence of a series is a stronger property than convergence of the series.

5.6. CONVERGENCE OF FUNCTIONAL SERIES

Suppose the terms $f_1(z), f_2(z), f_3(z), \dots$ of the series

$$f_1(z)+f_2(z)+f_3(z)+\dots$$

are functions defined in a set S and its nth partial sum is $s_n(z)$. Then the series is said to be *pointwise convergent* or simply *convergent* if the functional sequence $\{s_n(z)\}$ is convergent in S.

Sum function. When the series is convergent, i.e., when the sequence $\{s_n(z)\}$ is convergent, then the limit function $f(z)$ of the sequence is called the *sum function* of the series. The series is said to represent the sum function and this fact is written as

$$f(z) = f_1(z)+f_2(z)+f_3(z)+\dots \text{ for all } z \text{ in } S.$$

Criterion for convergence of a functional series. The series

$$f_1(z)+f_2(z)+f_3(z)+\dots$$

converges to $f(z)$ in S if, given an $\in > 0$, for each z in S, there exists an n_0 depending on $\in$ and z such that

$$|f_1(z)+f_2(z)+\dots+f_n(z)-f(z)| < \in \text{ for all } n > n_0.$$

Cauchy's criterian. The series $f_1(z)+f_2(z)+f_3(z)+\dots$ converges in S if, given an $\in > 0$, for each z in S, there exists an n_0 depending on $\in$ and z such that

$$|f_{n+1}(z)+f_{n+2}(z)+\dots+f_{n+p}(z)| < \in \text{ for all } n > n_0 \text{ and } p > 0.$$

5.7. UNIFORM CONVERGENCE OF A FUNCTIONAL SERIES

Suppose the series

$$f_1(z)+f_2(z)+f_3(z)+\dots$$

converges pointwise in a set S and its sum function is $f(z)$. The series is said to converge uniformly in S if, given an $\in > 0$, for all z's in S, there exists simultaneously an n_0 (i.e., an n_0 depending on $\in$ alone) such that

$$|f_1(z)+f_2(z)+\dots f_n(z)-f(z)| < \in \text{ for all } n > n_0.$$

Cauchy's criterian. The series $f_1(z) + f_2(z) + f_3(z) + \dots$ converges uniformly in S if, given an $\in > 0$, for all z's in S, there exists simultaneously an n_0 depending on $\in$ alone such that

$$|f_{n+1}(z)+f_{n+2}(z)+\dots+f_{n+p}(z)| < \in \text{ for all } n > n_0 \text{ and } p > 0.$$

Remark. Uniform convergence is a stronger property than convergence. That is, every convergent series need not be uniformly convergent. This is illustrated through the following example.

Example. In the open unit disc D, $|z| < 1$, the geometric series

$$1 + z + z^2 + z^3 + \dots$$

converges but not uniformly.

Proof: Let $s_n = 1 + z + z^2 + \dots + z^n$. Then

$$S_n = \frac{1 - z^{n+1}}{1 - z}.$$

In the example under theorem 5.3, we have shown that the sequence $\{s_n\}$ converges in D but not uniformly. So the given geometric series converges in D but not uniformly.

Next we study the case in which the terms of a uniformly convergent series are continuous. The case in which the terms are analytic functions will be taken up in chapter 6.

Theorem 5.6. If the terms $f_1(z), f_2(z), f_3(z), \dots$ of a uniformly convergent series

$$f_1(z) + f_2(z) + f_3(z) + \dots$$

are continuous in a set S, then the sum function $f(z)$ of the series is also continuous in S.

To establish continuity of $f(z)$ in S it is enough if we establish continuity of $f(z)$ at an arbitrary point z_0 of S. To do so, given an $\in > 0$, we have to show that it is possible to find a neighbourhood of z_0 such that

$$|f(z) - f(z_0)| < \in \text{ for all } z \text{ in the neighbourhood.}$$

Suppose an $\in$ is given. Due to uniform convergence of the series in S, for all z's in S, there exists an n_0 such that

$$|s_n(z) - f(z)| < \frac{1}{3}\in \quad \text{for all} \quad n > n_0.$$

Let N be a fixed number $> n_0$. Then

$$|s_N(z) - f(z)| < \frac{1}{3}\in \quad \text{for} \quad z \text{ in } S \tag{5.7.1}$$

and in particular, choosing $z = z_0$,

$$|s_N(z_0) - f(z_0)| < \frac{1}{3}\in. \tag{5.7.2}$$

Now $s_n(z) - s_n(z_0)$ is continuous in S because it is the sum of finite number of continuous functions. So there exists a $\delta > 0$ such that

$$|s_N(z) - s_N(z_0)| < \frac{1}{3}\in \qquad \text{for all } z \text{ in } |z - z_0| < \delta. \qquad (5.7.1)$$

$$\begin{aligned} \therefore \quad |f(z) - f(z_0)| &= |f(z) - s_N(z) + \{s_N(z) - s_N(z_0)\} + s_N(z_0) - f(z_0)| \\ &\leq |f(z) - s_N(z)| + |s_N(z) - s_N(z_0)| + |s_N(z_0) - f(z_0)| \\ &< \frac{1}{3}\in + \frac{1}{3}\in + \frac{1}{3}\in = \in \quad \text{for all } z \text{ in } |z - z_0| < \delta \end{aligned}$$

by (5.2.1), (5.2.3), (5.2.2) respectively. So $f(z)$ is continuous at $z = z_0$. Since z_0 is arbitrary in S, $f(z)$ is continuous in S.

Next we study a test to be used to ensure uniform convergence of a given series. This test is due to Weierstrass.

Theorem 5.7. (*Weierstrass' M-test for uniform convergence*).

Given a series,

$$f_1(z) + f_2(z) + f_3(z) + \dots$$

defined in a set S, if there exists a positive-termed convergent series

$$M_1 + M_2 + \dots + M_n + \dots$$

and an integer n_0 such that for all z in S and for all $n > n_0$,

$$|f_n(z)| < M_n$$

then the series is uniformly convergent in S.

Proof: Since the M-series is a convergent positive-termed number series, given an $\varepsilon > 0$, by Cauchy's criterian, there exists an n_1 such that

$$M_{n+1} + M_{n+2} + \dots + M_{n+p} < \in \text{ for all } n > n_1 \text{ and } p > 0$$

$$\therefore \quad |f_{n+1}(z) + \dots + f_{n+p}(z)| \le |f_{n+1}(z)| + \dots + |f_{n+p}(z)| \le M_{n+1} + \dots + M_{n+p}$$

for $n > n_0$ and for all $z < \in$ for all $n > n_1$ and $p > 0$.

So, given an $\in > 0$, simultaneously for all z's in S there exists an $N = \max(n_0, n_1)$ such that for all $n > N$ and $p > 0$

$$|f_{n+1}(z) + \dots + f_{n+p}(z)| < \in.$$

So the given series is uniformly convergent in S.

5.8. EXAMPLES

5.2 NUMBER SEQUENCE

1. **Solved sum.** Prove that, if the sequence $\{z_n\}$ converges to a, then the sequence is bounded (i.e., there is some constant M such that $|z_n| \le M$ for all n.

✍ Since $\{z_n\}$ converges to a, for any $\in > 0$ there exists an N such that

$$|z_n - a| < \in \text{ for all } n > N. \tag{1}$$

Now, from $z_n = a + (z_n - a)$ and the triangle inequality,

$$\begin{aligned} |z_n| &\le |a| + |z_n - a| \\ &< |a| + \in \quad \text{for all} \quad n > N, \text{ by (1).} \end{aligned} \tag{2}$$

Let M be max $(|z_1|, |z_2|, |z_3|, |z_n|, |a| + \varepsilon)$. Then by (2)

$$|z_n| \le M \qquad \text{for all } n$$

2. If the sequence $\{a_n\}$ and $\{b_n\}$ converge to A and B, show that the sequence $\{a_n + b_n\}$ converges to $A + B$.

[*Hint* : $|a_n + b_n - (A + B)| \le |a_n - A| + |b_n - A|$ and given an $\in > 0$, there exist integers n_1 and n_2 such that $|a_n - A| < \in$ for all $n > n_1$ and $|b_n - B| < \in$ for all $n > n_2$].

3. Set an example to show that the converse of the previous sum is not always true.

[*Hint* : $a_n = ni$, $b_n = (1/n) - ni$].

4. If $z_n \to a$, then prove that $|z_n| \to |a|$.

[*Hint* : By the triangle inequality $||z_n| - |a|| \le |z_n - a|$].

5. **Solved sum.** Check whether the sequence $\{i^n/n\}$ converges.

✍ By considering the distances of $i/1 + i^2/2 + i^3/3, \dots$ from the origin we suspect that the sequence may converge to 0. So consider

$$\left|\frac{i^n}{n} - 0\right| = \frac{|i|}{n} = \frac{1}{n}. \tag{1}$$

But, for any $\in > 0$, we get that, for all $n > 1/\in$, $1/n < \in$. Thus (1) becomes that

$$\left|\frac{i^n}{n} - 0\right| < \in \quad \text{for all} \quad n > \frac{1}{\in}$$

Thus by definition $i^n/n \to 0$ as $n \to \infty$.

6. Show that $\{z^n\}$ is divergent if $|z| > 1$.

[*Hint* : $|z| = r$; by triangle inequality $|z_{n+p} - z_n| \geq |z_{n+p}| - |z_n|$; So $|z^{n+p} - z^n| \geq |z^{n+p}| - |z^n|$ $= r^{n+p} - r^n = r^n(r^p - 1) = r^n(r-1)(r^{p-1} + r^{p-2} + ... + 1)$

$$> r^n (r-1)(1+1+1+...+1) = pr^n(r-1)$$

which increases with p infinitely].

7. Find the limit points of the set $\{1/m + i/n\}$, where m and n run through all the integers.

[Hint: When $n = m$, $(1 + i)/n \to 0$ as $n \to \infty$. When m is fixed, $(1/m) + (i/n) \to 1/m$ as $n \to \infty$. When n is fixed $(1/m) + (i/n) \to i/n$ as $m \to \infty$. So the limit points are zero and the points $1/m$, i/n, where m and n run through all integers].

8. Find the limit points of the set $\{p/m + iq/n\}$, where p, q, m and n run through all the integers.

[*Hint* : All the points of the complex plane are the limit points because the limit points of the set $\{p/m\}$ of rationals are all the real numbers, and similarly those of $\{q/n\}$ are all real numbers].

5.5 COMPLEX SERIES

9. If the series $z_1 + z_2 + z_3 + ...$ converges to a, show that the series $bz_1 + bz_2 + bz_3 + ...$ converges to ba.

10. If the series $z_1 + z_2 + z_3 + ...$, $z_1' + z_2' + z_3' + ...$ converge to a, a', show that the series $(z_1 + z_1') + (z_2 + z_2') + ...$ converges to $a + a'$.

11. Show that the series

$$1 + \frac{1}{1+i} + \frac{1}{(1+i)^2} + \frac{1}{(1+i)^3} + ...$$

converges and find its sum.

[*Hint* : Given series is a geometric series and converges since $|1/(1+i)| = 1/\sqrt{2} < 1$. Sum $= 1/(1-z)$, where $z = 1/(1+i)$].

5.6 CONVERGENCE OF A FUNCTIONAL SERIES

12. **Solved sum.** Show that the series

$$\frac{z}{1-z} + \frac{z^2}{1-z^2} + \frac{z^3}{1-z^3} + \cdots$$

converges in $|z| < 1$.

✍ Denote the nth term of the series by $f_n(z)$. Then

$$|f_n(z)| = \left|\frac{z^n}{1-z^n}\right| < \frac{|z|^n}{1-|z|^n} \text{ by triangle inequality}$$

$$= \frac{r^n}{1-r^n}, \text{ where } r = |z|.$$

Now we shall check by ratio test whether the positive-termed series

$$\frac{r}{1-r} + \frac{r^2}{1-r^2} + \frac{r^3}{1-r^3} + ...$$

converges.

Suppose $r < 1$. Then $$\lim_{n\to\infty} \frac{r^{n+1}}{1-r^{n+1}} \frac{1-r^n}{r^n} = r < 1$$

So the positive-termed series is convergent, and the given complex series is absolutely convergent in $|z| < 1$ and hence convergent in $|z| < 1$.

13. Show that the series whose nth term is $z^{2n}/(1 - z^{2n})$ is absolutely convergent in $|z| < 1$ and hence convergent in $|z| < 1$.

$$\left[Hint: \text{ If } |z| = r, \lim_{n\to\infty} \frac{r^{2n+2}}{1 - r^{2n+2}} \frac{1 - r^{2n}}{r^{2n}} = r^2 \right].$$

5.7 UNIFORM CONVERGENCE OF A FUNCTIONAL SERIES

14. Show that the series

(i) $\displaystyle\sum_{n=1}^{\infty} \frac{z^n}{1 - z^n}$

(ii) $\displaystyle\sum_{n=1}^{\infty} \frac{z^{2n}}{1 - z^{2n}}$

are uniformly convergent in the closed region $|z| = r$, where $r < 1$.

[*Hint* : Use M-test (*i*) $M_n = r^n/(1 - r^n)$; Refer to sum 12 (*ii*) $M_n = r^{2n}/(1 - r^{2n})$].

15. Show that the series

(i) $1 + z + \dfrac{z^2}{2!} + \dfrac{z^3}{3!} + \ldots$

(ii) $1 - \dfrac{z^2}{2!} + \dfrac{z^4}{4!} - \dfrac{z^4}{6!} + \ldots$

(iii) $z - \dfrac{z^3}{3!} + \dfrac{z^5}{5!} - \dfrac{z^7}{7!} + \ldots$

are all uniformly convergent in any bounded set.

[*Hint* : Use M-test]

16. Show that the series $\displaystyle\sum_{n=1}^{\infty} \frac{1}{n^2 + z^2}$ is uniformly convergent in $1 < |z| < 2$.

[*Hint* : For $n \geq 3$ and $1 \leq z \leq 2$,

$$\left|\frac{1}{n^2 + z^2}\right| = \frac{1}{|n^2 + z^2|} = \frac{1}{|n^2 - (-z^2)|} \leq \frac{1}{|n^2| - |z^2|}$$

$$< \frac{1}{n^2 - 4} = \frac{2}{2n^2 - 8} = \frac{2}{n^2 + (n^2 - 4)} < \frac{2}{n^2}; M_n = \frac{2}{n^2} \Bigg].$$

CHAPTER 6 POWER SERIES AND ELEMENTARY FUNCTIONS

6.1. POWER SERIES

In this section we consider the two power seriesin z and in z – a, namely,

$$a_0 + a_1 z + a_2 z^2 + a_3 z^3 + \ldots$$

$$a_0 + a_1 (z - a) + a_2 (z - a)^2 + \ldots$$

Geometric series. The power series with unit coefficients $1 + z + z^2 + z^3 + \ldots$, $1 + (z - a) + (z - a)^2 + \ldots$ are called geometric series.

Power series about z = a. The power series of the form $a_0 + a_1 (z - a) + a_2 (z - a)^2 + \ldots$ is said to be a series about $z = a$.

6.2. ABSOLUTE CONVERGENCE OF A POWER SERIES

The following theorem gives an open circular region in which a power series in z converges absolutely. The size of the region depends on the coefficients a_n of the series.

Theorem 6.1. The power series

$$a_0 + a_1 z + a_2 z^2 + a_3 z^3 + \ldots$$

converges absolutely in the open disc $|z| < R$, where

$$R = \frac{1}{l}, \quad l = \lim_{n \to \infty} \sup |a_n|^{1/n}.$$

Proof: It is evident that the series is convergent when $z = 0$. For $z \neq 0$, by Cauchy's root test, the positive-termed series

$$|a_0| + |a_1 z| + |a_2 z^2| + |a_3 z^3| + \ldots$$

is convergent if

$$\lim_{n \to \infty} \sup |a_n z^n|^{1/n} < 1 \quad \text{or} \quad \left(\lim_{n \to \infty} \sup |a_n|^{1/n}\right) |z| < 1,$$

i.e., if $l\,|z| < 1$ or if $|z| < 1/l = R$. Therefore the given series is absolutely converent in $|z| < R$.

Corollary 1. Since absolute convergence of a series implies its convergence (Theorem 5.5), the series

$$a_0 + a_1 z + a_2 z^2 + a_3 z^3 + \ldots$$

converges in the disc $|z| < R$.

Corollary 2. When $R = \infty$, the series is absolutely convergent in the entire finite plane.

Remark 1. The power series

$$a_0 + a_1 (z - a) + a_2 (z - a)^2 + \ldots$$

absolutely convergent in the disc $|z - a| < R$

(i.e., in the interior of the circle of convergence $|z-a|=R$) where

$$R=\frac{1}{l}, \quad l=\limsup_{n\to\infty}|a_n|^{1/n}$$

or $$R=\frac{1}{l}, \quad l=\limsup_{n\to\infty}\left|\frac{a_{n+1}}{a_n}\right|.$$

Remark 2. By *D'Alembert's ratio test*

The power series

$$a_0+a_1z+a_2z^2+a_3z^3+\dots$$

converges absolutely in the open disk

$$|z|<R,$$

where R is given by

$$l=\limsup_{n\to\infty}\left|\frac{a_{n+1}}{a_n}\right|, \quad R=\frac{1}{l}.$$

Circle of convergence. The circles

$$|z|=R, \quad |z-a|=R$$

are respectively called the *circles of convergence* of the series

$$a_0+a_1z+a_2z^2+a_3z^3+\dots,$$
$$a_0+a_1(z-a)+a_2(z-a)^2+\dots,$$

where R is called the *radius of convergence* of the series.

Example 1. For the geometric series

$$1+z+z^2+z^3+\dots$$

$R=1$ because l = limit superior of $1^{1/n}=1$. So the series converges absolutely in $|z|<1$. On the circle $|z|=1$ and outside it, $|z|\ge 1$, i.e., $|z^n|=|z|^n\ge 1$. So, on and outside the circle, the $(n+1)^{th}$ term z^n does not tend to zero as n tends to ∞. Therefore the necessary condition for convergence (Theorem 5.4) is not satisfied. Hence the series diverges in $|z|\ge 1$.

Example 2. The exponential series $1+z/1!+z^2/2!+z^3/3!+\dots$ is absolutely convergent in the entire finite plane because

$$l=\limsup_{n\to\infty}\left(\frac{1}{n!}\right)^{1/n}=0, \quad R=\infty.$$

Theorem 6.2. A power series is divergent in the exterior of its circle of convergence.

Proof: Consider the power series $a_0+a_1z+a_2z^2+\dots$ Its circle of convergence is $|z|=R$, where $R=1/l$ and

$$l=\limsup_{n\to\infty}|a_n|^{1/n}.$$

Since l is the limit superior of the sequence $\{|a_n|^{1/n}\}$, for any $\in>0$, only a finite number of terms are greater than $1+\in$ but an infinite number of terms are greater than $l-\in$. Thus, for a given $\in$, there exists an infinite number of a_n such that

$$|a_n|^{1/n}>l-\in. \tag{6.2.1}$$

Let z_1 be a point in the exterior of circle of convergence $|z|=1/l$. Then

$$|z_1|>1/l \quad \text{or} \quad l|z_1|-1>0.$$

Now choose ϵ to be $(l\,|\,z_1\,| - 1)/|\,z_1\,|$. Then $l - \varepsilon = 1/|\,z_1\,|$. Thus (6.2.1) becomes that, for an infinity of n,

$$|\,a_n\,|^{1/n} > 1/|\,z_1\,| \quad \text{or} \quad |\,a_n\,|^{1/n}\,|\,z_1\,| > 1$$

i.e.,

$$|\,a_n\,z_1{}^n\,|^{1/n} > 1 \quad \text{or} \quad |\,a_n\,z_1{}^n\,| > 1.$$

So the necessary condition for convergence that $a_n\,z_1{}^n$ should tend to zero as $n \to \infty$ (Theorem 5.4) is not satisfied. Therefore the given series is divergent at z_1. But z_1 is arbitrary. Hence the series diverges in the whole of the exterior of the circle of convergence.

Corollary 1. If $R = 0$, the series is divergent in the extended plane except at $z = 0$ where it is trivially convergent.

Corollary 2. If a power series is divergent at $z = z_1$ and if $|\,z_2\,| > |\,z_1\,|$, then the series is divergent at $z = z_2$ also because $|\,z_1\,|$ itself is either on the circle of convergence or in the exterior of the circle of convergence and consequently $|\,z_2\,|$ is in the exterior of the circle of convergence.

Corollary 3. The series

$$a_0 + a_1\,(z - a) + a_2\,(z - a)^2 + a_3\,(z - a)^3 + \ldots$$

diverges in the exterior of the circle of convergence $|\,z - a\,| = R$.

Remark. A power series converges absolutely in the interior of the circle of convergence and diverges in the exterior of the circle but at points on the circle the series converges or diverges.

Theorem 6.3. If the sequence $\{z_n\}$ converges, then the sequence is bounded (i.e., there exists a constant M such that $|\,z_n\,| \le M$ for all n).

Proof: Let $\{z_n\}$ converge to a. Then, for any $\varepsilon > 0$, there exists an N such that

$$|\,z_n - a\,| < \varepsilon \text{ for all } n > N. \tag{1}$$

Now from $z_n = a + (z_n - a)$ and the triangle inequality, we get

$$|\,z_n\,| \le |\,a\,| + |\,z_n - a\,|$$
$$< |\,a\,| + \varepsilon \text{ by (1), for all } n > N. \tag{2}$$

Let M be the maximum of the $N + 1$ numbers

$$|\,z_1\,|, |\,z_2\,|, |\,z_3\,|, \ldots |\,z_N\,|, |\,a\,| + \varepsilon.$$

then, by (2), we get that

$$|\,z_n\,| \le M, \text{ for all } n.$$

Corollary. If a series is convergent, then there exists a constant M such that

$$\{S_n\} \le M \text{ for all } n.$$

Theorem 6.4. (Abel's theorem 1). If a power series in z is convergent at $z = z_1$, then it converges absolutely in the circular open disc $|\,z\,| < |\,z_1|$.

Proof: Consider the power series

$$a_0 + a_1\,z + a_2\,z^2 + a_3\,z^3 + \ldots$$

which converges at $z = z_1$.Then $a_n\,z_1{}^n \to 0$ as $n \to \infty$ (Theorem 5.4). So, for any $\varepsilon > 0$, there exists an n_0 such that

$$|\,a_n\,z_1{}^n\,| < \varepsilon \qquad \text{for all } n > n_0.$$

Let $M = \max\,(|\,a_0\,|, |\,a_1\,z_1\,|, |\,a_2\,z_1{}^2\,| \ldots |\,a_{n_0}\,z_1{}^{n_0}\,|, \varepsilon)$. Then for all n

$$|\,a_n\,z_1{}^n\,| \le M \quad \text{or} \quad |\,a_n\,| \le \frac{M}{|\,z_1\,|^n}.$$

Now for any z whose modulus is less than that of z_1,

$$|\,a_0\,| + |\,a_1\,z\,| + |\,a_2\,z^2\,| + \ldots = |\,a_0\,| + |\,a_1\,|\,|\,z\,| + \ldots + |\,a_n\,|\,|\,z\,|^n + \ldots$$

$$\leq M + \frac{M}{|z_1|}|z| + \frac{M}{|z_1|^2}|z^2| + ...$$

$$= M(1 + \rho + \rho^2 + ...), \rho = \frac{|z|}{|z_1|}.$$

Since $\rho < 1$, the geometric series $M(1 + \rho + \rho^2 + ...)$ converges and consequently the series in question converges in $|z| < |z_1|$.

Remark. If the series $a_0 + a_1(z-a) + a_2(z-a)^2 + ...$ is convergent at $z = z_1$, then the series is absolutely convergent in $|z-a| < |z_1 - a|$.

6.3. UNIFORM CONVERGENCE OF A POWER SERIES

Theorem 6.5. (Abel's theorem 2). A power series in z converges uniformly in the closed circular disc

$$|z| = r,$$

whose r is less than the radius of convergence of the series.

Proof: Consider the power series.

$$a_0 + a_1 z + a_2 z^2 + a_3 z^3 + ...$$

Let its radius of convergence be R and let z_1 be a point lying in-between the circles $|z| = r$ and $|z| = R$. Then $r < |z_1| < R$ and z_1 is in the interior of the circle of convergence. So the series converges absolutely for $z = z_1$; that is,

$$|a_0| + |a_1 z_1| + |a_2 z_1^2| + ...$$

or

$$|a_0| + |a_1|\rho + |a_2|\rho^2 + ...,$$

where $\rho = |z_1|$, converges. But, for all z in $|z| \leq r$, $|a_n z^n| < |a_n|\rho^n$. Therefore, by Weierstrass M-test (Theorem 5.7), the series in question is uniformly convergent in $|z| \leq r$.

Remark 1. The series $a_0 + a_1(z-a) + a_2(z-a)^2 + ...$ converges uniformly in the closed circular disc

$$|z - a| \leq r < R.$$

Remark 2. A power series is not uniformly convergent in the interior of the circle of convergence. This is evident from the geometric series $1 + z + z^2 + z^3 + ...$ which is not uniformly convergent in the interior of the circle of convergence $|z| = 1$ (See example under Cauchy's criterion of uniform convergence of functional series).

6.4. ANALYTICITY OF THE SUM OF A POWER SERIES

In this section we study the nature of the sum of convergent power series in the interior of the circle of convergence and the term-by-term differentiability of the series therein. We begin with a lemma which will be used in theorem 6.5.

LEMMA. If a and b are two complex numbers, then, for $n \geq 2$,

$$\frac{b^n - a^n}{b - a} - na^{n-1} = (b-a)[b^{n-2} + 2ab^{n-3} + 3a^2 b^{n-4} + ... + (n-1)a^{n-2}]$$

Proof: Differentiating the finite geometric series

$$\frac{b^n - z^n}{b - z} = b^{n-1} + b^{n-2} z + b^{n-3} z^2 + ... + z^{n-1}$$

with respect to z and then multiplying by $(b - z)$,

$$\left\{\frac{(b-z)(-nz^{n-1})+(b^n-z^n)}{(b-z)^2}\right\}(b-z)$$

$$= (b-z)\,[b^{n-2}+2b^{n-3}z+\ldots+(n-1)\,z^{n-2}]$$

i.e.,
$$-nz^{n-1}+\frac{b^n-z^n}{b-z}=(b-z)\,[b^{n-2}+2b^{n-3}z+\ldots+(n-1)\,z^{n-2}].$$

Setting $z = a$, we get the result as stated.

Theorem 6.6. (*i*) The sum of a convergent power series in z is analytic in the interior of its circle of convergence.

(*ii*) A power series can be differentiated term by term in the interior of its circle of convergence; that is the derived series converges to the derivative of the sum of the series.

Proof: Consider the power series in z

$$a_0 + a_1 z + a_2 z^2 + a_3 z^3 + \ldots\ a_n z^n + \ldots = f(z).$$

Its first and second derived series are

$$a_1 + 2a_2 z + 3a_3 z^2 + \ldots + na_n z^{n-1} + \ldots$$
$$2a_2 + 6a_3 z + \ldots + n(n-1)\,a_n z^{n-2} + \ldots$$

Let the radii of convergence of the above three series be R, R_1 and R_2. Multiplication of the terms of a series by z does not alter its radius of convergence. So, multiplying the terms of the first derived series by z and then comparing the moduli of its terms with those of the given series,

$$|na_n z^n| > |a_n z^n| \quad \text{or} \quad (l_1 \geq l)$$

because of the additional n present on the left expression.

$$\therefore \qquad R_1 \leq R.$$

But
$$\limsup_{n\to\infty} (n|a_n|)^{1/n} \leq \left(\limsup_{n\to\infty} n^{1/n}\right)\times\left(\limsup_{n\to\infty} |a_n|^{1/n}\right).$$

$$\therefore \qquad l_1 \leq (1)\,(l)\ \left(\because \limsup_{n\to\infty} n^{1/n} = 1\right)$$

$$\therefore \qquad R_1 \geq R.$$

Now, considering both $R_1 \leq R$ and $R_1 \geq R$,

$$R_1 = R.$$

Similarly $R_2 = R_1$. Consequently $R = R_1 = R_2$; that is, the original series and its derived series have the same circle as the circle of convergence and they are absolutely convergent in $|z| < R$.

In particular, for the second derived series, we can find an M such that

$$\sum_{n=2}^{\infty} |a_n|\, n\,(n-1)\,|z|^{n-2} < M \text{ for all } z \text{ in } |z| < R. \qquad (6.4.1)$$

Let r be an arbitrary poitive number less than R. Take an arbitrary point z in the interior of the circle $|z| = r$ and consider a neighbourhood of z contained in this interior (Fig. 6.4.1). Let $z + a$ be a point in the neighbourhood. Then $|z + a| < r$. Suppose the first derived series converges to $\phi_1(z)$ in $|z| < R$, then

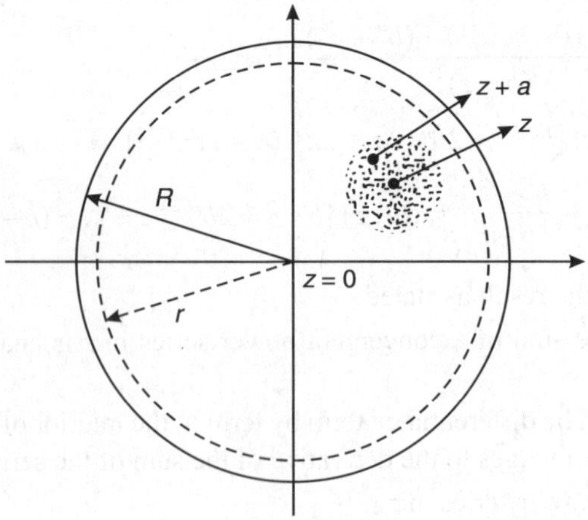

Fig. 6.4.1

$$\{f(z+a)-f(z)\} - a\,\phi_1(z) = [a_0 + a_1(z+a) + \sum_{n=2}^{\infty} a_n (z+a)^n]$$

$$- [a_0 + a_1 z + \sum_{n=2}^{\infty} a_n z^n] - [aa_1 + a \sum_{n=2}^{\infty} n a_n z^{n-1}]$$

$$= \sum_{n=2}^{\infty} a_n [(z+a)^n - z^n - an\,z^{n-1}]$$

$\therefore$
$$\frac{f(z+a)-f(z)}{a} - \phi_1(z) = \sum_{n=2}^{\infty} a_n \left[\frac{(z+a)^n - z^n}{a} - n\,z^{n-1}\right]$$

$$= \sum_{n=2}^{\infty} a_n \left[\frac{(z+a)^n - z^n}{(z+a)-z} - n\,z^{n-1}\right]$$

$$= \sum_{n=2}^{\infty} a_n\, a\,[(z+a)^{n-2} + 2z\,(z+a)^{n-3} + \ldots + (n-1)\,z^{n-2}] \text{ by lemma}$$

$\therefore$
$$\left|\frac{f(z+a)-f(z)}{a} - \phi_1(z)\right|$$

$$< |a| \sum_{n=2}^{\infty} |a_n|\,[r^{n-2} + 2r^{n-2} + 3r^{n-2} + \ldots + (n-1)\,r^{n-2}]$$

$$= |a| \sum_{n=2}^{\infty} |a_n|\,[1+2+3+\ldots+(n-1)]\,r^{n-2}$$

$$= |a| \sum_{n=2}^{\infty} |a_n| \,.\, \tfrac{1}{2}\,(n-1)\,n\,.\,r^{n-2} < |a| \,.\, \tfrac{1}{2} \,.\, M \qquad \text{by (6.4.1)}$$

$\therefore$
$$\lim_{a \to 0} \left|\frac{f(z+a)-f(z)}{a} - \phi_1(z)\right| = 0$$

or
$$\lim_{a \to 0} \left[\frac{f(z+a)-f(z)}{a} - \phi_1(z)\right] = 0$$

$$\therefore \qquad \left[\lim_{a \to 0} \frac{f(z+a) - f(z)}{a}\right] - \phi_1(z) = 0$$

or

$$\lim_{a \to 0} \frac{f(z+a) - f(z)}{a} = \phi_1(z).$$

Thus $f'(z)$ exists in $|z| < r$ and $f'(z) = \phi_1(z)$, that is, $f(z)$ is analytic in $|z| < r$. But r is arbitrary and less than R. Therefore $f(z)$ is analytic in $|z| < R$ and $f'(z) = \phi_1(z)$, that is, the derivative of the sum of a series is the sum of its derived series. Therefore, in conclusion,

(*i*) the sum $f(z)$ of a power series of analytic in $|z| < R$ and

(*ii*) the power series can be differentiated term by term in $|z| < R$, that is, the sum of the derived series is the derivative of the sum of the power series.

Corollary. The sum of a convergent power series in $z - a$ is analytic in the interior of the circle of convergence $|z - a| < R$ and the power series can be differentiated term by term and the sum of the derived series is the derivative of the sum of the original series.

6.5. UNIQUENESS OF REPRESENTATION OF A FUNCTION BY A POWER SERIES

Theorem 6.7. (Uniqueness theorem). If a power series in z converges to a function $f(z)$ in the interior of the circle of convergence, $|z| < R$, then no other power series converges to the same function $f(z)$ in $|z| < R$.

Proof: Suppose the series

$$a_0 + a_1 z + a_2 z^2 + a_3 z^3 + \ldots$$

converges to $f(z)$. Then, setting $z = 0$,

$$a_0 = f(0).$$

Similarly, setting $z = 0$, in the first, second, etc. derived series,

$$a_1 = f'(0), \quad 2.1.a_2 = f''(0), 3.2.1.a_3 = f'''(0), \ldots$$

i.e.,

$$a_0 = f(0), \quad a_1 = \frac{f'(0)}{1!}, a_2 = \frac{f''(0)}{2!}, a_3 = \frac{f'''(0)}{3!}, \ldots$$

Similarly, if another series also converges to the same function $f(z)$, then its coefficients also are

$$f(0), \frac{f'(0)}{1!}, \frac{f''(0)}{2!}, \frac{f'''(0)}{3!}, \ldots$$

in the same manner. This completes the proof.

The theorem establishes the uniqueness of representations of functions by power series.

6.6. ELEMENTARY FUNCTIONS

In chapter 4 we defined polynomials and rational functions. Now we define some more elementary functions namely exponential, logarithmic, circular and hyperbolic functions.

6.7. EXPONENTIAL FUNCTIONS

The function $f(z)$ which is the solution of the differential equation

$$f'(z) = f(z), \quad f(0) = 1$$

is called the *exponential function*. Suppose the series

$$a_0 + a_1 z + a_2 z^2 + a_3 z^3 + \ldots$$

converges to $f(z)$. Then, in the interior of its circle of convergence, it can be differentiated term by term. So, from $f'(z) = f(z)$, comparing the coefficients of

$$a_1 + 2a_2 z + 3a_3 z^2 + ..., \quad a_0 + a_1 z + a_2 z^2 + a_3 z^3 + ...,$$

$$a_1 = a_0, 2a_2 = a_1, 3a_3 = a_2, ..., na_n = a_{n-1}, ...$$

$$\therefore \quad a_n = \frac{a_{n-1}}{n} = \frac{a_{n-2}}{n(n-1)} = ... = \frac{a_0}{n(n-1)...3.2.1} = \frac{a_0}{n!} = \frac{1}{n!}$$

since $a_0 = 1$ due to the fact that $f(0) = 1$. Hence

$$f(z) = 1 + \frac{z}{1!} + \frac{z^2}{2!} + \frac{z^3}{3!} + ... \text{ is the exponential function } e^z.$$

Non-vanishing of e^z. For all finite $z = h + ik$,

$$| e^z | = | e^{h+ik} | = e^h | \cos k + i \sin k | = e^h > 0$$

since the real number e^h is always positive. Hence e^z does not vanish for any z. Also it does not become ∞ because, if $e^{z_1} = \infty$, then $e^{-z_1} = 0$ which is a contraidction.

Region of analyticity. The radius of convergence of this series is ∞ because $R = 1/l$, where

$$l = \limsup_{n \to \infty} (1/n!)^{1/n} = 0.$$

So the exponential series is convergent in the entire finite plane and its sum is analytic there. As such, it is an entire function, that is, a function analytic in the whole of the finite plane. This function is denoted by the symbol

$$e^z \quad \text{or} \quad \exp z.$$

Theorem 6.8. Addition theorem. $e^{z_1} e^{z_2} = e^{z_1 + z_2}$. That is,

$$(\exp z_1)(\exp z_2) = \exp(z_1 + z_2).$$

Proof: Let a be a complex constant. Then for all z in the finite plane

$$\frac{d}{dz}(e^z e^{a-z}) = e^z e^{a-z} + e^z(-e^{a-z}) = 0.$$

$$\therefore \quad e^z e^{a-z} = c$$

a constant. Setting $z = 0$ and remembering $e^0 = 1$, $e^a = c$. Thus

$$e^z e^{a-z} = e^a.$$

Now, setting $z_1 = z$ and $z_2 = a - z$, we get $z_1 + z_2 = a$ and so

$$e^{z_1} e^{z_2} = e^{z_1 + z_2}.$$

Theorem 6.9. (Euler's relation). $e^{x+iy} = e^x(\cos y + i \sin y)$.

Proof: By addition theorem

$$e^{x+iy} = e^x e^{iy} = e^x \left(1 + \frac{iy}{1!} + \frac{i^2 y^2}{2!} + \frac{i^3 y^3}{3!} + ...\right)$$

$$= e^x \left[\left(1 - \frac{y^2}{2!} + \frac{y^4}{4!} - ...\right) + i\left(y - \frac{y^3}{3!} + \frac{y^5}{5!} - ...\right)\right]$$

$$= e^x(\cos y + i \sin y).$$

Remark. The polar form $r(\cos\theta + i\sin\theta)$ of a complex number can be written in exponential as $re^{i\theta}$.

Theorem 6.10. e^z is not defined at $z = \infty$.

Proof: e^z is a single-valued function defined in the finite plane. To check whether e^z is defined at ∞, we have to check whether $e^{1/z}$ is defined at $z = 0$. Now, as z approaches 0 from the right travelling along the real axis,

$$\lim_{z \to 0} |e^{1/z}| = \lim_{h \to 0} e^{1/h} = \infty,$$

where h is positive and, as z approaches 0 from the left travelling along the real axis,

$$\lim_{z \to 0} |e^{1/z}| = \lim_{h \to 0} e^{-1/h} = 0.$$

Therefore the limit of $e^{1/z}$, as z tends to zero, does not exist because it is not unique. Hence e^z is not defined at ∞.

Remark. For all finite $z = h + ik$

$$|e^z| = |e^{h+ik}| = e^h |\cos k + i \sin k| = e^h = e^h > 0$$

since the real number e^h is always positive. Hence e^z does not vanish for any z. Also it does not become ∞ because, if $e^{z_1} = \infty$, then $e^{-z_1} = 0$ which is a contradiction.

Singular point. Since e^z is not defined at $z = \infty$, e^z is not analytic at $z = \infty$ and hence $z = \infty$ is a singular point of e^z.

6.8. LOGARITHMIC FUNCTION AND FUNCTION a^z

The inverse function of the exponential function is the *logarithmic function.* To define this inverse consider the mapping

$$w = e^z.$$

Under this mapping the point $z = x + iy$ goes to w, where

$$w = e^{x+iy} = e^x (\cos y + i \sin y)$$

whose argument is y. This implies that points on the straight line $y = y_0$ of the z-plane go onto the points on the ray arg $w = y_0$ of the w-plane and the mapping is one-to-one. If $w = u + iv$, the ray arg $w = y_0$ is the line $\tan^{-1}(v/u) = y_0$ or $v = (\tan y_0)\, u$ emerging from $w = 0$. Thus, when y_0 takes values from 0 to 2π, the line $y = y_0$ in the z-plane moves upward from the real axis $y = 0$ to $y = 2\pi$ and correspondingly the ray in the w-plane rotates about the point $w = 0$ completely once starting from the positive real semi-axis (Fig. 6.8.1). This implies that the mapping is one-to-one and onto, and the strip $0 \le y < 2\pi$ in the z-plane is mapped onto $0 \le \arg w < 2\pi$ which is the whole of w-plane excluding the two points $w = 0$ and $w = \infty$ since arguments of 0 and ∞ are not defined. But

$$e^{x+iy+2\pi ni} = e^{x+iy}\, e^{2\pi ni} = e^{x+iy}, \; n = 0, \pm 1, \pm 2, \ldots$$

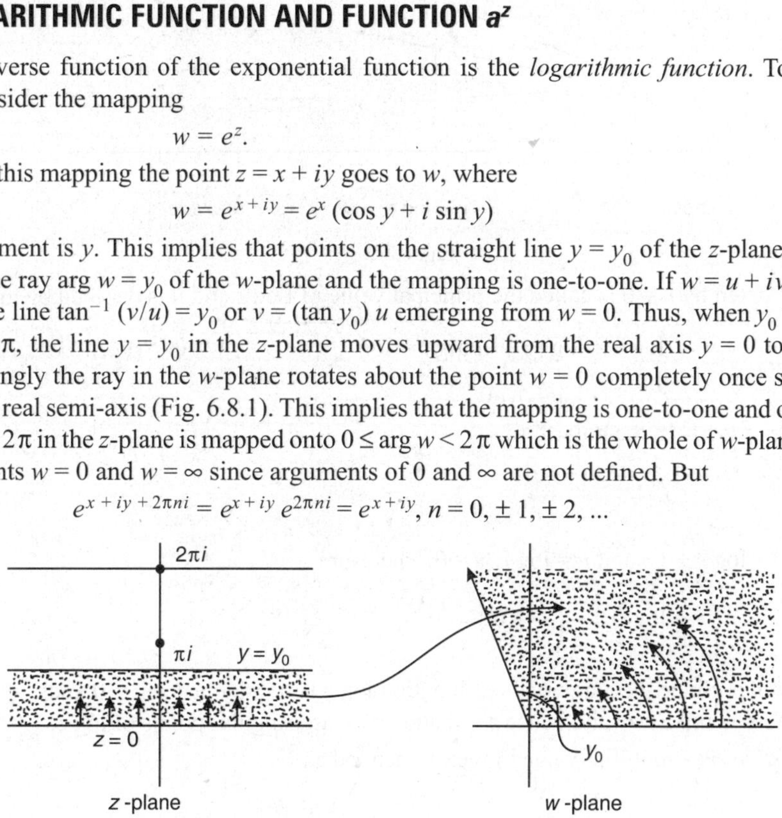

Fig. 6.8.1

So each of the following strips in the z-plane

$$\ldots, -2\pi \le y < 0, \; 0 \le y < 2\pi, \; 2\pi \le y < 4\pi, \ldots \tag{6.8.1}$$

goes onto the w-plane once. That is, for each w there are many z's satisfying $e^z = w$ and each of the strips (6.8.1) contains one z satisfying $e^z = w$. Further, for a given w, the respective values of z of the strips are $\log |w| + i \arg w + n(2\pi i)$ because

$$e^{\log |w| + i \arg w + n(2\pi i)} = e^{\log |w|} e^{i \arg w} e^{2\pi i n}$$

$$= |w| \{\cos \arg w + i \sin \arg w\} = w.$$

Now we are in a position to define the many-valued function, the logarithmic function, its values being those of z satisfying $e^z = w$.

Logarithmic function. Logarithmic function is a multiple-valued function defined in the whole of z-plane (Fig. 6.8.2) except at $z = 0$ and $z = \infty$ such that

$$\text{Log } z = \log |z| + i \arg z + n(2\pi i), \quad n = 0, \pm 1, \pm 2, \ldots$$

$$= \log \sqrt{x^2 + y^2} + i \tan^{-1}(y/x) + n(2\pi i), \quad n = 0, \pm 1, \pm 2, \ldots$$

$$= \log r + i\theta + n(2\pi i), \quad n = 0, \pm 1, \pm 2, \ldots$$

The set of values of Log z given by one value of n is called a branch of Log z. As such, Log z has infinite number of branches.

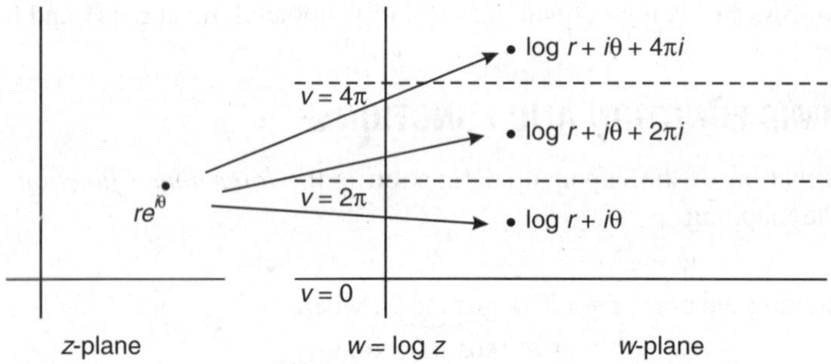

Fig. 6.8.2

The value given by $n = 0$ is called the principal value of Log z and it is denoted by $\log z$ so that

$$\log z = \log |z| + i \arg z = \log \sqrt{x^2 + y^2} + i \tan^{-1}(y/x)$$

$$= \log r + i\theta.$$

Derivative of log z. In the polar form

$$f'(z) = \frac{r}{z}(u_r + iv_r).$$

Now $\log z = \log r + i\theta$ and $u = \log r$, $v = \theta$. Therefore

$$\frac{d}{dz} \log z = \frac{r}{z}\left(\frac{1}{r} + 0\right) = \frac{1}{z}.$$

Singular point. Since the single-valued function $\log z$ is not defined at $z = 0$ and $z = \infty$, $\log z$ is not analytic at $z = 0$ and $z = \infty$ and hence $z = 0$ and $z = \infty$ are singular points of $\log z$.

Function a^z. a^z is a multiple-valued function defined as

$$a^z = e^{z \log a},$$

where a is a complex constant.

6.9. BRANCH POINT

Suppose $w = f(z)$ is a multiple-valued function defined in a region D and having single-valued branches. Suppose there exists a point $z = a$ in D such that, as z travels once along a simple closed curve C in D encircling $a, f(z)$ travels in w plane, stretching across a branch once and, as z travels again along $C, f(z)$ travels in the same manner, stretching through the neighbouring branch once and so on. Then $z = a$ is said to be a branch point of $f(z)$.

Example 1. The function $w = z^{1/2}$ has two branches. One branch is the upper half w-plane and the other branch is the lower half w plane. As z travels along the unit circle C, $z = e^{i\theta}$, $0 \le \theta < 2\pi$ once, the point $z^{1/2}$ travels along the semicircle $w = e^{i\phi}$, $0 \le \phi < \pi$ once, and, as z travels again along C, $z = e^{i\theta}$, $2\pi \le \theta < 4\pi$, $z^{1/2}$ travels along the lower semicircle $w = e^{i\phi}$, $\pi \le \phi < 2\pi$. So $z = 0$ is a branch point of $f(z) = z^{1/2}$.

Now $z = \infty$ is also a branch point of $z^{1/2}$ because every circuit around $z = \infty$ along an arbitrarily large circle $|z| = R$ is also simultaneously a circuit around $z = 0$.

Example 2. $z = 0$ and $z = \infty$ are the branch points of $z^{1/n}$, where n is a positive integer. These branch points are said to be of order $n - 1$.

Example 3. $z = 1$ and $z = \infty$ are the branch points of $(z - 1)^{1/2}$, each of order 1.

Example 4. $z = 1$ and $z = 2$ are the branch points of the function $\{(z - 1)/(z - 2)\}^{1/2}$, each of order 1.

Example 5. $z = 0$ is a branch point of log z because, as z travels along the unit circle $z = e^{i\theta}$, $0 \le \theta < 2\pi$, once, Log z travels along the straight line joining (0, 0) and (0, 2π) stretching across the branch $0 < v < 2\pi$ once. The order of this branch point is ∞. Also $z = \infty$ is a branch point of Log z because $z = 0$ is branch point of Log $(1/z)$.

6.10. TRIGONOMETRIC FUNCTIONS

First we define sine and cosine of a complex number z as the functions

$$\cos z = \frac{e^{iz} + e^{-iz}}{2}, \quad \sin z = \frac{e^{iz} - e^{-iz}}{2i}.$$

Like e^z, the functions sin z and cos z are defined everywhere in the finite plane and not at $z = \infty$. They are differentiable in the entire finite plane and so they are entire functions. Now

$$\frac{d}{dz} \sin z = \cos z, \quad \frac{d}{dz} \cos z = -\sin z.$$

Also tan z, cot z, sec z and cosec z are defined as

$$\tan z = \frac{\sin z}{\cos z}, \quad \cot z = \frac{\cos z}{\sin z},$$

$$\sec z = \frac{1}{\cos z}, \quad \text{cosec } z = \frac{1}{\sin z}.$$

tan z and sec z are analytic in a bounded region in which sin $z \ne 0$ and cot z and cosec z are analytic in a bounded region in which cos $z \ne 0$.

6.11. HYPERBOLIC FUNCTIONS

Hyperbolic functions are defined as

$$\cosh z = \frac{e^z + e^{-z}}{2}, \quad \sinh z = \frac{e^z - e^{-z}}{2},$$

$$\tanh z = (\sinh z)/\cosh z, \quad \coth z = (\cosh z)/\sinh z,$$

$$\text{sech } z = 1/\cosh z, \quad \text{cosech } z = 1/\sinh z.$$

With these definitions, we see that

$$\sin iz = i \sinh z, \quad \cos iz = \cosh z.$$

and $\cos^2 z + \sin^2 z = 1$, $\cosh^2 z - \sinh^2 z = 1$, ...

$$\sinh iz = i \sin z, \quad \cosh iz = \cos z.$$

6.12. HARMONIC FUNCTIONS

Harmonic function. A real-valued function $u(x, y)$ defined in a region is said to be harmonic

(*i*) if its second partials $u_{xx}, u_{yy}, u_{xy}, u_{yx}$ exist and are continuous, and

(*ii*) if it satisfies the Laplace equation

$$\frac{\partial^2 u}{\partial x^2} + \frac{\partial^2 u}{\partial y^2} = 0.$$

i.e., $$\nabla^2 u = 0, \text{ where } \nabla^2 = \frac{\partial^2}{\partial x^2} + \frac{\partial^2}{\partial y^2}.$$

Conjugate harmonic functions. Two harmonic functions are said to be conjugate harmonic functions if they satisfy the C.R. equations.

Harmonic conjugate. If $u + iv$ is analytic, then u and v are harmonic functions and of them, u is said to be the harmonic conjugate of u.

Of two conjugate harmonic functions, one is called the harmonic conjugate of the other.

Remark. Suppose u and v are conjugate harmonic functions. Then the C.R. equations are true. That is,

$$u_x = v_y, \quad u_y = -v_x.$$

But v and u are not conjugate harmonic functions because they do not satisfy the C.R. equations because

$$v_x \neq u_y, \quad v_y \neq -u_x.$$

However, $-v$ and u are conjugate harmonic functions. Thus in the conjugate harmonic functions the order in which they are written cannot be changed.

Theorem 6.11. A function $f(z) = u(x, y) + iv(x, y)$ defined in a region D is analytic in it if and only if $u(x, y)$ and $v(x, y)$ are congugate harmonic functions.

Proof: *Necessity part.* Suppose $f(z)$ is analytic in D. Then $f'(z), f''(z), f'''(z)$, ... are analytic in D (vide result 8.9.4). Therefore u_x, u_y, v_x, v_y exist and

$$u_x = v_y, \quad u_y = -v_x.$$

Also $u_{xx}, u_{yy}, u_{xy}, u_{yx}, v_{xx}, v_{yy}, v_{xy}, v_{yx}$ exist and are continuous. Due to continuity, in u_{xy}, u_{yx} and v_{xy}, v_{yx} the order of differentiation is immaterial. That is,

$$u_{xy} = u_{yx}, \quad v_{xy} = v_{yx}.$$

Differentiating $u_x = v_y$ and $u_y = -v_x$ with respect to x and y respectively and then adding,

$$u_{xx} + u_{yy} = 0.$$

Again differentiating them with respect to y and x respectively and then subtracting,

$$v_{xx} + v_{xy} = 0.$$

Hence $u(x, y)$ and $v(x, y)$ are conjugate harmonic functions.

Remark 1. Suppose $u + iv$ is analyic. Then it may be noted that $v + iu$ is not analytic since v, u do not satisfy the C-R equations. So u is not the conjugate of v.

For example, considering z^2, we can see that

$$u = x^2 - y^2, \quad v = 2xy$$

satisfy the C-R equations but

$$u = 2xy, \quad v = x^2 - y^2$$

do not satisfy the C-R equations.

Remark 2. Now we shall show that, if u and v are any two harmonic functions, then $u + iv$ may or may not be analytic.

If $u = x^2 - y^2$, $v = 4xy$, then

$$u_{xx} + u_{yy} = 0, \quad v_{xx} + v_{yy} = 0.$$

So u and v are harmonic functions. But

$$u_x = 2x, \quad v_y = 4x.$$

Hence the first equation of the C-R equations, $u_x = v_y$, is not satisfied. Therefore $u + iv$ is not analytic.

Here note that $x^2 - y^2$ is the real part of z^2 and $4xy$ is the imaginary part of a different function $2z^2$.

Remark 3. If u or v is not harmonic, then $u + iv$ is not analytic.

Sufficiency part. Now the hypothesis is that $u(x, y)$ and $v(x, y)$ are conjugate harmonic functions. Therefore

(*i*) the second partialsof u and v are continuous and

(*ii*) u and v satisfy the C.R. equations.

Since the second partials exist, the first partials u_x, u_y, v_x, v_y exist and are continuous. Now u and v satisfy the C.R. equations. Therefore, by theorem 4.9, $f(z)$ is analytic in D.

HARMONIC FUNCTIONS IN POLAR COORDINATES

The Laplacian operator in cartesian coordinates is

$$\nabla^2 = \frac{\partial^2}{\partial x^2} + \frac{\partial^2}{\partial y^2}.$$

In polar coordinates, it is

$$\nabla^2 = \frac{\partial^2}{\partial r^2} + \frac{1}{r}\frac{\partial}{\partial r} + \frac{1}{r^2}\frac{\partial^2}{\partial \theta^2}.$$

Thus, if $\phi(r, \theta)$ is harmonic function, then

$$\phi_{rr} + \frac{1}{r}\phi_r + \frac{1}{r^2}\phi_{\theta\theta} = 0.$$

Theorem 6.12. If $f(z) = u(r, \theta) + iv(r, \theta)$ is an analytic function, then $u(r, \theta)$ and $v(r, \theta)$ are harmonic functions.

Proof: The C-R equations in polar coordinates are

$$u_r = \frac{1}{r} v_\theta, \quad u_\theta = -r v_r.$$

Now differentiating u_r and u_θ partially w.r.t θ and r,

$$u_{\theta r} = \frac{1}{r} v_{\theta\theta}, \quad u_{r\theta} = -(v_r + r v_{rr}).$$

Since $u_{\theta r} = u_{r\theta}$, we get

$$\frac{1}{r} v_{\theta\theta} = -(v_r + r\, v_{rr}).$$

Dividing this by r,

$$v_{rr} + \frac{1}{r} v_r + \frac{1}{r^2} v_{\theta\theta} = 0.$$

Thus v is a harmonic function. Again, from the C-R equations,

$$v_r = -\frac{1}{r} u_\theta, \quad v_\theta = r u_r,$$

differentiating v_r and v_θ partially w.r.t θ and r, and equating $v_{\theta r}$ and $v_{r\theta}$, we get

$$-\frac{1}{r} u_{\theta\theta} = u_r + r u_{rr}$$

$$\therefore \qquad u_{rr} + \frac{1}{r} u_r + \frac{1}{r^2} u_{\theta\theta} = 0.$$

So u also is a harmonic function.

ANALYTIC FUNCTION WHEN ITS REAL OR IMAGINARY PART IS GIVEN

When $f(z) = u(x, y) + iv(x, y)$ is analytic in a region and when one of $u(x, y)$ and $v(x, y)$ is given, we can find the other and then $f(z)$. In other words, when a harmonic function $u(x, y)$ is given we can find another harmonic function $v(x, y)$ so that $u + iv$ is analytic.

Method 1. (Exact differential method). Suppose the harmonic function $u(x, y)$ is given. Now

$$dv = v_x\, dx + v_y\, dy$$

is an exact differential where v_x and v_y are known from u through the C.R. equations. Therefore

$$v = \int v_x\, dx \text{ treating } y \text{ as a constant}$$

$$+ \int (\text{terms of } v_y \text{ not containing } x)\, dy.$$

Illustration 1. Consider the harmonic function $u = x^2 - y^2$. Then $v_x = -u_y = 2y$, $v_y = u_x = 2x$.

$$\therefore \qquad v = \int 2y\, dx + \int (0)\, dy = 2xy + 0 = 2xy.$$

$$\therefore \qquad f(z) = x^2 - y^2 + i(2xy) + C = (x + iy)^2 + C = z^2 + C.$$

Method 2. (Substitution method). This method provides us with a formal formula to write down at once $f(z)$ itself when u or v or $u + v$ or $u - v$ or $2u + v$ or $u - 2v$ is given. We know that, if $f(z)$ is analytic, then $\partial f / \partial \bar{z} = 0$, that is, $f(z)$ is independent of $\bar{z}$. So, if we write

$$f(z) = u\left(\frac{z+\bar{z}}{2}, \frac{z-\bar{z}}{2i}\right) + iv\left(\frac{z+\bar{z}}{2}, \frac{z-\bar{z}}{2i}\right) \tag{6.12.1}$$

then $f(z)$ is independent of $\bar{z}$ which means that $\bar{z}$ can be replaced by any number without altering the value of $f(z)$. Also, conjugation of (6.12.1) gives,

$$\overline{f(z)} = u\left(\frac{z+\bar{z}}{2}, \frac{z-\bar{z}}{2i}\right) - iv\left(\frac{z+\bar{z}}{2}, \frac{z-\bar{z}}{2i}\right) \tag{6.12.2}$$

since $(z + \bar{z})/2$, $(z - \bar{z})/2i$, being real, remain unchanged. Now addition of (6.12.1) and (6.12.2) gives

$$f(z) = 2u\left(\frac{z+\bar{z}}{2}, \frac{z-\bar{z}}{2i}\right) - \overline{f(z)}. \tag{6.12.3}$$

If $\bar{z}$ is replaced by any number a, then (6.12.3) becomes

$$f(z) = 2u\left(\frac{z+a}{2}, \frac{z-a}{2i}\right) - \overline{f(a)},$$

provided $\overline{f(a)}$ exists or becomes

$$f(z) = 2u\left(\frac{z+a}{2}, \frac{z-a}{2i}\right) \tag{6.12.4}$$

apart from an arbitrary additive constant. This formula can be simplified as

$$f(z) = 2u\left(\frac{z}{2}, \frac{z}{2i}\right) = 2u\left(\frac{1}{2}z, -\frac{1}{2}iz\right) \tag{6.12.5}$$

by choosing $a = 0$ provided $\overline{f(0)}$ exists, that is, $u(0, 0) - iv(0, 0)$ or, in particular, $u(0, 0)$ exists, apart from an arbitrary additive constant.

Remark. If $u(0, 0)$ is either infinity or undefined, then (6.12.5) fails to give $f(z)$. In such a case (6.12.4) may be used with a particular choice for a other than 0.

Illustration 2. If $u(x, y) = x^2 - y^2$, from (6.12.5), setting $\frac{1}{2}z$ for x and $-\frac{1}{2}iz$ for y,

$$u\left(\tfrac{1}{2}z, -\tfrac{1}{2}iz\right) = \left(\tfrac{1}{2}z\right)^2 - \left(-\tfrac{1}{2}iz\right)^2 = \tfrac{1}{2}z^2.$$

$$\therefore \qquad f(z) = 2u\left(\tfrac{1}{2}z, -\tfrac{1}{2}iz\right) = z^2.$$

Illustration 3. Consider the harmonic function $u = \frac{1}{2}\log(x^2 + y^2)$. In this case $u(0, 0) = -\infty$. So we cannot use the formula (6.12.5). Alternatively, using (6.12.4) with a choice $a = 1$,

$$f(z) = 2 \cdot \tfrac{1}{2} \cdot \log\left[\left\{\tfrac{1}{2}(z+1)\right\}^2 + \left\{-\tfrac{1}{2}i(z-1)\right\}^2\right] = \log z. \qquad \textbf{(1978, 80)}$$

Corollary to method 2. Suppose $f(z) = u + iv$ is an analytic function. Now

$$-if(z) = v + i(-u),$$
$$(1-i)f(z) = (u+v) + i(v-u),$$
$$(1+i)f(z) = (u-v) + i(u+v).$$

So, if v, $u+v$, $u-v$ are given, we can find $(-u)$, $v-u$, $u+v$, the imaginary parts on the left, using (6.12.4) if $v(a)$, $u(a)+v(a)$, $u(a)-v(a)$ exist. Thus the corresponding analytic functions $f(z)$'s are

$$f(z) = 2iv\left(\frac{z+a}{2}, \frac{z-a}{2i}\right) \tag{6.12.6}$$

$$f(z) = \frac{2}{1-i}\left[(u+v)\left(\frac{z+a}{2}, \frac{z-a}{2i}\right)\right] \tag{6.12.7}$$

$$f(z) = \frac{2}{1+i}\left[(u-v)\left(\frac{z+a}{2}, \frac{z-a}{2i}\right)\right]. \tag{6.12.8}$$

Illustration 4. If $u + v = \dfrac{\sin 2x}{\cosh 2y - \cos 2x}$, in (6.12.7) we cannot choose $a = 0$ because $u(0, 0) + v(0, 0)$ is not defined. So, choosing $a = \frac{1}{2}\pi$,

$$f(z) = \frac{2}{1-i} \cdot \frac{\sin 2\left\{\frac{1}{2}\left(z+\frac{1}{2}\pi\right)\right\}}{\cosh 2\left\{-\frac{1}{2}i\left(z-\frac{1}{2}\pi\right)\right\} - \cos 2\left\{\frac{1}{2}\left(z+\frac{1}{2}\pi\right)\right\}}$$

$$= \frac{2}{1-i} \cdot \frac{\sin\left(z+\frac{1}{2}\pi\right)}{\cos\left(z-\frac{1}{2}\pi\right) - \cos\left(z+\frac{1}{2}\pi\right)}$$

$$= \frac{2}{1-i} \cdot \frac{\cos z}{2 \sin z} = \frac{\cot z}{1-i}.$$

Method 3. (Milne-Thomson method). Now $f'(z)$ is analytic and it is

$$f'(z) = u_x(x, y) + iv_x(x, y) = u_x(x, y) - iu_y(x, y)$$

$$= u_x\left[\tfrac{1}{2}(z+\bar{z}), -\tfrac{1}{2}i(z-\bar{z})\right] - iu_y\left[\tfrac{1}{2}(z+\bar{z}), -\tfrac{1}{2}i(z-\bar{z})\right].$$

Replacing $\bar{z}$ by z,

$$f'(z) = u_x(z, 0) - iu_y(z, 0).$$

$$\therefore \qquad f(z) = \int [u_x(z, 0) - iu_y(z, 0)]\, dz.$$

Similarly, if v is known,

$$f(z) = \int [v_y(z, 0) + iv_x(z, 0)]\, dz.$$

Illustration 5. Let $u(x, y) = x^2 - y^2$. Then

$$u_x(x, y) = 2x,\ u_y(x, y) = -2y,\ u_x(z, 0) = 2z,\ u_y(z, 0) = 0.$$

$$\therefore \qquad f(z) = \int [2z - i(0)]\, dz = z^2.$$

VELOCITY POTENTIAL AND STREAM FUNCTION

Suppose the complex function

$$f(z) = u(x, y) + iv(x, y)$$

is analytic. Then the real part $u(x, y)$ is the velocity potential and the imaginary part $v(x, y)$ is the steam function in some two dimensional steady state irrotational flows in Fluid Dynamics, Thermodynamics and Electronics. In this, $f(z)$ is called the complex potential of the flow.

The family of equipotential lines

$$u(x, y) = c_1$$

are orthogonal to the family of stream lines

$$v(x, y) = c_2,$$

where c_1 and c_2 are parameters.

Notation. In general, the velocity potential and the stream function are denoted by

$$\phi(x, y),\ \psi(x, y)$$

so that, in two-dimensional steady state flows, the complex potential

$$f(z) = \phi(x, y) + i\psi(x, y)$$

is analytic.

Illustration 5. In a two dimensional flow, the stream function is $\psi = \tan^{-1} y/x$. Find the velocity potential ϕ.

✍ We shall denote ϕ, ψ by u, v. Then $f(z) = u + iv$ is an analytic function.

$$v_y = \frac{1}{1 + y^2/x^2} \cdot \frac{1}{x} = \frac{x}{x^2 + y^2},$$

$$v_x = \frac{1}{1 + y^2/x^2}\left(-\frac{y}{x^2}\right) = -\frac{y}{x^2 + y^2}.$$

Consider the exact equation,

$$du = u_x\, dx + u_y\, dy$$
$$= v_y\, dx + (-v_x)\, dy \text{ by C-R equations.}$$

$\therefore$
$$u = \int v_y\, dx \text{ treating } y \text{ as a constant}$$
$$+ \int [\text{Terms of } (-v_x) \text{ not containing } x]\, dy$$
$$= \int \frac{x}{x^2 + y^2}\, dx + \int 0\, dy = \frac{1}{2} \log (x^2 + y^2).$$

So the velocity potential ϕ is

$$\phi = 1/2 \log (x^2 + y^2).$$

6.13. EXAMPLES

6.2. Absolute Convergence of a Power Series

1. Obtain the radii of convergence of the following series (wherever necessary, use limit of $n^{1/n}$ as $n \to \infty$ is 1 and limit of $(n^n/n\,!)^{1/n}$ as $n \to \infty$ is e):

(*i*) $\sum\limits_{n=1}^{\infty} \frac{n}{2^n} z^n$ (*ii*) $\sum\limits_{n=1}^{\infty} \frac{n\,!}{n^n} z^n$

(*iii*) $\sum\limits_{n=1}^{\infty} n^2 z^n$ (*iv*) $\sum\limits_{n=1}^{\infty} n^k z^n$

(*v*) $\sum\limits_{n=1}^{\infty} n^n z^n$ (*vi*) $\sum\limits_{n=1}^{\infty} n\,! z^n$

(*vii*) $\sum\limits_{n=1}^{\infty} \frac{n^k}{n\,!} z^n$ (*viii*) $\sum\limits_{n=1}^{\infty} \frac{2^n}{n\,!} z^n$ **(1983)**

(*ix*) $\sum \frac{2 + in}{2^n} z^n$ (*x*) $\sum\limits_{n=1}^{\infty} \frac{1}{(1 + in^2)\, 2^n} z^n$

(*xi*) $\sum \left(1 - \frac{1}{n}\right)^{n^2} z^n$ **(1981)** (*xii*) $\sum 2^{\sqrt{n}} z^n$

(*xiii*) $\sum \left(1 + \frac{1}{n}\right)^{n^2} z^n$ (*xiv*) $\sum \frac{1}{n^p} z^n$

[*Hint* : Use $\frac{1}{R} = \lim\limits_{n \to \infty} |a_n|^{1/n}$ or $\lim\limits_{n \to \infty} \left|\frac{a_{n+1}}{a_n}\right|$.

(i) $\frac{1}{R} = \lim_{n \to \infty} \left(\frac{n}{2^n} \right)^{1/n} = \lim_{n \to \infty} \left(\frac{n^{1/n}}{2} \right) = \frac{1}{2};\ R = 2$

(ii) $\frac{1}{R} = \lim_{n \to \infty} \left(\frac{n!}{n^n} \right)^{\frac{1}{n}} \frac{1}{e};\ R = e$

(iii) $\frac{1}{R} = \lim_{n \to \infty} \frac{(n+1)^2}{n^2} = \lim_{n \to \infty} \left(1 + \frac{1}{n} \right)^2 = 1;\ R = 1$

(iv) $\frac{1}{R} = \lim_{n \to \infty} \left(\frac{n+1}{n} \right)^k = 1;\ R = 1$

(v) $\frac{1}{R} = \lim_{n \to \infty} (n^n)^{1/n} = \lim_{n \to \infty} n = \infty;\ R = 0$

(vi) $\frac{1}{R} = \lim_{n \to \infty} \left(\frac{n!}{n^n} n^n \right)^{1/n} = \left(\lim_{n \to \infty} \left(\frac{n!}{n^n} \right)^{1/n} \right) \left(\lim_{n \to \infty} n \right) = \infty;\ R = 0$

(vii) $\frac{1}{R} = \lim_{n \to \infty} \left(\frac{n^k}{n^n} \frac{n^n}{n!} \right)^{1/n} = \lim_{n \to \infty} \left(\frac{1}{n^{n-k}} \frac{n^n}{n!} \right)^{1/n} = 0;\ R = \infty$

(viii) $\frac{1}{R} = \lim_{n \to \infty} \left(\frac{2^n}{n^n} \frac{n^n}{n!} \right)^{1/n} = 0;\ R = \infty$

(ix) $\frac{1}{R} = \lim_{n \to \infty} \left\{ \frac{2 + i(n+1)}{2^{n+1}} \cdot \frac{2^n}{2 + in} \right\} = \frac{1}{2};\ R = 2$

(x) $\frac{1}{R} = \lim_{n \to \infty} \frac{(1 + in^2)\, 2^n}{\{1 + i(n+1)^2\}\, 2^{n+1}} = \frac{1}{2};\ R = 2$

(xi) $\frac{1}{R} = \lim_{n \to \infty} \left\{ \left(1 - \frac{1}{n} \right)^{n^2} \right\}^{1/n} = \lim_{n \to \infty} \left(1 - \frac{1}{n} \right)^n = \frac{1}{e};\ R = e$

(xii) $\frac{1}{R} = \lim_{n \to \infty} \left(2^{\sqrt{n}} \right)^{\frac{1}{n}} = \lim_{n \to \infty} 2^{\sqrt{\frac{1}{n}}} = 1;\ R = 1$

(xiii) $\frac{1}{R} = \lim_{n \to \infty} \left(1 + \frac{1}{n} \right)^n = e;\ R = \frac{1}{e}$

(xiv) $\frac{1}{R} = \lim_{n \to \infty} \frac{1}{(n+1)^p} \cdot n^p = \lim_{n \to \infty} \frac{1}{\left(1 + \frac{1}{n}\right)^p} = 1;\ R = 1$]

2. Show that $\sum_{n=1}^{\infty} \frac{z^n}{n(n+1)}$ converges absolutely in the disc $|z| \leq 1$. **(1981)**

[*Hint* : $\frac{1}{R} = \lim_{n\to\infty} \frac{n(n+1)}{(n+1)(n+2)} = 1$; $R = 1$. On the circle $|z| = 1$, $\left|\frac{z^n}{n(n+1)}\right| = \frac{1}{n(n+1)} < \frac{1}{n^2}$,

where $\sum \frac{1}{n^2}$ is convergent.]

3. Solved sum. Find the domain in which the following geometric series converge:

(*i*) $\sum_{n=1}^{\infty} \left(\frac{z+i}{2i}\right)^n$ (*ii*) $\sum_{n=1}^{\infty} \left(\frac{iz-1}{2+i}\right)^n$ (iii) $\sum_{n=1}^{\infty} \frac{1}{(1+z^2)^n}$

(*i*) The geometric series converges in

$$\left|\frac{z+i}{2i}\right| < 1 \quad \text{or} \quad |z-(-i)| < 2$$

That is, the series converges in the interior of the circle whose centre is at $z = -i$ and radius is 2.

(*ii*) The geometric series converges in

$$\left|\frac{iz-1}{2+i}\right| < 1 \quad \text{or} \quad \frac{|i(z+i)|}{\sqrt{5}} < 1 \quad \text{or} \quad |z-(-i)| < \sqrt{5}.$$

That is, the series converges in the interior of the circle whose centre is at $z = -i$ and radius is $\sqrt{5}$.

(*iii*) The geometric series converges in

$$\left|\frac{1}{z^2+1}\right| < 1 \quad \text{or} \quad |z^2+1| > 1 \quad \text{or} \quad |z+i||z-i| > 1$$

6.10. Trigonometric Functions

4. Prove that

(*i*) $\sin(z_1 + z_2) = \sin z_1 \cos z_2 + \cos z_1 \sin z_2$

(*ii*) $\cos(z_1 + z_2) = \cos z_1 \cos z_2 - \sin z_1 \sin z_2$

(*iii*) $\sin(\frac{1}{2}\pi - z) = \cos z, \cos(\frac{1}{2}\pi - z) \ \sin z.$

[*Hint* : (*i*) $\frac{e^{iz_1} - e^{-iz_1}}{2i} \frac{e^{iz_2} + e^{-iz_2}}{2} + \frac{e^{iz_1} + e^{-iz_1}}{2} \frac{e^{iz_2} - e^{-iz_2}}{2i} = \{e^{i(z_1+z_2)} - e^{-i(z_1+z_2)}\}/2i.$]

5. Show that $\sin 2z = 2 \sin z \cos z$.

6.11. Hyperbolic Function

6. Establish the following results and deduce that the functions $\sin z$ and $\cos z$ are not bounded:

(*i*) $|\sin z|^2 = \sin^2 x + \sinh^2 y$ (*ii*) $|\cos z|^2 = \cos^2 x + \sinh^2 y$.

[*Hint* : $|\sin z|^2 = |\sin x \cosh y + i \cos x \sinh y|^2$; use $\cosh^2 y = 1 + \sinh^2 y$, $\cos^2 x = 1 - \sin^2 x$ to get (*i*). Now $\sinh y = \frac{1}{2}(e^y - e^{-y})$ is unbounded]

7. From $\sin^2 z + \cos^2 z = 1$ it appears that $\sin z$ and $\cos z$ are bounded. Explain the illusion.

[*Hint* : $\cos^2 z = \frac{1}{2} + \frac{1}{2}\cos 2z$, $\sin^2 z = \frac{1}{2} - \frac{1}{2}\cos 2z$. But $\cos 2z$ is unbounded (vide previous sum)].

6.12. Harmonic Functions

8. Prove that both real and imaginary parts of an analytic function are harmonic.

9. Give the Laplace equation in polar coordinates.

[**Ans.** $\nabla^2 = \frac{\partial^2}{\partial r^2} + \frac{1}{r}\frac{\partial}{\partial r} + \frac{1}{r^2}\frac{\partial^2}{\partial \theta^2}$]

10. Give an example of a function such that u and v are harmonic but $u + iv$ is not analytic.

11. True or false: If u and v are harmonic, then $u + iv$ is an analytic function.

[**Ans.** False]

12. If u and v are harmonic, then can we say that $u + v$ is analytic?

[**Ans.** No. The function is $(u + v) + i0$]

13. Prove that $x^2 - y^2 - 2xy - 2x + 3y$ is harmonic.

[*Hint* : $u_{xx} = 2$, $u_{yy} = -2$. So $u_{xx} + u_{yy} = 0$]

14. The value of m such that $2x - x^2 + my^2$ may be harmonic is

1. 3 **2.** 1 **3.** 2 **4.** 3

[*Hint* : $u_{xx} + u_{yy} = 0 \Rightarrow (-2) + 2m = 0$. So $m = 1$]

15. Prove that, if $u = x^2 - y^2$, $v = \frac{y}{x^2 + y^2}$, then u and v satisfy the Laplace equation but $u + iv$ is not analytic.

[*Hint* : Show that $\nabla^2 u = 0$, $\nabla^2 v = 0$ and $u_x \neq v_y$]

16. Is there an analytic function $f(z) = u + iv$ for which $v = e^{y/x}$?

[*Hint* : $v_{xx} = \frac{ye^{y/x}(y + 2x)}{x^4}$, $v_{yy} = \frac{e^{y/x}}{x^2}$. Since $v_{xx} + v_{yy} \neq 0$, v is not harmonic and $f(z)$ is not analytic]

17. If $f(z)$ is analytic, prove that

(*i*) $\nabla^2 \log |f(z)| = 0$

(*ii*) $\nabla^2 |f(z)|^p = p^2 |f(z)|^{p-2} |f'(z)|^2$

[*Hint* : (*ii*) $\nabla^2 |f(z)|^p = 4\frac{\partial}{\partial z}\frac{\partial}{\partial \bar{z}}[f(z) \,.\, \overline{f(z)}]^{p/2}$]

18. Given the function $w = z^3$, where $w = u + iv$, show that u and v satisfy the C-R equations and the Laplace equation and that the families of curves $u = c_1$ and $v = c_2$ are orthogonal to each other.

[*Hint* : $u = x^3 - 3xy^2$, $v = 3x^2y - y^3$; The slopes of the curves are $\frac{x^2 - y^2}{2xy}$ and $-\frac{2xy}{x^2 - y^2}$]

19. Solved sum. If $e^z = u + iv$, then show that u and v are harmonic functions.

✍ Since $e^z = e^{x+iy} = e^x e^{iy} = e^x(\cos y + i \sin y)$,

$u = e^x \cos y$	$v = e^x \sin y$
$u_x = e^x \cos y$	$v_x = e^x \sin y$
$u_{xx} = e^x \cos y$	$v_{xx} = e^x \sin y$
$u_y = -e^x \sin y$	$v_y = e^x \cos y$
$u_{yy} = -e^x \cos y$	$v_{yy} = -e^x \sin y$
$u_{xx} + u_{yy} = 0$	$v_{xx} + v_{yy} = 0.$

Further the second partials of u and v are continuous. So u and v are harmonic functions.

20. Solved sum. If $z^3 = u + iv$, then show that u and v are harmonic functions.

Now $z^3 = (x + iy)^3 = x^3 + 3x^2 (iy) + 3x (iy)^2 + (iy)^3$,

$u = x^3 - 3xy^2$	$v = 3x^2y - y^3$
$u_{xx} = 6x$	$v_{xx} = 6y$
$u_{yy} = -6x$	$v_{yy} = -6y$
$u_{xx} + u_{yy} = 0$	$v_{xx} + v_{yy} = 0.$

Further the second partials of u and v are continuous. So u and v are harmonic functions.

21. Solved sum. If u and v are harmonic functions, prove that

$$(u_y - v_x) + i\,(u_x + v_y)$$

is an analytic function.

Since u and v are harmonic functions,

$$u_{xx} + u_{yy} = 0, \quad v_{xx} + v_{yy} = 0, \tag{1}$$

$$u_{xy} = u_{yx}, \quad v_{xy} = v_{yx}. \tag{2}$$

Let $U = u_y - v_x$, $V = u_x + v_y$. Then

$$\begin{aligned} U_x - V_y &= (u_{xy} - v_{xx}) - (u_{yx} + v_{yy}) \\ &= (u_{xy} - u_{yx}) - (v_{xx} + v_{yy}) \\ &= 0 \text{ by (1), (2).} \end{aligned}$$

$$\therefore \quad U_x = V_y.$$

$$\begin{aligned} U_y + V_x &= (u_{yy} - v_{yx}) + (u_{xx} + v_{xy}) \\ &= (u_{xx} + u_{yy}) - (v_{yx} - v_{xy}) \\ &= 0 \text{ by (1), (2).} \end{aligned}$$

So $U_x = V_y$, $U_y = -V_x$. Therefore U and V satisfy the C-R equations. Further, since u and v are harmonic functions, their second partials are continuous and consequently the first partials of U and V are continuous. Thus $U + iV$ satisfies the sufficiency conditions for differentiability and hence $U + iV$ is analytic.

22. Solved sum. Show that

$$\frac{\partial^2}{\partial x^2} + \frac{\partial^2}{\partial y^2} = 4\frac{\partial^2}{\partial z\, \partial \bar{z}}.$$

Let f be a function of x, y, where x, y are functions of z, $\bar{z}$, that is,

$$x = \frac{z + \bar{z}}{2}, \quad y = \frac{z - \bar{z}}{2i}.$$

$$\frac{\partial f}{\partial z} = \frac{\partial f}{\partial x}\frac{\partial x}{\partial z} + \frac{\partial f}{\partial y}\frac{\partial y}{\partial z} = \frac{\partial f}{\partial x}\frac{1}{2} + \frac{\partial f}{\partial y}\frac{1}{2i} = \frac{1}{2}\left(\frac{\partial}{\partial x} + \frac{1}{i}\frac{\partial}{\partial y}\right) f.$$

$$\therefore \quad 2\frac{\partial}{\partial z} = \frac{\partial}{\partial x} + \frac{1}{i}\frac{\partial}{\partial y}.$$

$$\frac{\partial f}{\partial \bar{z}} = \frac{\partial f}{\partial x}\frac{\partial x}{\partial \bar{z}} + \frac{\partial f}{\partial y}\frac{\partial y}{\partial \bar{z}} = \frac{\partial f}{\partial x}\frac{1}{2} - \frac{\partial f}{\partial y}\frac{1}{2i} = \frac{1}{2}\left(\frac{\partial}{\partial x} - \frac{1}{i}\frac{\partial}{\partial y}\right) f.$$

$$\therefore \quad 2\frac{\partial}{\partial \bar{z}} = \frac{\partial}{\partial x} - \frac{1}{i}\frac{\partial}{\partial y}.$$

$$\therefore \quad \frac{\partial^2}{\partial x^2} + \frac{\partial^2}{\partial y^2} = \left(\frac{\partial}{\partial x} + \frac{1}{i}\frac{\partial}{\partial y}\right)\left(\frac{\partial}{\partial x} - \frac{1}{i}\frac{\partial}{\partial y}\right)$$

$$= \left(2\frac{\partial}{\partial z}\right)\left(2\frac{\partial}{\partial \bar{z}}\right) = 4\left(\frac{\partial^2}{\partial z\, \partial \bar{z}}\right).$$

23. Solved sum. If $f(z)$ is analytic, show that

(*i*) $\nabla^2 \,|\, \text{Re}\, f(z) \,|^2 = 2\,|\, f'(z) \,|^2$

(*ii*) $\nabla^2 \,|\, \text{Im}\, f(z) \,|^2 = 2\,|\, f'(z) \,|^2$

(*iii*) $\nabla^2 \,|\, f(z) \,|^2 = 4\,|\, f'(z) \,|^2$.

✍ (*i*) Let $f(z) = u + iv$, then

$$\frac{\partial (u^2)}{\partial x} = 2uu_x; \quad \frac{\partial^2 (u^2)}{\partial x^2} = 2\,[u_x^2 + uu_{xx}]$$

$$\frac{\partial (u^2)}{\partial y} = 2uu_y; \quad \frac{\partial^2 (u^2)}{\partial y^2} = 2\,[u_y^2 + uu_{yy}]$$

$$\therefore \quad \frac{\partial^2 u^2}{\partial x^2} + \frac{\partial^2 u^2}{\partial y^2} = 2\,(u_x^2 + u_y^2) + 2u\,(u_{xx} + u_{yy})$$

$$= 2\,(u_x^2 + u_y^2) + 2u \times 0 \text{ since } u \text{ is harmonic}$$

$$= 2\,|\, u_x + iu_y \,|^2 = 2\,|\, f'(z) \,|^2.$$

$$\therefore \quad \nabla^2 \,|\, \text{Re}\, f(z) \,|^2 = 2\,|\, f'(z) \,|^2. \qquad (1)$$

(*ii*) In (*i*), replacing u by v, we get that

$$\frac{\partial^2 v^2}{\partial x^2} + \frac{\partial^2 v^2}{\partial y^2} = 2\,|\, f'(z) \,|^2.$$

$$\therefore \quad \nabla^2 \,|\, \text{Im}\, f(z) \,|^2 = 2\,|\, f'(z) \,|^2. \qquad (2)$$

(*iii*) **First method.** Adding (1) and (2), we get

$$\nabla^2 (u^2 + v^2) = 4\,|\, f'(z) \,|^2.$$

Second method

$$\nabla^2 \,|\, f(z) \,|^2 = \nabla^2 \,[f(z)\, \overline{f(z)}\,] \quad \text{since } |\, z \,|^2 = z\bar{z}$$

$$= 4\,\frac{\partial}{\partial z}\frac{\partial}{\partial \bar{z}}\,[f(z)\, \overline{f(z)}]$$

$$= 4\left[\frac{\partial}{\partial z}\, f(z)\right]\left[\frac{\partial}{\partial \bar{z}}\, \overline{f(z)}\right]$$

since $f(z)$ is independent of $\bar{z}$ and $\overline{f(z)}$ is independent of z

$$= 4\left[\frac{\partial}{\partial z}\, f(z)\right]\left[\overline{\frac{\partial}{\partial z}\, f(z)}\right]$$

$$= 4f'(z)\, [\overline{f'(z)}] = 4\,|\, f'(z) \,|^2 \text{ since } z\bar{z} = |\, z \,|^2.$$

24. Solved sum. Show that

$$\left(\frac{\partial^2}{\partial x^2}+\frac{\partial^2}{\partial y^2}\right)\log|f(z)|=0$$

✍

$$\text{LHS} = \nabla^2 [1/2 \log|f(z)|^2] = 1/2\, \nabla^2 \log [f(z)\, \overline{f(z)}\,]$$

$$= 1/2\, \nabla^2 [\log f(z) + \log \overline{f(z)}\,]$$

$$= \frac{1}{2}\,.\,4\,.\,\frac{\partial}{\partial z}\frac{\partial}{\partial \bar{z}}\,[\log f(z) + \log \overline{f(z)}\,].$$

$$= 2\frac{\partial}{\partial z}\left[\frac{\partial}{\partial \bar{z}}\log f(z)\right] + 2\frac{\partial}{\partial z}\left[\frac{\partial}{\partial \bar{z}}\log \overline{f(z)}\right]$$

$$= 2\frac{\partial}{\partial z}\left[\frac{\partial}{\partial \bar{z}}\log f(z)\right] + 2\frac{\partial}{\partial z}\overline{\left[\frac{\partial}{\partial z}\log f(z)\right]}.$$

Now $\log f(z)$ is analytic. Therefore $\log f(z)$ is independent of $\bar{z}$ and

$$\overline{\frac{\partial}{\partial z}[\log f(z)]}$$

is independent of z. Therefore

$$\text{LHS} = 2\frac{\partial}{\partial z}\,[0] + 2\,[0] = 0.$$

25. Check whether $xy\,(x + y)$ is a harmonic function. Can it be the real part of an analytic function? **(1981)**

[*Hint* : $u_{xx} + u_{yy} = 2y + 2x \neq 0$. So u is not harmonic and hence it cannot be the real part of an analytic function].

26. Show that $u\,(x, y)$ is a harmonic function in the following cases and find the analytic function $f(z)$ having it as its real part:

(*i*) $u = 2x\,(1 - y)$

(*ii*) $u = x^3 - 3xy^2$ **(1974, 77, 80)**

(*iii*) $u = x^3 - 3xy^2 + x$

(*iv*) $u = y^3 - 3x^2y$ **(1981)**

(*v*) $u = y^3 - 3x^2y + 2y$

(*vi*) $u = -x^3 + 3xy^2 + 2x$ **(1981)**

(*vii*) $u = x^3 - 3xy^2 + 3x^2 - 3y^2 + 1$ **(1975, 79)**

(*viii*) $u = x^2 - y^2 - 2xy - 2x + 3y$ **(1982)**

(*ix*) $u = 2x^3 - 5x^2 - 6xy^2 + 3x + 5y^2 - 7.$ **(1980)**

[*Hint* : Get $u_{xx} + u_{yy} = 0$. To use (6.12.5) set $x = \frac{1}{2}z$, $y = -\frac{1}{2}iz$ in u.

(*i*) $f(z) = 2\,[2(\tfrac{1}{2}z)(1 + \tfrac{1}{2}iz)] = 2x + iz^2$

(*ii*) $f(z) = 2\,[(\tfrac{1}{2}z)^3 - 3(\tfrac{1}{2}z)(-\tfrac{1}{2}iz)^2] = z^3$

(*iii*) $f(z) = z^3 + z$

(*iv*) $f(z) = iz^3$

(v) $f(z) = iz^3 - 2iz$
(vi) $f(z) = z^3 + 2z$
(vii) $f(z) = z^3 + 3z^2 + 1$
(viii) $f(z) = z^2 + iz^2 - 2z - 3iz$
(ix) $f(z) = 2z^3 - 5z^2 + 3z - 7$]

27. Find the analytic function whose real part is $x/(x^2+y^2)$.

[*Hint* : $u(0, 0)$ is not defined; use (6.12.4) with $a = 1$.

$$f(z) = 2\frac{\frac{1}{2}(z+a)}{\{\frac{1}{2}(z+1)\}^2 + \{-\frac{1}{2}i(z-1)\}^2} = \frac{z+1}{z} = 1 + \frac{1}{z}$$

Therefore $f(z) = 1/z$ since the real part of $1/z$ is $x/(x^2+y^2)$]

28. Solved sum. Show that the function $u(x, y) = \sin x \cosh y$ is harmonic. Find its harmonic conjugate $v(x, y)$ and the analytic function $f(z) = u + iv$. **(1975, 79, 83)**

✍ Now $u_{xx} = -\sin x \cosh y$, $u_{yy} = \sin x \cosh y$. $\therefore u_{xx} + u_{yy} = 0$. So u is a harmonic function.

Method 1.
$$v_x = -u_y = -\sin x \sinh y,$$
$$v_y = u_x = \cos x \cosh y.$$

$$\therefore \quad v(x, y) = -\sinh y \int \sin x \, dx + \int (0)\, dy = \cos x \sinh y.$$

$$\therefore \quad f(z) = u + iv = \sin x \cosh y + i \cos x \sinh y$$
$$= \sin x \cos iy + \cos x \sin iy = \sin(x + iy) = \sin z$$

Method 2.
$$f(z) = 2\sin(\tfrac{1}{2}z)\cosh(-\tfrac{1}{2}iz) \qquad \text{by (6.12.5)}$$
$$= 2\sin(\tfrac{1}{2}z)\cos(\tfrac{1}{2}z) = \sin z = \sin(x+iy)$$
$$= \sin x \cos iy + \cos x \sin iy$$
$$= \sin x \cosh y + i\cos x \sinh y.$$
$$\therefore \quad v(x, y) = \cos x \sinh y.$$

29. Solved sum. Find the analytic function $f(z)$ if its real part is $u(x, y) = e^x(x\cos y - y\sin y)$. **(1979, 80, 81)**

✍ As in method 2 setting $\frac{1}{2}z$ for x and $-\frac{1}{2}iz$ for y in $u(x, y)$,

$$f(z) = 2e^{\frac{1}{2}z}[(\tfrac{1}{2}z)\cos(-\tfrac{1}{2}iz) - (-\tfrac{1}{2}iz)\sin(-\tfrac{1}{2}iz)]$$
$$= e^{\frac{1}{2}z} z[\cos(\tfrac{1}{2}iz) - i\sin(\tfrac{1}{2}iz)] = e^{\frac{1}{2}z} z[\cosh\tfrac{1}{2}z + \sinh\tfrac{1}{2}z]$$
$$= e^{\frac{1}{2}z} z[\tfrac{1}{2}(e^{\frac{1}{2}z} + e^{-\frac{1}{2}z}) + \tfrac{1}{2}(e^{\frac{1}{2}z} - e^{-\frac{1}{2}z})] = e^{\frac{1}{2}z} z e^{\frac{1}{2}z} = ze^z.$$

30. Solved sum. Find the analytic function $f(z)$ if its real part is $u(x, y) = e^{-x}[(x^2-y^2)\cos y + 2xy\sin y]$. **(1979)**

✍ By method 2, setting $\frac{1}{2}z$ for x and $-\frac{1}{2}iz$ for y in $u(x, y)$,

$$f(z) = 2e^{-\frac{1}{2}z}[\{(\tfrac{1}{2}z)^2 - (\tfrac{1}{2}iz)^2\}\cos(-\tfrac{1}{2}iz) + 2(\tfrac{1}{2}z)(-\tfrac{1}{2}iz)\sin(-\tfrac{1}{2}iz)]$$
$$= 2e^{-\frac{1}{2}z}[\tfrac{1}{2}z^2\cos(-\tfrac{1}{2}iz) - i\tfrac{1}{2}z^2\sin(-\tfrac{1}{2}iz)]$$
$$= e^{-\frac{1}{2}z} z^2[\cosh\tfrac{1}{2}z - \sinh\tfrac{1}{2}z] = e^{-\frac{1}{2}z} z^2(e^{-\frac{1}{2}z}) = z^2 e^{-z}$$

31. Find the analytic function if $e^{-x}(x \sin y - y \cos y)$ is its real part. **(1982)**
(**Ans.** ize^{-z})

32. Find the analytic function if $e^{x}(x \sin y + y \cos y)$ is its imaginary part. **(1980)**

[*Hint* : $f(z) = 2ie^{\frac{1}{2}z}[(\frac{1}{2}z)\sin(-\frac{1}{2}iz) + (-\frac{1}{2}iz)\cos(-\frac{1}{2}iz)]$ by (6.12.6)

$= ie^{\frac{1}{2}z}\, z\{-i \sinh \frac{1}{2}z - i\cosh\frac{1}{2}z\} = ze^{\frac{1}{2}z}(e^{\frac{1}{2}z}) = ze^{z}$]

33. Find the analytic function whose real part is

$$\sin 2x/(\cosh 2y + \cos 2x).$$

(**Ans.** $\tan z$)

34. Solved sum. Verify whether $v = e^{x}(x \cos y - y \sin y)$ is a harmonic function and construct the analytic function $f(z)$.

$$v = (e^{x}x)\cos y - e^{x} y \sin y$$

$$\left.\begin{aligned} v_x &= e^{x}(x+1)\cos y - e^{x} y \sin y \\ v_y &= e^{x} x(-\sin y) - e^{x}(\sin y + y\cos y) \end{aligned}\right\}$$

$$\left.\begin{aligned} v_{xx} &= e^{x}(x+2)\cos y - e^{x} y \sin y \\ v_{yy} &= -e^{x} x\cos y - e^{x}(\cos y + \cos y - y \sin y) \\ &= -e^{x}(x+2)\cos y + e^{x} y\sin y \end{aligned}\right\}$$

$\therefore$ $$v_{xx} + v_{yy} = 0$$

and so $v(x, y)$ is harmonic.

To find u, consider the exact equation

$$\begin{aligned} du &= u_x\, dx + u_y\, dy \\ &= (v_y)\, dx + (-v_x)\, dy \text{ by C-R equations} \end{aligned}$$

$\therefore$ $$\begin{aligned} u &= \int v_y\; dx \text{ treating } y \text{ as a constant} \\ &\qquad + \int [\text{Terms of } (-v_x) \text{ not containing } x]\, dy \\ &= \int [-e^{x} x\sin y - e^{x}(\sin y + y\cos y)]\, dx + \int 0\, dy \\ &= [(-e^{x}x - e^{x})\sin y - e^{x}(\sin y + y\cos y)] \\ &= e^{x}[-x\sin y - y\cos y]. \end{aligned}$$

$$\begin{aligned} f(z) &= u + iv \\ &= e^{x}[(-x\sin y - y\cos y) + i(x\cos y - y\sin y)] \\ &= e^{x}[x(-\sin y + i\cos y) - y(\cos y + i\sin y)] \\ &= e^{x}[ix(\cos y + i\sin y) - y(\cos y + i\sin y)] \\ &= e^{x}(ix - y)(\cos y + i\sin y) \\ &= e^{x} i(x + iy)e^{iy} = ie^{x+iy}(x+iy) = ei^{z}z. \end{aligned}$$

35. Solved sum. Find the analytic function $f(z) = u + iv$, given that

$$u - v = e^{x}(\cos y - \sin y).$$

To get $u - v$, we have

$$u + iv = f(z). \qquad (1)$$

$i \times (1)$: $$iu - v = if(z)$$

Adding these two, we get

$$u - v + i(u + v) = (1 + i) f(z)$$

i.e., $$e^x (\cos y - \sin y) + i(u + v) = (1 + i) f(z).$$

So $e^x (\cos y - \sin y)$ is the real part of the analytic function

$$(1 + i) f(z) = U + iV \text{ say.}$$

Then we have

$$U(x, y) = e^x (\cos y - \sin y).$$

$\therefore$ $$U_x = e^x (\cos y - \sin y), \quad U_y = e^x (-\cos y - \sin y).$$

Considering the differential of V,

$$dV = V_x\, dx + V_y\, dy$$
$$= (-U_y)\, dx + U_x\, dy \text{ by C-R equations}$$

$\therefore$ $$V = -\int U_y\, dx \text{ treating } y \text{ as a constant}$$
$$+ \int (\text{Terms of } U_x \text{ not containing } x)\, dy$$
$$= -\int e^x (-\sin y - \cos y)\, dx + \int 0\, dy$$
$$= e^x (\sin y + \cos y).$$

$\therefore$ $$(1 + i) f(z) = U + iV$$
$$= e^x [(\cos y - \sin y) + i(\sin y + \cos y)]$$
$$= e^x [(\cos y + i \sin y) + i(\cos y + i \sin y)]$$
$$= e^x (1 + i)[\cos y + i \sin y] = e^x (1 + i) e^{iy}$$
$$= e^{x + iy} (1 + i) = (1 + i) e^z$$

$\therefore$ $$f(z) = e^z.$$

SUBSTITUTION METHOD

36. Solved sum. Construct the analytic function $f(z) = u + iv$, given that

$$u(x, y) = e^x (\cos y - \sin y).$$

✍ Now $u(x, y)$ is the real part of $f(z)$ and

$$u(0, 0) = e^0 (\cos 0 - \sin 0) = 1.$$

Thus $u(0, 0)$ exists. Therefore, by substitution method,

$$f(z) = 2[u(z/2, -iz/2)] + C,$$

where C is an arbitrary constant. But

$$u(x, y) = e^x (\cos y - \sin y).$$

$\therefore$ $$f(z) = 2[e^{z/2} \{\cos(-iz/2 - \sin(-iz/2)\}] + C$$
$$= 2[e^{z/2} (\cosh z/2 + i \sinh z/2)] + C$$
$$= 2\left[e^{z/2}\left(\frac{e^{z/2} + e^{-z/2}}{2} + i\frac{e^{z/2} - e^{-z/2}}{2}\right)\right] + C$$
$$= (e^z + 1) + i(e^z - 1) + C = (1 + i) e^z + 1 - i + C$$
$$= (1 + i) e^z + C'.$$

37. Solved sum. Find the analytic functions $f(z) = u + iv$, given that

(i) $-2v(x, y) = e^x (\cos y - \sin y)$

(ii) $u + v = e^x (\cos y - \sin y)$

(iii) $u - v = e^x (\cos y - \sin y)$

(iv) $2u + v = e^x (\cos y - \sin y)$

(v) $u - 2v = e^x (\cos y - \sin y)$.

✍ **(i)** Now it is given that

$$u + iv = f(z). \quad (1)$$

Multiplying both sides by $2i$, we get

$$-2v + 2iu = 2if(z).$$

Let $2if(z) = U + iV$. Then

$$U = -2v = e^x (\cos y - \sin y).$$

Now $U(0, 0) = 1$. So $U(0, 0)$ exists. Thus, by substitution method, $2if(z) = 2U(x, y)$, where $x = z/2, y = -iz/2$. So

$$(2i)f(z) = 2[e^{z/2} \{\cos(-iz/2) - \sin(-iz/2)\}] \quad (2)$$

$$= 2[e^{z/2} (\cosh z/2 + i \sinh z/2)] + C$$

$$= 2\left[e^{z/2}\left(\frac{e^{z/2} + e^{-z/2}}{2} + i\,\frac{e^{z/2} - e^{-z/2}}{2}\right)\right] + C$$

$$= (e^z + 1) + i(e^z - 1) + C = (1 + i)e^z + (1 - i) + C$$

$$= (1 + i)e^z + C'. \quad (3)$$

$$\therefore \quad f(z) = \frac{1+i}{2i} e^z + C'' = \frac{-i(1+i)}{2} e^z + C''$$

$$= \frac{1-i}{2i} e^z + C''$$

(ii) Now we should consider the analytic function whose real part is $u + v$.

(1): $\quad u + iv = f(z)$.

$(-i) \times (1)$: $\quad v - iu = -if(z)$.

Adding these two, we get the analytic function as

$$u + v + i(v - u) = (1 - i)f(z).$$

Now let $(1 - i)f(z) = U + iV$. Then we have

$$U = u + v = e^x (\cos y - \sin y).$$

By the same working as in (i), corresponding to (2), we get

$$(1 - i)f(z) = 2[e^{z/2} \{\cos(-iz/2) - \sin(-iz/2)\}]$$

$$= (1 + i)e^z + C.$$

$$\therefore \quad f(z) = \frac{1+i}{1-i} e^z + C' = ie^z + C'.$$

(iii) (1): $\quad u + iv = f(z)$

$i \times (1)$: $\quad -v + iu = if(z)$.

Adding these two, we get the analytic function as

$$(u - v) + i(u + v) = (1 + i)f(z).$$

In this sum, for (3), we have

$$(1 + i)f(z) = (1 + i)e^z + C.$$

$$\therefore \quad f(z) = e^z + C'.$$

(iv) (1): $\quad u + iv = f(z)$

$2 \times (1)$: $\quad 2u + i2v = 2f(z)$

$(-i) \times (1)$: $\quad v - iu = -if(z).$

Adding the last two, we get

$$(2u + v) + i(2v - u) = (2 - i) f(z).$$

In this sum, for (3), we have

$$(2 - i) f(z) = (1 + i) e^z + C.$$

$$\therefore \quad f(z) = \frac{1+i}{2-i} e^z + C' = \frac{1+3i}{5} e^z + C'.$$

(v) (1): $\quad u + iv = f(z).$

$(2i) \times (1)$: $\quad -2v + 2iu = 2i f(z).$

Adding these two, we get

$$u - 2v + i(2u + v) = (1 + 2i) f(z).$$

In this sum, for (3), we have

$$(1 + 2i) f(z) = (1 + i) e^z + C.$$

$$\therefore \quad f(z) = \frac{1+i}{1+2i} e^z + C' = \frac{3-i}{5} e^z + C'.$$

38. Solved sum. Find the analytic function $f(z) = u + iv$, given that

$$2u + v = e^{2x} [(2x + y) \cos 2y + (x - 2y) \sin 2y].$$

✍ Given that

$$u + iv = f(z). \tag{1}$$

$2 \times (1)$: $\quad 2u + i2v = 2f(z)$

$(-i) \times (1)$: $\quad v - iu = -if(z).$

Adding these two, we get

$$2u + v + i(2v - u) = (2 - i) f(z).$$

Thus $2u + v$ is the real part of the analytic function $(2 - i) f(z)$. Now $2u(0, 0) + v(0, 0)$ exists. Hence, by substitution method, replacing x by $z/2$, y by $-iz/2$, we get

$$(2 - i) f(z) = 2e^z \left[\left(z - \frac{iz}{2}\right) \cos(-iz) + \left(\frac{z}{2} + iz\right) \sin(-iz)\right]$$

$$= 2e^z z \left[\left(1 - \frac{i}{2}\right) \cosh z - i\left(\frac{1}{2} + i\right) \sinh z\right]$$

$$= 2e^z z \left[\left(1 - \frac{i}{2}\right) \cosh z + \left(\frac{-i}{2} + 1\right) \sinh z\right]$$

$$= 2e^z z \left(1 - \frac{i}{2}\right) [\cosh z + \sinh z]$$

$$= (2 - i)\, e^z z \left[\frac{e^z + e^{-z}}{2} + \frac{e^z - e^{-z}}{2}\right]$$

$$= (2 - i)\, e^z z\, [e^z] = (2 - i)\, e^{2z} z.$$

$$\therefore \quad f(z) = e^{2z} z.$$

39. Solved sum. Find the analytic function $f(z) = u + iv$ given that

$$u - v = \frac{\sin 2x}{\cosh 2y - \cos 2x}.$$

Now we have

$$u + iv = f(z). \tag{1}$$

$i \times (1)$: $\qquad -v + iu = if(z).$

Adding these two, we get

$$u - v + i(u + v) = (1 + i) f(z).$$

Thus $u - v$ is the real part of the analytic function $(1 + i) f(z)$. Now $u(0, 0) - v(0, 0)$ does not exist but $u(\pi/2, 0) - v(\pi/2, 0)$ exists. Hence, by the general substitution method, replacing

$$x \text{ by } \frac{z + \pi/2}{2}, \; y \text{ by} - i\,\frac{z - \pi/2}{2},$$

$$(1 + i) f(z) = \frac{\sin 2\left(\dfrac{z + \pi/2}{2}\right)}{\cosh 2\left(-i\,\dfrac{z - \pi/2}{2}\right) - \cos 2\left(\dfrac{z + \pi/2}{2}\right)}$$

$$= \frac{\sin(\pi/2 + z)}{\cosh[i(\pi/2 - z)] - \cos(\pi/2 + z)}$$

$$= \frac{\cos z}{\cos(\pi/2 - z) - (-\sin z)}$$

$$= \frac{\cos z}{\sin z + \sin z} = \frac{1}{2}\,\frac{\cos z}{\sin z} = \frac{1}{2}\cot z.$$

$$\therefore \qquad f(z) = \frac{1}{2(1+i)}\cot z.$$

40. Solved sum. Construct the analytic function $f(z) = u + iv$ given that

$$u - v = \frac{1}{2}\,\frac{\cos x + \sin x - e^{-y}}{(\cos x - \cosh y)}$$

and $f(\pi/2) = 0$.

Now we have

$$u + iv = f(z) \tag{1}$$

$i \times (1)$: $\qquad -v + iu = if(z).$

Adding these two, we get

$$(u - v) + i(u + v) = (1 + i) f(z).$$

Thus $u - v$ is the real part of the analytic function $(1 + i) f(z)$. Now $u(\pi, 0) - v(\pi, 0)$ exists. Hence, by substitution method, replacing

$$x \text{ by } \frac{z + \pi}{2}, \; y \text{ by} - \frac{i(z - \pi)}{2},$$

$$(1+i)f(z) = 2\left[\frac{1}{2}\frac{\cos\dfrac{z+\pi}{2} + \sin\dfrac{z+\pi}{2} - e^{i(z-\pi)/2}}{\cos\dfrac{z+\pi}{2} - \cosh\dfrac{i(\pi - z)}{2}}\right]$$

$$= \frac{-\sin z/2 + \cos z/2 - e^{iz/2}\, e^{-i\pi/2}}{-\sin z/2 - \cos(\pi/2 - z/2)}$$

$$= \frac{-\sin z/2 + \cos z/2 + i\, e^{iz/2}}{-\sin z/2 - \sin z/2}$$

$$= \frac{-\sin z/2 + \cos z/2 + i(\cos z/2 + i \sin z/2)}{-2\sin z/2}$$

$$= \frac{-2\sin z/2 + (1+i)\cos z/2}{-2\sin z/2} = 1 - \frac{1+i}{2}\cot\frac{z}{2}.$$

$$\therefore \qquad f(z) = \frac{1}{1+i} - \frac{1}{2}\cot\frac{z}{2} = C - \frac{1}{2}\cot\frac{z}{2}.$$

But it is given that $f(\pi/2) = 0$. Therefore

$$0 = C - 1/2 \cot \pi/4 \quad \text{or} \quad C = 1/2.$$

$$\therefore \qquad f(z) = 1/2 - 1/2 \cot z/2.$$

MILNE-THOMSON METHOD

41. Solved sum. If $f(z) = u + iv$ is analytic, and if

$$u - v = e^x(\cos y - \sin y),$$

find $f(z)$.

✍ It is given that

$$u + iv = f(z). \tag{1}$$

$i \times (1)$: $\quad -v + iu = if(z).$

Adding these two, we get

$$u - v + i(u + v) = (1+i)f(z).$$

We shall denote the analytic function $(1+i)f(z)$ as

$$(1+i)f(z) = U + iV,$$

where

$$U = e^x(\cos y - \sin y).$$

Considering the derivaive of $(1+i)f(z)$,

$$(1+i)f'(z) = U_x + iV_x$$
$$= U_x + i(-U_y) \text{ since } U_y = -V_x.$$

But $U_x = e^x(\cos y - \sin y)$, $U_y = e^x(-\sin y - \cos y)$.

$$\therefore \qquad (1+i)f'(z) = e^x(\cos y - \sin y) + ie^x(\sin y + \cos y)$$
$$= e^z(\cos 0 - \sin 0) + ie^z(\sin 0 + \cos 0)$$

by Milne-Thomson method

$$= e^z + ie^z = (1+i)e^z.$$

$\therefore \qquad f'(z) = e^z.$

So, integrating w.r.t z,

$$f(z) = e^z.$$

42. Solved sum. Find the analytic functions $f(z) = u + iv$, given that

(i) $-2v(x, y) = e^x(\cos y - \sin y)$

(ii) $2u + v = e^x(\cos y - \sin y)$

(iii) $u - 2v = e^x(\cos y - \sin y)$

✍ $\qquad u + iv = f(z). \qquad (1)$

$-i \times (1)$: $\qquad v - iu = -if(z). \qquad (2)$

Using these two we shall obtain $-2v$, $2u + v$, $u - 2v$.

$-2 \times (2)$: $\qquad -2v + i2u = 2if(z). \qquad (3)$

$2 \times (1) + (2)$: $\qquad 2u + v + i(2v - u) = (2 - i)f(z). \qquad (4)$

$(1) - 2 \times (2)$: $\qquad u - 2v + i(2u + v) = (1 + 2i)f(z). \qquad (5)$

(i) Taking (3), we have

$$2if(z) = -2v + i(2u) \qquad (6)$$
$$= U + iV, \text{ say.}$$

Then, from the given result,

$$U = e^x(\cos y - \sin y).$$

But the derivative of $2if(z)$ is

$$2if'(z) = U_x + iV_x$$
$$= U_x + i(-U_y) \text{ since } U_y = -V_x$$
$$= e^x(\cos y - \sin y) + ie^x(\sin y + \cos y)$$
$$= e^z(\cos 0 - \sin 0) + ie^z(\sin 0 + \cos 0)$$

by Milne-Thomson method

$$= e^z + ie^z = (1 + i)e^z.$$

Integrating w.r.t z, we get

$$2if(z) = (1 + i)e^z + C. \qquad (7)$$

$$\therefore \qquad f(z) = \frac{1+i}{2i}e^z + C' = \frac{1-i}{2}e^z + C'.$$

(ii) From (4) we have

$$(2 - i)f(z) = 2u + v + i(2v - u).$$

Proceeding in the manner in which (7) was obtained

$$(2 - i)f(z) = (1 + i)e^z + C.$$

$$\therefore \qquad f(z) = \frac{1+i}{2-i}e^z + C' = \frac{1+3i}{5}e^z + C'.$$

(iii) From (5) we have

$$(1 + 2i)f(z) = u - 2v + i(2u + v).$$

Proceeding in the manner in which (7) was obtained

$$(1 + 2i)f(z) = (1 + i)e^z + C.$$

$$\therefore \qquad f(z) = \frac{1+i}{1+2i}e^z + C' = \frac{3-i}{5}e^z + C'.$$

CHAPTER 7 ELEMENTARY AND CONFORMAL MAPPINGS

7.1. BILINEAR TRANSFORMATION (MOBIUS TRANSFORMATION)

In chapter 2 we studied the elementary transformations namely, translation, rotation, homothetic transformation, inversion and reflection. In this section we study bilinear transformation.

The transformation

$$w = \frac{az+b}{cz+d}, \quad \begin{vmatrix} a & b \\ c & d \end{vmatrix} \neq 0, \tag{7.1.1}$$

where a, b, c, d are complex numbers, is called a bilinear transformation. This transformation was first studied by A, F, Mobius.

Nomenclature, When the fraction is cleared,

$$cwz + dw - az - b = 0$$

which is linear in w and also linear in z. Hence the transformation is called a bilinear transformation.

Linear fractional transformation. A bilinear transformation is also called a linear fractional transformation because $(az + b)/(cz + d)$ is a fraction formed by linear function $az + b$ and $cz + d$.

Situation when the determinant $ad - bc$ of transformation is zero. Suppose, under the bilinear transformation, z_1 and z_2 $(z_1 \neq z_2)$ go to w_1 and w_2. Then

$$w_1 - w_2 = \frac{az_1+b}{cz_1+d} - \frac{az_2+b}{cz_2+d}$$

$$= \frac{(ad-bc)(z_1-z_2)}{(cz_1+d)(cz_2+d)} = 0.$$

That is, $w_1 = w_2$. This implies that, if $ad - bc = 0$, then the entire z plane goes to a single point in the w plane. So the transformation is trivial. The single point is $w = b/d$ because $z = 0$ goes into b/d.

Elementary transformations. Certain elementary transformations are particular cases of bilinear transformation as detailed below.

Particularity	*Transformation*	*Nature of transformation*
$c = 0, \frac{a}{d} = 1$	$w = z + \frac{b}{d}$	Translation
$c = 0, b = 0$	$w = \frac{a}{d} z$	Rotation and homothetic transformation
$c = 0$	$w = \frac{a}{d} z + \frac{b}{d}$	Rotation and homothetic transformation followed by translation
$a = d = 0, b = c$	$w = \frac{1}{z}$	Inversion

Theorem 7.1. Under a bilinear transformation no two points in z plane go to the same point in w plane.

Proof : Suppose z_1 and z_2 go to the same point in the w plane under the transformation $w = (az + b)/(cz + d)$. Then

$$\frac{az_1 + b}{cz_1 + d} = \frac{az_2 + b}{cz_2 + d}$$

i.e., $\quad (az_1 + b)(cz_2 + d) - (az_2 + b)(cz_1 + d) = 0.$

$\therefore \quad (ac - bd)(z_1 - z_2) = 0$ or $z_1 = z_2$.

This implies that no two distinct points in the z plane go to the same point in the w plane. So each point in the z plane goes to a unique point in the w plane.

Remark 1. The point $z = -d/c$, given by $cz + d = 0$, goes to $w = \infty$. The point $z = \infty$ goes to $w = a/c$ because

$$w = \frac{a + (b/\infty)}{c + (d/\infty)} = \frac{a}{c}.$$

Remark 2. The inverse transformation from w plane to z plane is also bilinear because it is

$$z = \frac{(-d)\,w + b}{cw + (-a)}, \quad \begin{vmatrix} -d & b \\ c & -a \end{vmatrix} \neq 0.$$

So under the inverse transformation each w goes into a unique z. Thus the original transformation sets up a *one to one correspondence* between the points in the extended z plane and the points in the extended w plane.

Fixed points. Points which are invariant under a transformation are called fixed points or constant points or invariant points of the transformation. The fixed points of the transformation

$$w = (az + b)/(cz + d)$$

is obtained from

$$z = (az + b)/(cz + d) \quad \text{or} \quad cz^2 + (d - a)z - b = 0.$$

In general, these points are two in number unless the discriminant is zero in which case the number of points is one. Certain particular cases are as follows :

Particular cases	*Fixed Points*
$b = 0$	$z = 0, z = (a - d)/c$
$c = 0$	$z = \infty, z = b/(d - a)$
$(d - a)^2 + 4bc = 0$	$z = \frac{1}{2}(a - d)/c$ (coincident points)

Fixed points of the elementary transformations are as follows :

Transformation	Fixed points
$w = z + b$	$z = \infty$
$w = az$	$z = 0, z = \infty$
$w = 1/z$	$z = -1, z = 1.$

Theorem 7.2. The set of all bilinear transformations is a group under composition.

Proof: Consider the element

$$w = T(z) = \frac{az + b}{cz + d}.$$

A complex number can be expressed as the quotient of two complex numbers as $z = (z_1/z_2)$. Thereby

$$w = \frac{a(z_1/z_2)+b}{c(z_1/z_2)+d} = \frac{az_1+bz_2}{cz_1+dz_2}$$

Now, denoting $az_1 + bz_2$ by w_1 and $cz_1 + dz_2$ by w_2, the equations of the transformation are

$$w = \frac{w_1}{w_2}, \begin{pmatrix} w_1 \\ w_2 \end{pmatrix} = \begin{pmatrix} a & b \\ c & d \end{pmatrix} \begin{pmatrix} z_1 \\ z_2 \end{pmatrix}$$

Therefore the non-singular matrix

$$M = \begin{pmatrix} a & b \\ c & d \end{pmatrix}$$

can be considered as the matrix of transformation. Let

$$M_1 = \begin{pmatrix} a_1 & b_1 \\ c_1 & d_1 \end{pmatrix}, \quad M_2 = \begin{pmatrix} a_2 & b_2 \\ c_2 & d_2 \end{pmatrix}$$

be the matrices of the transformations T_1, T_2. Then the composite transformation in which T_1 is carried out first and T_2 is carried out next, is

$$\begin{pmatrix} w_1 \\ w_2 \end{pmatrix} = M_2 M_1 \begin{pmatrix} z_1 \\ z_2 \end{pmatrix}.$$

It is also a bilinear transformation and its matrix is M_2M_1. So the set is closed under composition. Further composition is associative since matrix multiplication is associative. The identity element under the operation is the transformation whose matrix is

$$\begin{pmatrix} 1 & 0 \\ 0 & 1 \end{pmatrix}$$

Also, the inverse of a transformation with matrix M is the transformation with matrix M^{-1}. So the set is a group under composition.

Remark. Matrix multiplication is not commutative. Therefore the above group is not commutative.

Though a bilinear transformation involves a, b, c, d effectively it involves only three constants because az + b and cz + d can be divided by one of a, b, c, d so that one of them is 1. This shows that there is just one bilinear transformation that maps three given distinct points z_1, z_2, z_3 into three specified points w_1, w_2, w_3.

If $a \neq 0$ and $c \neq 0$ then the transformation can be assumed to be

$$w = A\frac{z+B}{z+C}.$$

Theorem 7.3. The bilinear transformation which transforms z_1, z_2, z_3 into w_1, w_2, w_3 is

$$\frac{(w-w_1)(w_2-w_3)}{(w-w_3)(w_2-w_1)} = \frac{(z-z_1)(z_2-z_3)}{(z-z_3)(z_2-z_1)} \tag{7.1.2.}$$

Note. In this result w, w_1, w_2, w_3 are in an order in the numerator and are in the anti-order in the denominator.

Proof: If the required transformation is

$$w = \frac{az+b}{cz+d},$$

$$w - w_1 = \frac{az+b}{cz+d} - \frac{az_1+b}{cz_1+d} = \frac{(ad-bc)(z-z_1)}{(cz+d)(cz_1+d)}$$

$$(cz + d)(cz_1 + d)(w - w_1) = (ad - bc)(z - z_1).$$

Similarly,

$$(cz_2 + d)(cz_3 + d)(w_2 - w_3) = (ad - bc)(z_2 - z_3),$$
$$(cz + d)(cz_3 + d)(w - w_3) = (ad - bc)(z - z_3),$$
$$(cz_2 + d)(cz_1 + d)(w_2 - w_1) = (ad - bc)(z_2 - z_1)$$

These four results lead to the required result.

Remark 1. The result (7.1.2) can be written as $\dfrac{(w-w_1)(w_2-w_3)}{(w_1-w_2)(w_3-w)} = \dfrac{(z-z_1)(z_2-z_3)}{(z_1-z_2)(z_3-z)}$ which is easier for remembering.

Remark 2. Sometimes one of $z_1, z_2, z_3, w_1, w_2, w_3$ may be ∞. Suppose $z_3 = \infty$. Then the z expression on the right becomes

$$\frac{(z-z_1)\{(z_2/z_3)-1\}}{\{(z/z_3)-1\}(z_2-z_1)} = \frac{(z-z_1)(-1)}{(-1)(z_2-z_1)} = \frac{z-z_1}{z_2-z_1}.$$

*Cross ratio***.** Given four points z_1, z_2, z_3, z_4 in this order, the ratio

$$\frac{(z_1-z_3)(z_2-z_4)}{(z_1-z_4)(z_2-z_3)}$$

is called the cross ratio of the points.

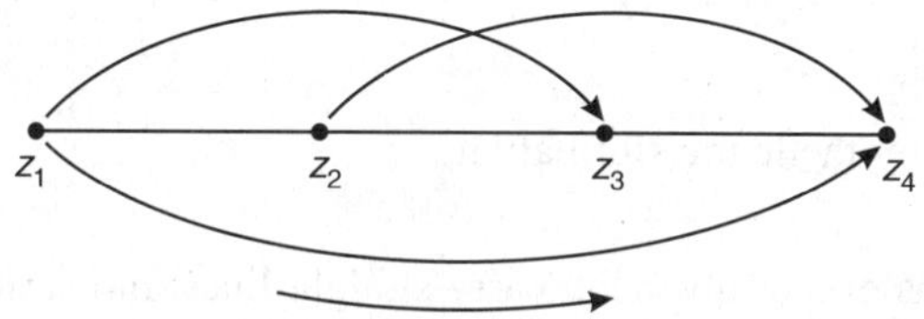

Fig. 7.1.1

Theorem 7.4. The cross ratio is preserved by a bilinear transformation.

Let the given points be z_1, z_2, z_3, z_4. Consider the bilinear transformation

$$w = \frac{az+b}{cz+d}.$$

As in theorem 3,

$$(cz_1 + d)(cz_3 + d)(w_1 - w_3) = (ad - bc)(z_1 - z_3)$$
$$(cw_2 + d)(cw_4 + d)(w_2 - w_4) = (ad - bc)(z_2 - z_4)$$
$$(cz_1 + d)(cz_4 + d)(w_1 - w_4) = (ad - bc)(z_1 - z_4)$$
$$(cz_2 + d)(cz_3 + d)(w_2 - w_3) = (ad - bc)(z_2 - z_3)$$

These four results lead to

$$\frac{(z_1-z_3)(z_2-z_4)}{(z_1-z_4)(z_2-z_3)} = \frac{(w_1-w_3)(w_2-w_4)}{(w_1-w_4)(w_2-w_3)}.$$

This completes the proof.

Remark. Figure 7.1.1 suggests how one can write down the cross ratio of z_1, z_2, z_3, z_4.

Theorem 7.5. A bilinear transformation maps straight lines and circles into straight lines and circles.

Proof : A bilinear transformation is composed of the elementary transformations, translation, rotation and homothetic transformation and inversion.

Case (*i*) : Suppose c = 0. Then the transformation is

$$w = \frac{az+b}{d} \quad \text{or} \quad w = \frac{a}{d}z + \frac{b}{d}.$$

So it can be thought of as the transformations

$$w_1 = \frac{a}{d}z, \quad w = w_1 + \frac{b}{d}$$

to be carried out in the same order. Hence the transformation is a rotation and homothetic transformation followed by a translation.

Case (*ii*) : Suppose $c \neq 0$. Then

$$w = \frac{bc-ad}{c^2} \cdot \frac{1}{z+(d/c)} + \frac{a}{c}.$$

This can be written as

$$w_1 = z + \frac{d}{c}, \quad w_2 = \frac{1}{w_1}, \quad w_3 = \frac{bc-ad}{c^2} w_2, \quad w = w_3 + \frac{a}{c}.$$

So the bilinear transformation is composed of the following transformations to be effected in the same order :

(*i*) Translation,

(*ii*) Inversion,

(*iii*) Rotation and homothetic transformation,

(*iv*) Translation.

These transformations except inversion carry straight lines into straight lines and circles into circles. So it is enough if we show that inversion $w = 1/z$ carries straight lines and circles into straight lines and circles. The equation

$$\alpha(x^2+y^2) + 2gx + 2fy + \lambda = 0$$

represents a straight line when $\alpha = 0$ and represents a circle when $\alpha \neq 0$, where α, g, f, λ are real. Now the inversion is given by

$$X + iY = \frac{1}{x+iy} \quad \text{or} \quad x = \frac{X}{X^2+Y^2}, \quad y = \frac{-Y}{X^2+Y^2}$$

The equation of the image of the above straight line or a circle is

$$\alpha\left(\frac{X^2}{(X^2+Y^2)^2} + \frac{Y^2}{(X^2+Y^2)^2}\right) + 2g\frac{X}{X^2+Y^2} - 2f\frac{Y}{X^2+Y^2} + \lambda = 0$$

i.e.,
$$\alpha + 2gX - 2fY + \lambda(X^2+Y^2) = 0.$$

Thus the image is a circle if $\lambda \neq 0$ and a straight line if $\lambda = 0$. This completes the proof.

Theorem 7.5. (*i*) If *p, q* are the the fixed points (i.e., invariant points) under a bilinear transformation, then the transformation can be expressed in the form

$$\frac{w-p}{w-q} = k\frac{z-p}{z-q}.$$

(*ii*) If *p* is the coincident fixed point, then

$$\frac{1}{w-p} = k + \frac{1}{z-p}.$$

Proof: Let the bilinear transformation be $w = (az + b)/(cz + d)$. Then, as in theorem 7.3.

$$(cz + d)(cp + d)(w - p) = (ad - bc)(z - p),$$
$$(cz + d)(cq + d)(w - q) = (ad - bc)(z - q).$$

$$\therefore \quad \frac{cp + d}{cq + d} \cdot \frac{w - p}{w - q} = \frac{z - p}{z - q}$$

This is of the form

$$\frac{w - p}{w - q} = k\frac{z - p}{z - q} \tag{7.1.3}$$

(*ii*) The most general bilinear form is

$$(\ldots)\, wz + (\ldots)\, w + (\ldots)\, z + (\ldots) = 0.$$

Now, this form should reduce to the form

$$(z - p)^2 = 0$$

when w is replaced by z. So the form should be

$$k(w - p)(z - p) + (w - z) = 0$$

i.e., $$k(w - p)(z - p) + (w - p) - (z - p) = 0$$

$$\therefore \quad k + \frac{1}{z - p} - \frac{1}{w - p} = 0 \quad \text{or} \quad \frac{1}{w - p} = k + \frac{1}{z - p} \tag{7.1.4}$$

7.2. SPECIAL BILINEAR TRANSFORMATIONS

In this section we obtain bilinear transformations which map.

(*i*) Upper half plane **Im** $z \geq 0$ onto the unit circular disc $w| \leq 1$,

(*ii*) Right half plane **Re** $z \geq 0$ onto the unit circular disc $|w| \leq 1$.

(*iii*) Unit circular disc $|z| \leq 1$ onto the unit circular disc $|w| \leq 1$.

Theorem 7.6. The bilinear transformations which map **Im** $z \geq 0$ onto $|w| \leq 1$ are of the form

$$w = e^{i\lambda}\frac{z - z_1}{z - \bar{z}_1},$$

where λ is a real number (Fig. 7.2.1)

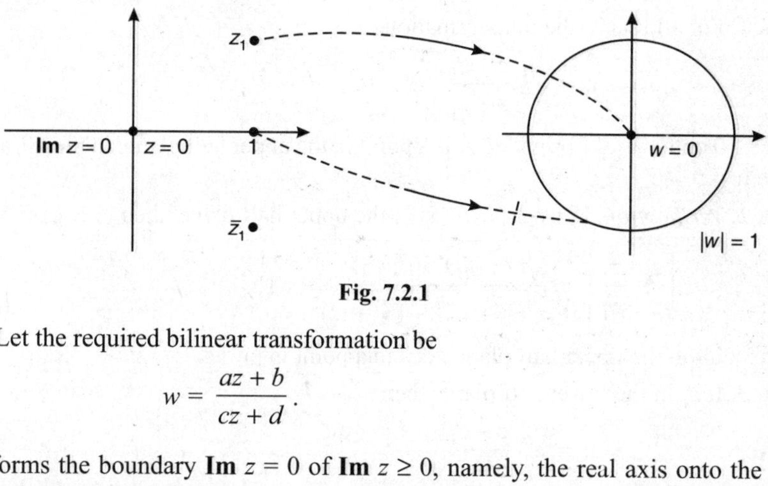

Fig. 7.2.1

Proof: Let the required bilinear transformation be

$$w = \frac{az + b}{cz + d}.$$

It transforms the boundary **Im** $z = 0$ of **Im** $z \geq 0$, namely, the real axis onto the boundary of $|w| \leq 1$, namely, the unit circle $|w| = 1$.

Let us take the points $z = 0$, $z = 1$, $z = \infty$ on the real axis. They go to three points on the circle $|w| = 1$. Considering, the images of $z = 0$ and $z = \infty$,

$$\left|\frac{a\cdot 0+b}{c\cdot 0+d}\right| = 1, \quad \left|\frac{a+(b/\infty)}{c+(d/\infty)}\right| = 1 \quad \text{or} \quad \left|\frac{b}{d}\right| = 1, \quad \left|\frac{a}{c}\right| = 1.$$

The second one, $|a/c| = 1$, implies that $a/c = e^{i\lambda}$, where λ is real. Hence the transformation becomes

$$w = \frac{a}{c}\cdot\frac{z+(b/a)}{z+(d/c)} = e^{i\lambda}\,\frac{z-z_1}{z-z_2}, \tag{7.2.1}$$

where $z_1 = -b/a$, $z_2 = -d/c$. Now

$$\left|\frac{z_1}{z_2}\right| = \left|\frac{b}{a}\cdot\frac{c}{d}\right| = \left|\frac{c}{a}\right| \left|\frac{b}{d}\right| = (1)\,(1) = 1.$$

$\therefore$ $|z_1| = |z_2|$

From (7.2.1), since $z = 1$ goes to a point on the circle $|w| = 1$,

$$\left|e^{i\lambda}\,\frac{1-z_1}{1-z_2}\right| = 1.$$

i.e., $|1 - z_1| = |1 - z_2|$ or $|1 - z_1|^2 = |1 - z_2|^2$

$\therefore$ $(1 - z_1)(1 - \bar{z}_1) = (1 - z_2)(1 - \bar{z}_2)$

But, expanding and using $z_1\bar{z}_1 = |z_1|^2$, $z_2\bar{z}_2 = |z_2|^2$, $|z_1| = |z_2|$,

$$z_1 + \bar{z}_1 = z_2 + \bar{z}_2$$

$\therefore$ $2\,\mathbf{Re}\,z_1 = 2\,\mathbf{Re}\,z_2$ or $\mathbf{Re}\,z_1 = \mathbf{Re}\,z_2$.

But $|z_1| = |z_2|$. Hence, either $z_2 = z_1$ or $z_2 = \bar{z}_1$. But, if $z_2 = z_1$, then the transformation (7.2.1) becomes $w = e^{i\lambda}$ which maps the entire z plane into a single point $w = e^{i\lambda}$, Thus $z_2 = \bar{z}_1$ and the required transformation is

$$w = e^{i\lambda}\,\frac{z-z_1}{z-\bar{z}_1}$$

However, we have to still check whether the upper half plane goes onto the interior of the circle $|w| = 1$. If z_1 lies in the upper half plane, then it goes to $w = 0$, a point in the interior of $|w| = 1$. So the above transformation maps $\mathbf{Im}\,z \geq 0$ onto $|w| \leq 1$ if z_1 is a point in the upper half plane.

Remark 1. For all real λ, the transformations

$$w = e^{i\lambda}\,\frac{z-z_1}{z-\bar{z}_1}$$

map $\mathbf{Im}\,z \geq 0$ onto $|w| \leq 1$ provided z_1 is a point in the upper half plane. In these transformations $z = z_1$ goes to $w = 0$.

Remark 2. *Verification.* If $z_1 = x_1 + iy_1$ is in the upper half plane, then y_1 is positive and

$$\left|\frac{z-z_1}{z-\bar{z}_1}\right| = \frac{\sqrt{(x-x_1)^2+(y-y_1)^2}}{\sqrt{(x-x_1)^2+(y+y_1)^2}} < 1$$

So every point in the upper half plane goes to a point in $|w| < 1$.

Remark 3. If z_1 in the lower half plane, then

$$w = e^{i\lambda}\,\frac{z-z_1}{z-\bar{z}_1}$$

transforms the lower half plane $\mathbf{Im}\,z \leq 0$ onto $|w| \leq 1$.

Note. For different values of λ and z_1, the transformation sends the real axis onto $|w| = 1$. But under each transformation the points in the z plane go to different points in the w plane.

Theorem 7.7. The bilinear transformation which map $\textbf{Re } z \geq 0$ onto $|w| \leq 1$ are of the form

$$w = e^{i\lambda}\frac{z - z_1}{z + \bar{z}_1}$$

where λ is a real number (Fig. 7.2.2)

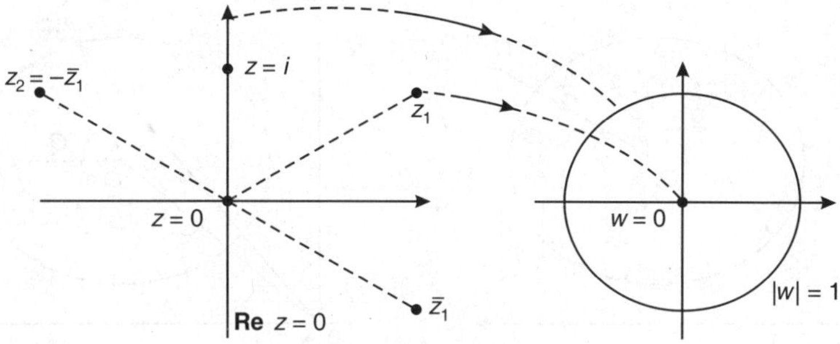

Fig. 7.2.2

Proof : Let the required bilinear transformation be

$$w = \frac{az + b}{cz + d}$$

Now the imaginary axis, $\textbf{Re } z = 0$, goes onto the circle $|w| = 1$. Taking the particular points $z = 0$, $z = i$, $z = \infty$ on the imaginary axis and first considering $z = 0$ and $z = \infty$, we get, as in the previous theorem,

$$w = e^{i\lambda}\frac{z - z_1}{z - z_2} \tag{7.2.1}$$

and $\qquad |z_1| = |z_2|.$

Further, since $z = i$ goes to a point on the circle $|w| = 1$,

$$\left| e^{i\lambda}\frac{i - z_1}{i - z_2} \right| = 1$$

i.e., $\qquad |i - z_1| = |i - z_2| \;\text{ or }\; |i - z_1|^2 = |i - z_2|^2.$

$\therefore \qquad (i - z_1)(-i - \bar{z}_1) = (i - z_2)(-i - \bar{z}_2)$

But, expanding and using

$$z_1\bar{z}_1 = |z_1|^2, z_2\bar{z}_2 = |z_2|^2, |z_1| = |z_2|,$$

$$z_1 - \bar{z}_1 = z_2 - \bar{z}_2$$

$\therefore \qquad 2\,\textbf{Im } z_1 = 2\,\textbf{Im } z_2 \;\text{ or }\; \textbf{Im } z_1 = \textbf{Im } z_2.$

Now, since $|z_1| = |z_2|$ and $\text{Im } z_1 = \text{Im } z_2$ (Fig. 7.2.2),

$$z_2 = -\bar{z}_1$$

Hence the required bilinear transformation is

$$w = e^{i\lambda}\frac{z - z_1}{z + \bar{z}_1}$$

where λ is real and $z = z_1$ is a point in the right half plane $\textbf{Re } z \geq 0$. The point $z = z_1$ goes to $w = 0$.

7.3. CIRCLES AND INVERSE POINTS

Inverse points. Suppose C is the centre of a circle and ρ, its radius (Fig. 7.3.1). If P and Q are points such that

(*i*) C, P, Q are collinear and P, Q are on the same side of C and

(*ii*) $CP \cdot CQ = \rho^2$, then P and Q are said to be inverse with respect to the circle. It is conventional to denote P and Q by the complex numbers p and q.

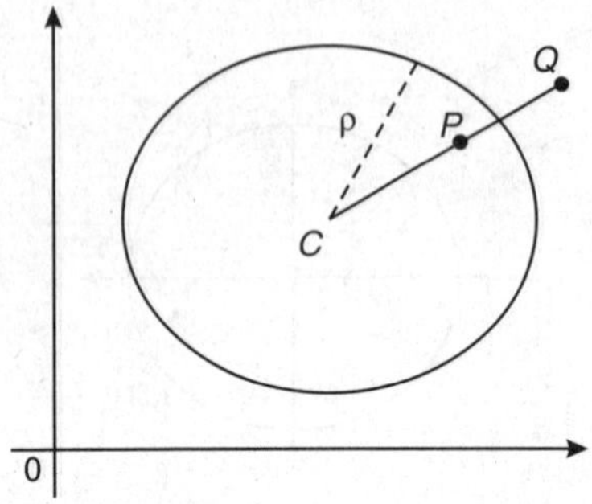

Fig. 7.3.1

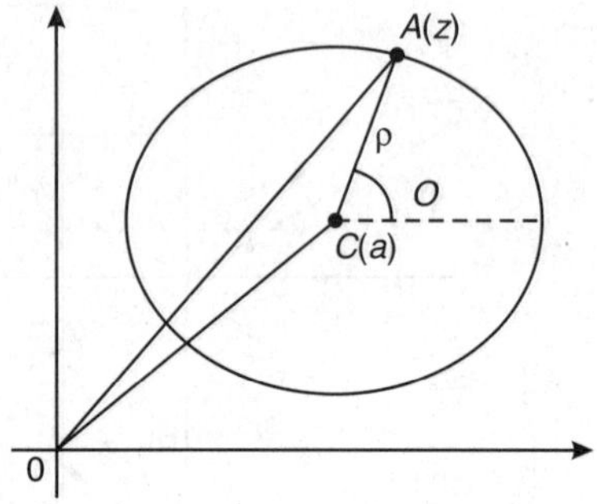

Fig. 7.3.2

Theorem 7.8. If $z = p$ and $z = q$ are inverse points with respect to a circle, then the equation of the circle can be put in the form

$$\frac{z-p}{z-q} = k, \quad k \neq 1.$$

Proof: Let us make the following assumptions (Fig. 7.3.2).

(*i*) P, Q, l : Inverse points p, q and the length CP,

(*ii*) $C(a), \rho$: Centre and radius of the circle,

(*iii*) $A(z)$: Any point on the circle.

Then $$z = \overline{OA} = \overline{OC} + \overline{CA} = a + \rho e^{i\theta}$$

Hence the equation of the circle is

$$z = a + \rho e^{i\theta}, \quad 0 \le \theta < 2\pi$$

Also $$P = \overline{OP} = \overline{OC} + \overline{CP} = a + CP\, e^{i\lambda} = a + le^{i\lambda},$$

$$q = \overline{OQ} = \overline{OC} + \overline{CQ} = a + CQ\, e^{i\lambda} = a + \frac{\rho^2}{l} e^{i\lambda},$$

where λ is real. Consequently, for any point z on the circle,

$$z - p = (a + \rho e^{i\theta}) - (a + l\, e^{i\lambda}) = \rho e^{i\theta} - l\, e^{i\lambda},$$

$$z - q = (a + \rho\, e^{i\theta}) - \left(a + \frac{\rho^2}{l} e^{i\lambda}\right) = \frac{\rho}{l}(l\, e^{i\theta} - \rho e^{i\lambda}).$$

Thus, for any point z on the circle,

$$\left|\frac{z-p}{z-q}\right| = \frac{l}{\rho}\left|\frac{\rho e^{i\theta} - l\, e^{i\lambda}}{l\, e^{i\theta} - \rho e^{i\lambda}}\right| = \frac{l}{\rho} = k, \text{ say.}$$

So this is the equation of the circle. Hence the result.

Particular case : Suppose $k = 1$. Then the equation is

$$|z - p| = |z - q|.$$

That is, z is equidistant from p and q. So the equation represents the perpendicular bisector of the segment joining p and q.

Converse. It can be established that the equation

$$\left|\frac{z-p}{z-q}\right| = k, \quad k \neq 1$$

represents a circle with p and q as points inverse with reference to it and that the equation represents, when $k = 1$, a straight line about which p and q are symmetrical.

Theorem 7.9. A bilinear transformation transforms a circle into a circle and inverse points into inverse points. In the particular case in which the circle transforms into a straight line the inverse points transform into points symmetrical about the line.

Proof: Let p, q be the inverse points. Then the equation of the circle is of the form

$$\left|\frac{z-p}{z-q}\right| = k, \ k \neq 1.$$

Let the given bilinear transformation be

$$w = \frac{az+b}{cz+d}$$

and let p', q' be the images of p, q. Then

$$(cp+d)(w-p') = (ad-bc)(z-p),$$
$$(cq+d)(w-q') = (ad-bc)(z-q).$$

$$\therefore \quad \frac{z-p}{z-q} = \frac{cp+d}{cq+d} \cdot \frac{w-p'}{w-q'}.$$

Hence the equation of the image of the circle is

$$\left|\frac{cp+d}{cq+d}\right| \left|\frac{w-p'}{w-q'}\right| = k \quad \text{or} \quad \left|\frac{w-p'}{w-q'}\right| = k \left|\frac{cq+d}{cp+d}\right|$$

This implies the truth of the first result. The particular case occurs, when $k\,|(cq+d)/(cp+d)| = 1$, in which case the equation of the image is

$$|w-p'| = |w-q'|$$

This equation represents a straight line about which p' and q' are symmetrical.

Illustration. As an application of inverse points we now find the bilinear transformation which maps the imaginary axis **Re** $z = 0$ onto the unit circle $|w| = 1$, sending

$$z = 1, 0 \text{ into } w = 0, -1.$$

Now $z = 1$, $z = -1$ are inverse points with respect to the imaginary axis and $w = 0$, $w = \infty$ are inverse points with respect to $|w| = 1$. So the transformation sends

$$z = 1, 0, -1 \text{ into } w = 0, -1, \infty.$$

Suppose the required bilinear transformation is

$$\frac{(w-w_1)(w_2-w_3)}{(w-w_3)(w_2-w_1)} = \frac{(z-z_1)(z_2-z_3)}{(z-z_3)(z_2-z_1)}.$$

When ∞ is set for w_3, the left side expression becomes

$$\frac{(w-w_1)\{(w_2/w_3)-1\}}{\{(w/w_3)-1\}(w_2-w_1)} = \frac{w-w_1}{(-1)} \cdot \frac{(-1)}{(w_2-w_1)} = \frac{w-w_1}{w_2-w_1}$$

$$\therefore \qquad \frac{w - w_1}{w_2 - w_1} = \frac{(z - z_1)(z_2 - z_3)}{(z - z_3)(z_2 - z_1)}$$

Now, setting $z_1 = 1, z_2 = 0, z_3 = -1, w_1 = 0, w_2 = -1$ in this, we get the required transformation as

$$w + 1 = \frac{(z)(2)}{(z+1)(1)} \quad \text{or} \quad w = \frac{z-1}{z+1}.$$

Theorem 7.10. The bilinear transformations which map the disc $|z| \leq 1$ onto the disc $|w| \leq 1$ are of the form

$$w = e^{i\lambda} \frac{z - z_1}{\bar{z}_1 z - 1},$$

where λ is real.

Proof: Suppose the points,

$$z = z_1, \quad z = \frac{1}{\bar{z}_1},$$

which are inverse points with respect to the circle $|z| = 1$, go to the points which are

$$w = 0, \quad w = \infty$$

inverse with respect to the circle $|z| = 1$. Then

$$0 = \frac{az_1 + b}{cz_1 + d}, \quad \infty = \frac{(a/\bar{z}_1) + b}{(c/\bar{z}_1) + d}$$

$$\therefore \qquad z_1 = -\frac{b}{a} \qquad \bar{z}_1 = -\frac{c}{d}$$

$$\therefore \qquad w = \frac{a}{c} \cdot \frac{z + (b/a)}{z + (d/c)} = \frac{a}{c} \cdot \frac{z - z_1}{z - (1/\bar{z}_1)} = \frac{a\bar{z}_1}{c} \cdot \frac{z - z_1}{\bar{z}_1 z - 1}.$$

Using the fact that $z = 1$ goes to a point on $|w| = 1$,

$$\left|\frac{a\bar{z}_1}{c}\right| \left|\frac{1 - z_1}{\bar{z}_1 - 1}\right| = 1$$

In this $|1 - z_1| = |\bar{z}_1 - 1|$ because, if $z_1 = x + iy$,

$$|1 - z_1| = \sqrt{(1-x)^2 + y^2}, |\bar{z}_1 - 1| = \sqrt{(x-1)^2 + y^2}$$

$$\therefore \qquad \left|\frac{a\bar{z}_1}{c}\right| = 1 \quad \text{or} \quad \frac{a\bar{z}_1}{c} = e^{i\lambda},$$

where λ is real.

$$w = e^{i\lambda} \frac{z - z_1}{\bar{z}_1 z - 1}.$$

Now $z = z_1$ goes to $w = 0$. So, if $z = z_1$ is in $|z| \leq 1$, then the interior of $|z| = 1$ goes onto the interior of $|w| = 1$.

Corollary. If $z_1 = 0$, i.e., if $z = 0$ goes to $w = 0$, then the bilinear transformation is $w = ze^{i\lambda}$ which is a pure rotation about the origin through the angle λ.

7.4. TRANSFORMATION $w = z^2$

If $z = re^{i\theta}$ with radius vector r and argument θ goes to $w = Re^{i\phi}$ under the transformation $w = z^2$, then

$$R = r^2, \quad \phi = 2\,\theta.$$

That is, the transformation squares the distance of a point from the origin and doubles the polar angle. It is evident that the image of the half of the circle $|z| = a$ which lies above the real axis is the full circle $|w| = a^2$. In particular, the circle $|z| = \frac{1}{2}$ is contracted to the circle $|w| = \frac{1}{4}$ and the circle $|z| = 2$ is magnified to the circle $|w| = 4$. The entire first quadrant region of the z plane goes onto the upper half plane in the w plane. The upper half z plane alone goes onto the entire w plane once and the lower half z plane alone goes again onto the entire w plane. In particular,

(*i*) the infinite sectorial domain $\alpha \le \arg z \le \beta$ is mapped onto the infinite sectorial domain $2\alpha \le \arg w \le 2\beta$ and

(*ii*) the circle $r = \lambda$ in the z plane is mapped on to the circle $R = \lambda^2$ in the w plane twice.

(*iii*) the infinite sector bounded by $\arg z = 0$ and $\arg z = \pi/4$ goes onto the entire first quadrant of the w plane.

If $z = x + iy$ and $w = u + iv$, then $w = z^2$ gives

$$u = x^2 - y^2, \quad v = 2xy$$

So the transformation $w = z^2$ maps the rectangular hyperbolas

$$x^2 - y^2 = k_1, \quad xy = k_2$$

onto the straight lines

$$u = k_1, \quad v = 2k_2$$

7.5. TRANSFORMATION $w = z^{1/2}$

The function $f(z) = z^{1/2}$ is a double-valued function. For example, if $z = 4\,(\cos\alpha + i\sin\alpha)$, then the values of $f(z)$ are

$$w = 2\text{ cis}\left(\tfrac{1}{2}\alpha\right), \quad w = 2\text{ cis}\left(\pi + \tfrac{1}{2}\alpha\right)$$

Suppose $\qquad z = r\,(\cos\theta + i\sin\theta)$

Then under the transformation $w = z^{1/2}$ the image of z is

$$w = r^{1/2}\,(\cos\theta + i\sin\theta)^{1/2}$$

$$= r^{1/2}\left(\cos\frac{2k\pi + \theta}{2} + i\sin\frac{2k\pi + \theta}{2}\right), \quad k = 0, 1$$

Taking $z^{1/2}$ as the single valued function given by $k = 0$,

$$w = r^{1/2}\,(\cos\theta/2 + i\sin\theta/2)\,.$$

Then the transformation changes the radius vector of z into its square root and halves the argument. Thus the upper half plane goes onto the first quadrant, etc.

Theorem 7.11. Under the transformation $w = z^{1/2}$, the concentric circles $|z - 1| = \rho$ in the w plane go onto leminscates with focal points $w = \pm 1$.

The given transformation is $w^2 = z$. That is,

$$w^2 - 1 = z - 1 \quad \text{or} \quad (w - 1)(w + 1) = z - 1$$

$$\therefore \qquad |w - 1|\,|w + 1| = |z - 1|$$

Hence the circles $|z - 1| = \rho$ go onto the lemniscates

$$|w - 1|\,|w + 1| = \rho$$

with foci at $w = \pm 1$. (Lemniscate is the locus of a point which moves such that the product of its distances from two fixed points (called foci) is constant.)

7.6. TRANSFORMATION $w = z + \frac{1}{z}$

The function $z + 1/z$ is called Zhukosky's function.

Let

$$z = re^{i\theta} = r(\cos\theta + i\sin\theta)$$

$w = u + iv$ under $w = z + \frac{1}{z}$. Then

$$u + iv = r(\cos\theta + i\sin\theta) + \frac{1}{r(\cos\theta + i\sin\theta)}$$

$$= r(\cos\theta + i\sin\theta) + \frac{1}{r}(\cos\theta - i\sin\theta)$$

$$\therefore \quad u = \left(r + \frac{1}{r}\right)\cos\theta, \quad v = \left(r - \frac{1}{r}\right)\sin\theta \tag{1}$$

Now we shall show the following two cases:

Case 1. A circle whose centre is the origin goes onto an ellipse whose centre is the origin under the zhukosky's transformation.

Case 2. A ray emanating from the origin goes onto a hyperbola whose centre is the origin.

Case 1. Consider the circle in the z plane whose polar equation is $r = a$. Its image is obtained by eliminating r, θ from

$$u = \left(r + \frac{1}{r}\right)\cos\theta, \quad v = \left(r - \frac{1}{r}\right)\sin\theta, \quad r = a$$

Eliminating r from these three, we get

$$u = \left(a + \frac{1}{a}\right)\cos\theta, \quad v = \left(a - \frac{1}{a}\right)\sin\theta .$$

To eliminate θ, find $\cos\theta$ and $\sin\theta$, and square and add. Then

$$\frac{u^2}{\left(a + \frac{1}{a}\right)^2} + \frac{v^2}{\left(a - \frac{1}{a}\right)^2} = 1 \tag{2}$$

This is the required image of the circle. It is an ellipse. Its centre is the origin. Its foci are (2, 0), (– 2, 0) which can be obtained from the fact that the foci of the ellipses.

$$\frac{x^2}{\alpha^2} + \frac{y^2}{\beta} = 1 \text{ are } \left(\pm\sqrt{\alpha^2 - \beta^2}, 0\right).$$

So the image of the circles $x^2 + y^2 = a^2$, where a is a parameter are the confocal ellipses (2), having the same foci (± 2, 0).

It is evident that the image ellipses of the two circles

$$x^2 + y^2 = a^2, \quad x^2 + y^2 = \frac{1}{a^2}$$

of which one circle lies inside the unit circle $x^2 + y^2 = 1$ and the other lies outside the unit circle, are same ellipses.

Image of $|z| = 1$ or $r = 1$. The image of circle, $r = 1$, is given by the eliminant of r, θ, from

$$u = \left(r + \frac{1}{r}\right)\cos\theta, \quad v = \left(r - \frac{1}{r}\right)\sin\theta, \quad r = 1.$$

Eliminating r by substituting $r = 1$,

$$u = 2\cos\theta, \quad v = 0.$$

Here θ varies from 0 to 2π. But $\cos\theta$ varies from – 1 to 1. Thus the image is the straight line segment joining (2, 0), (– 2, 0).

Case 2. Consider the ray in the z plane whose equation is

$$\arg z = \alpha.$$

This ray emanates from the origin. Its image is given by the eliminant of r, θ from

$$u = \left(r + \tfrac{1}{r}\right)\cos\theta, \quad v = \left(r - \tfrac{1}{r}\right)\sin\theta, \quad \theta = \alpha\,.$$

Eliminating θ by substituting $\theta = \alpha$, we get

$$u = \left(r + \tfrac{1}{r}\right)\cos\alpha, \quad v = \left(r - \tfrac{1}{r}\right)\sin\alpha$$

i.e., $$r + \frac{1}{r} = \frac{u}{\cos\alpha}, \quad r - \frac{1}{r} = \frac{v}{\sin\alpha}$$

Squaring these and subtracting, we get the eliminant of r as

$$\frac{u^2}{\cos^2\alpha} - \frac{v^2}{\sin^2\alpha} = \left(r + \frac{1}{r}\right)^2 - \left(r - \frac{1}{r}\right)^2$$

i.e., $$\frac{u^2}{\cos^2\alpha} - \frac{v^2}{\sin^2\alpha} = 4 \quad \text{or} \quad \frac{u^2}{4\cos^2\alpha} - \frac{v^2}{4\sin^2\alpha} = 1\,.$$

So the image is a hyperbola if $\alpha \neq 0, \pi/2, \pi, 3\pi/2$. Its foci are

$$(2, 0) \quad \text{and} \quad (-2, 0)$$

Since the foci of the hyperbola $\dfrac{x^2}{\alpha^2} - \dfrac{y^2}{\beta^2} = 1$ are $\left(\pm\sqrt{\alpha^2 + \beta^2}, 0\right)$.

7.7. TRANSFORMATIONS $w = e^z$, $w = \text{Log } z$

To begin with we shall take z as

$$z = x + iy$$

in cartesian coordinates and its image under $w = e^z$ as

$$w = Re^{i\phi}$$

in polar coordinates. Then

$$Re^{i\phi} = e^z = e^{x+iy} = e^x e^{iy}$$

$$\therefore \quad R = e^x, \quad \phi = y$$

i.e., $$|w| = |Re^{i\phi}| = |R(\cos\phi + i\sin\phi)| = |R|\sqrt{\cos^2\phi + \sin^2\phi} = R = e^x,$$

$$\arg w = \arg(Re^{i\phi}) = \arg(R\cos\phi + iR\sin\phi) = \tan^{-1}\left(\frac{R\sin\phi}{R\cos\phi}\right)$$

$$= \tan^{-1}(\tan\phi) = \phi = y. \qquad (1)$$

In section 6.8 we saw that the lines

$$y = y_0, \quad y = y_0 + 2n\pi, \quad y = y_0 + 4n\pi, \ldots$$

in the z plane go onto the same ray

$$v = (\tan y_0)\, u$$

in the w plane emanating from $w = 0$ and that each of the strips

$$0 \le y < 2\pi,\ 2\pi \le y < 4\pi,\ 4\pi \le y < 6\pi, \ldots$$

in the z plane goes onto the w plane once. In particular, a horizontal strip $y_1 < y < y_2$ $(y_2 - y_1 = \pi)$ goes onto a half plane. Thus, to map the horizontal strip $-\frac{1}{2}\pi < y < \frac{1}{2}\pi$ onto $|w| < 1$, we should first map the strip onto the half plane Re $z > 0$ by the transformation $w = e^z$, and then onto $|w| < 1$ by $w = (z - 1)/(z + 1)$.

Thus the resultant transformation is

$$w = \frac{e^z - 1}{e^z + 1} \quad \text{or} \quad w = \tanh \frac{z}{2}.$$

Image of the line $y = y_0$. For all points on the line $y = y_0$, the x coordinates vary from $-\infty$ to ∞ and the y coordinates are the constant y_0. Thus, from (1), for the corresponding points, the radius vectors and the arguments are such that

$$0 \le R < \infty, \quad \phi = y_0$$

because e^x varies from 0 to ∞ as x varies from $-\infty$ to ∞. Thus the radius vectors of the w points vary from 0 to ∞ while the arguments are the constant y_0.

Since the image of the entire line $y = y_0$ is the entire ray, emanating from the origin which makes an angle y_0

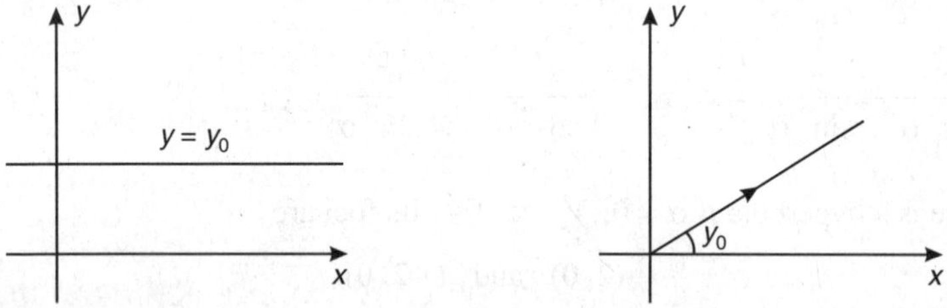

with the real axis. As particular cases of the transformation, we have the following :

Preimage in z plane	*Image in w plane*
$y = 0$	Right half of the real axis
$y = \frac{\pi}{4}$	Right emanating from the origin and making an angle $\pi/4$ with the real axis.
$y = \frac{\pi}{2}$	Upper half of the imaginary axis

Image of the infinite rectangular strip bounded by the lines $y = 0, y = 2\pi$. As y_0 takes values from 0 to 2π, the line $y = y_0$ moves upward parallely from the real axis $y = 0$ to the line $y = 2\pi$ and correspondingly the image ray in the w plane rotates about the point w = 0 completely once starting

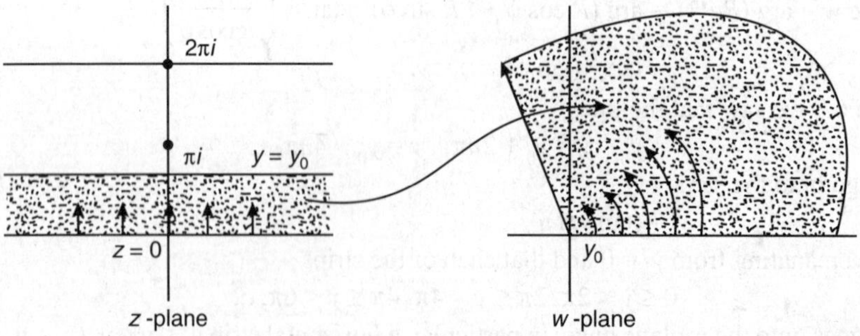

from the right half of the real axis. This implies that the mapping is one to one and onto, and the strip $0 \le y < 2\pi$ in the z plane is mapped onto the region $0 \le \arg w < 2\pi$ which is the whole of w plane. But

$$e^{x + iy + 2\pi ni} = e^{x + iy}\, e^{2\pi ni} = e^{x + iy}, \; n = 0, \; \pm 1, \; \pm 2, \ldots$$

So each of the following strips in the z plane

$$\ldots, -2\pi \le y < 0, \quad 0 \le y < 2\pi, 2\pi \le y < 4\pi, \ldots$$

goes onto the w plane once. That is, for each w there are many z's satisfying $e^z = w$ and each of the strips contains one z satisfying $e^z = w$. Further, for a given w, the respective values of z of the strips are $\log|w| + i \arg w + n(2\pi i)$ because

$$e^{\log|w| + i\arg w + n(2\pi i)} = e^{\log|w|}\, e^{i\arg w}\, e^{2\pi i n}$$

$$= |w| \{\cos(\arg w) + i \sin(\arg w)\} = w.$$

Now we are in a position to define the many-valued function, the logarithmic function, its values being those of z satisfying $e^z = w$.

Logarithmic function. From the above discussion we see that the logarithmic function is a multiple-valued function defined in the whole of z plane except at $z = 0$ and $z = \infty$.

We shall denote this function by Log z so that

$$\text{Log } z = \log|z| + i \arg z + n(2\pi i), \quad n = 0, \ \pm 1, \ \pm 2, \ldots$$

$$= \log\sqrt{x^2 + y^2} + i\tan^{-1}\left(y/x\right) + n(2\pi i), \quad n = 0, \ \pm 1, \ \pm 2, \ldots$$

$$= \log r + i\,\theta + n(2\pi i), \quad n = 0, \ \pm 1, \ \pm 2, \ldots$$

The set of values of Log z given by one value of n is called a branch of Log z. As such, Log z has infinite number of branches.

The value given by $n = 0$ is called the principal value of log z and it is denoted by log z so that

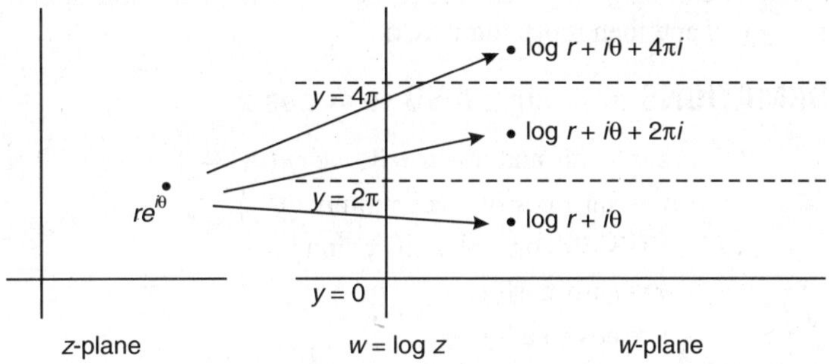

$$\text{Log } z = \log|z| + i \arg z = \log\sqrt{x^2 + y^2} + i\tan^{-1}\left(y/x\right)$$

$$= \log r + i\theta.$$

Images of $x = a$ and $x = b$ under $w = e^z$. If the image of $z = x + iy$ is $w = Re^{i\phi}$, then

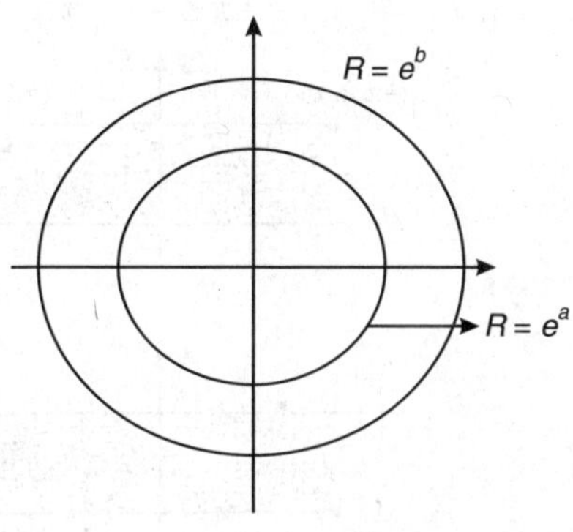

$$R = e^x, \quad \phi = y$$

The images of the lines

$$x = a, \quad x = b$$

are the circles

$$R = e^a, \quad R = e^b$$

whose centres are at the origin and radii are e^a and e^b.

A particular case. The image of the y axis, $x = 0$, is

$$R = e^0 \quad \text{or} \quad R = 1$$

which is the unit circle $|w| = 1$.

Images of $x = h, y = k$. Using $w = Re^{i\phi}$ and $z = x + iy$, the transformation $w = e^z$ is

$$Re^{i\phi} = e^x e^{iy} \quad \text{or} \quad R = e^x, \ \phi = y.$$

So the transformation maps the line $x = h$ onto the circle $R = e^h$ and the line $y = k$ $(0 \le k \le 2\pi)$ onto the ray $\phi = k$.

The rectangular domain

$$h_1 \le x \le h_2, \ k_1 \le y \le k_2$$

is mapped onto the domain bounded by the circles

$$R = e^{h_1}, \ \text{R} = e^{h_2}$$

and the rays (Fig. 7.7.1)

$$\phi = k_1, \ \phi = k_2.$$

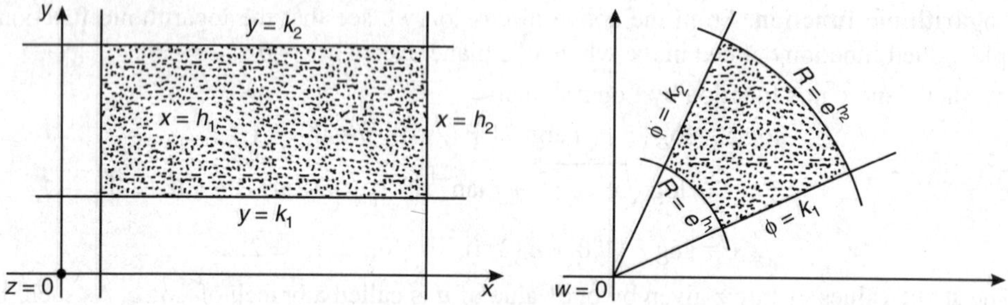

Fig. 7.7.1

Image of y = mx. The line $y = mx$ goes onto $R = e^{(1/m)\phi}$ is a logarithmic spiral. This is obtained by first replacing x by y/m and then replacing y by ϕ.

7.8. TRANSFORMATIONS $w = \sin z$ AND $w = \cos z$

Setting $\qquad z = x + iy$ and $w = u + iv$, we get

$$u + iv = \sin x \cos iy + \cos x \sin iy$$

$$= \sin x \cosh y + i \cos x \sinh y$$

$$\therefore \qquad u = \sin x \cosh y,$$

$$y = \cos x \sinh y.$$

Image of a line segment parallel to x axis. We shall consider the images of line segments

(*i*) $y = c, \quad -\frac{1}{2}\pi \le x \le \frac{1}{2}\pi$

(*ii*) $y = -c, \quad -\frac{1}{2}\pi \le x \le \frac{1}{2}\pi$

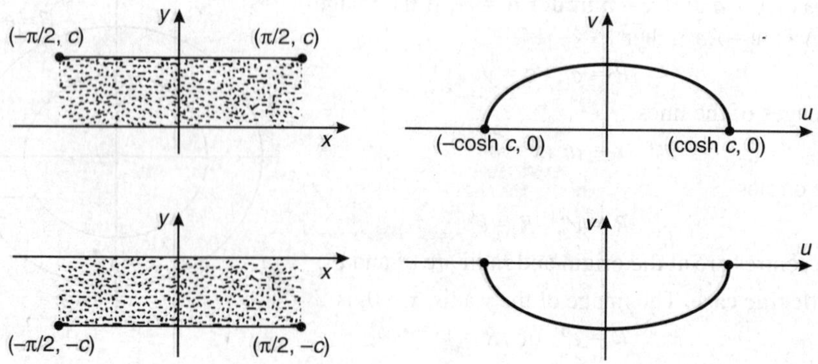

Fig. 7.8.1

(Fig. 7.8.1). The first segment goes onto the curve whose parametric equation are

$$u = (\cosh c) \sin x$$

$$\therefore \quad v = (\sinh c) \cos x$$

It is the upper semielliptic arc

$$\frac{u^2}{(\cosh c)^2} + \frac{v^2}{(\sinh c)^2} = 1, \; v \geq 0$$

As c increases from zero, the line segment moves upwards and its image also increases in size. As c varies from 0 to ∞, the infinite upper semistrip

$$-\tfrac{1}{2}\pi \leq x \leq \tfrac{1}{2}\pi, \quad y \geq 0$$

goes onto the entire upper half w plane and similarly the infinite lower semistrip

$$-\tfrac{1}{2}\pi \leq x \leq \tfrac{1}{2}\pi, \quad y \leq 0$$

goes onto the entire lower half w plane. Thus $w = \sin z$ sets up a one to one correspondence between the points in the infinite strip

$$-\tfrac{1}{2}\pi \leq x \leq \tfrac{1}{2}\pi$$

and the entire w plane. The portion of the real axis

$$-\tfrac{1}{2}\pi \leq x \leq \tfrac{1}{2}\pi, \quad y = 0$$

goes onto the line segment joining the points $w = -1$, $w = 1$ which are the common focus of all the above ellipses.

Image of a line segment parallel to y axis. The line segment parallel to the y axis

$$x = c, \quad -\tfrac{1}{2}\pi < y \leq \tfrac{1}{2}\pi$$

maps onto the right-hand branch of the hyperbola

$$\frac{u^2}{\sin^2 c} - \frac{v^2}{\cos^2 c} = 1,$$

when $c > 0$ and the left-hand branch of the hyperbola when $c < 0$. The foci of the hyperbola are $w = -1$, $w = 1$.

Transformation $w = \cos z$. Since $\sin\left(\tfrac{1}{2}\pi + z\right) = \cos z$, the transformation $w = \cos z$ is a composition of the transformations

$$w_1 = \tfrac{1}{2}\pi + z, \quad w = \sin w_1$$

As such, it is a translation which shifts all the points to the right through a distance $\tfrac{1}{2}\pi$ followed by the transformation $w = \sin z$.

7.9. CONFORMAL MAPPING

Isogonal mapping. Let $f(z)$ be a single-valued function defined in a region D. Then the mapping.

$$w = f(z)$$

is said to be *isogonal* if it preserves the magnitude of the angle between every two curves. It is immaterial whether the sense of orientation of the angle is preserved or not.

Conformal mapping. The mapping

$$w = f(z)$$

is said to be conformal if it preserves the magnitude of the angle between every two curves and also if it preserves the sense of orientation of the angle.

Remark. Conformal mapping is a particular case of isogonal mapping.

Example 1. The mapping $w = \bar{z}$ is a reflection in the real axis. As such, it preserves the magnitudes of angles but not the sense of their orientations. So it is isogonal but not conformal.

Example 2. If $w = f(z)$ is a conformal mapping, then the mapping $w = f(\bar{z})$ is isogonal because it is the isogonal mapping $w_1 = \bar{z}$ followed by the conformal mapping $w = f(w_1)$.

Example 3. If $w = u + iv = f(z)$ is a conformal mapping, then $w = u - iv = \overline{f(z)}$ is isogonal but not conformal because $w = \overline{f(z)}$ is the mapping $w = f(z)$ followed by a reflection in the real axis.

Angle made by a tangent to a curve with the real axis. Let C be a continuous curve and z_0 and z be points on it. Then the angle made by the chord joining z_0 and z with the real axis is

$$\arg (z - z_0)$$

As z travels along the curve and coincides with z_0, the chord becomes the tangent to C at z_0. So the angle made by the tangent with the real axis is

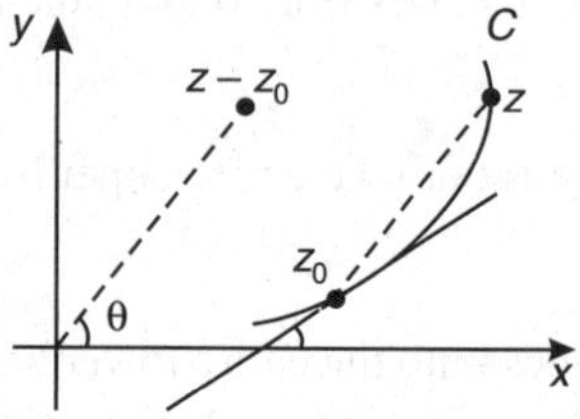

$$\lim_{z \to z_0} (z - z_0)$$

Theorem 7.12. If $f(z)$ is analytic in a region D and if $f'(z) \neq 0$ in D, then the mapping

$$w = f(z)$$

is conformal in D.

Proof: Let $z = a$ be in D. Then, from the definition of derivative,

$$\arg \left(\lim_{z \to a} \frac{f(z) - f(a)}{z - a} \right) = \arg f'(a)$$

i.e.,
$$\lim_{z \to a} \left(\arg \frac{f(z) - f(a)}{z - a} \right) = \arg f'(a)$$

But $\arg z_1/z_2 = \arg z_1 - \arg z_2$. Therefore, for any particular mode of approach of z to a,

$$\lim_{z \to a} \arg \{f(z) - f(a)\} - \lim_{z \to a} \arg (z - a) = \arg f'(a)$$

$$\therefore \quad \lim_{z \to a} \arg \{f(z) - f(a)\} = \lim_{z \to a} \arg (z - a) + \arg f'(a) \tag{7.8.1}$$

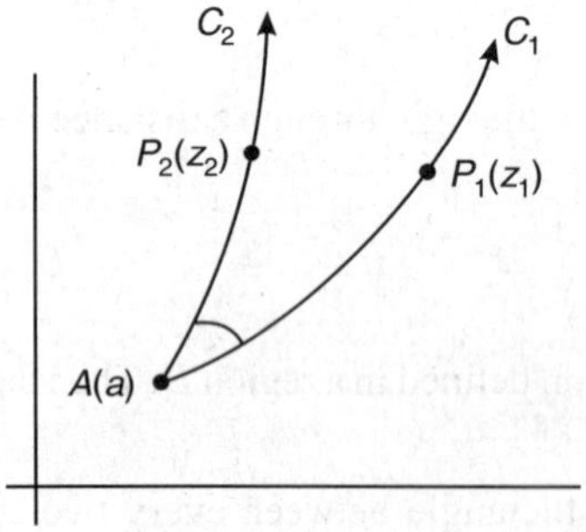

Fig. 7.9.1

Let two arcs C_1 and C_2 emanate from A (a) (Fig. 7.9.1). If P_1 is a point on C_1 and P_2, a point on C_2, then, with the notation used in section 2.4, if θ is the angle between the arcs C_1 and C_2, then

$$\theta = \lim \left[\text{angle} \left(\overline{AP_1}, \overline{AP_2} \right) \right]$$

as $P_1 \to A$ along C_1 and $P_2 \to A$ along C_2. Therefore

$$\theta = \lim_{z_2 \to a} [\arg (z_2 - a)] - \lim_{z_1 \to a} [\arg (z_1 - a)]$$

Similarly, if Γ_1 and Γ_2 are the images of C_1 and C_2 and if ϕ is the angle between Γ_1 and Γ_2, then

$$\phi = \lim_{f(z_2) \to f(a)} [\arg \{f(z_2) - f(a)\}] - \lim_{f(z_1) \to f(a)} [\arg \{f(z_1) - f(a)\}]$$

$$= \lim_{z_2 \to a} [\arg (z_2 - a)] + \arg f'(a) - \lim_{z_1 \to a} [\arg (z_1 - a)] - \arg f'(a) \quad \text{by} \quad (7.8.1)$$

$$= \lim_{z_2 \to a} [\arg (z_2 - a)] - \lim_{z_1 \to a} [\arg (z_1 - a)] = \theta .$$

So the mapping preserves the magnitude and the sense of orientation of the angle. Hence the mapping is conformal.

Remark. From (7.9.1) it can be seen that the angle of rotation at $z = a$ is $\arg f'(a)$.

Coefficient of magnification. In a conformal mapping, we shall show that at a point the magnification in all directions is the same.

Theorem 7.13. Suppose $f(z)$ is analytic in a region D. If l is the length of a small arc C emanating from $z = a$ and if L is the length of its image, then

$$\lim_{z \to a} \frac{L}{l} = |f'(a)|,$$

where $z \to a$ along C,

Proof: Let z be a point on C and $z \to a$. Then

$$\lim_{z \to a} \frac{L}{l} = \lim_{z \to a} \frac{|f(z) - f(a)|}{|z - a|} = \lim_{z \to a} \left| \frac{f(z) - f(a)}{z - a} \right|$$

$$= \left| \lim_{z \to a} \frac{f(z) - f(a)}{z - a} \right| = |f'(a)|.$$

Remark. If $|f'(a)| > 1$, then the transformation has a constant magnification at $z = a$ and the constant of magnification is $|f'(a)|$.

Critical point of the transformation. A point z at which $f'(z) = 0$ is called a critical point of the transformation $w = f(z)$.

At a critical point the mapping is not conformal. The following theorem embodies this result.

Theorem 7.14. If $f(z)$ is analytic in a region D and if $z = a$ is a point in it such that $f'(a) = 0$, then the transformation $w = f(z)$ is not conformal at $z = a$.

Proof: Using Taylor's series (Chapter 9) for $f(z)$ about $z = a$,

$$f(z) = f(a) + \frac{(z - a)}{1!} f'(a) + \frac{(z - a)^2}{2!} f''(a) + ...$$

Now $\quad f'(a) = 0.$

Therefore $\quad f(z) = f(a) + a_2 (z - a)^2 + a_3 (z - a)^3 + ...$

or, in general,

$$f(z) = f(a) + a_n (z - a)^n + a_{n+1} (z - a)^{n+1} + ...$$

$$\therefore \quad f(z) - f(a) = (z - a)^n \phi(z),$$

where $\phi(z)$ is the sum of an absolutely convergent series and $\phi(a) \neq 0$.

$\therefore \quad \arg\{f(z)-f(a)\} = n \arg(z-a) + \arg \phi(z).$

Thus, with the notations used in theorem 7.12.

$$\phi = \lim_{f(z_2)\to f(a)} [\arg\{f(z_2)-f(a)\}] - \lim_{f(z_1)\to f(a)} [\arg\{f(z_1)-f(a)\}]$$

$$= \lim_{z_2\to a} [n \arg(z_2-a) + \arg \phi(z_2)] - \lim_{z_1\to a} [n \arg(z_1-a) + \arg \phi(z_1)]$$

$$= n\left[\lim_{z_2\to a} \{\arg(z_2-a)\} - \lim_{z_1\to a} \{\arg(z_1-a)\}\right] = n\theta.$$

Since $\phi \neq \theta$, the mapping is not conformal.

Corollary. If $z = a$ is a zero of $f(z)$ of order n, that is, if $f(z) = (z-a)^n \phi(z)$, $\phi(a) \neq 0$, then $\phi = n\theta$, that is, the angle between Γ_1 and Γ_2 at $w = f(a)$ is n times the angle between C_1 and C_2 at $z = a$.

Remark. The mapping at critical points and the singular points, is not conformal.

Theorem. 7.15.

(*i*) If the transformation $w = f(z)$ is conformal in a region D,

(*ii*) If u_x, u_y, v_x, v_y exist and are continuous in D and

(*iii*) If the magnification at each point in D is a constant $\neq 0$, then $f(z)$ is analytic in D.

Proof : Let ds and $d\sigma$ be lengths of elementary arc in the z plane at $z = a$ and the image arc in the w plane. Then there exists a positive k such that $d\sigma = k\, ds$ or $d\sigma^2 = k^2 ds^2$. Hence

$$d\sigma^2 = k^2 (dx^2 + dy^2) \qquad (\because ds^2 = dx^2 + dy^2)$$

Also $\quad d\sigma^2 = du^2 + dv^2$, where $du = u_x d_x + u_y d_y$, $dv = v_x d_x + v_y d_y$,

So $\quad d\sigma^2 = (u_x^2 + v_x^2)\, dx^2 + (u_y^2 + v_y^2)\, dy^2 + 2(u_x u_y + v_x v_y)\, d_x d_y$

$\therefore \quad u_x^2 + v_x^2 = k^2, \; u_y^2 + v_y^2 = k^2, \; u_x u_y + v_x v_y = 0$

since dx and dy are arbitrary. From the first and the second, we see that there exist α and β such that

$$u_x = k \cos\alpha, \quad v_x = k \sin\alpha,$$
$$u_y = k \cos\beta, \quad v_y = k \sin\beta,$$

and, on substitution, in the third,

$$\cos(\alpha-\beta) = 0 \quad \text{or} \quad \alpha - \beta = \pm \tfrac{1}{2}\pi$$

of which the result $\alpha - \beta = -\tfrac{1}{2}\pi$ or $\alpha = \beta - \tfrac{1}{2}\pi$ leads to the C.R. equations

$$u_x = v_y, \quad u_y = -v_x.$$

Since the C.R. equations are true and since u_x, u_y, v_x, v_y exist and are continuous, $f(z)$ is analytic.

The result $\alpha - \beta = \tfrac{1}{2}\pi$ has to be neglected because it leads to

$$u_x = -v_y, \quad u_y = v_x$$

which lead to the result that $\overline{f(z)}$ is analytic and hence $w = \overline{f(z)}$ is conformal which is a contradiction to the hypothesis that $w = f(z)$ is conformal. This completes the proof.

7.10. EXAMPLES

7.1. BILINEAR TRANSFORMATION

1. Show that the linear transformation $w = az + b$, where a and b are complex constants, is a composition of homothetic transformation, rotation and translation. Which are fixed points under this transformation ?

[**Ans.** ∞, if $b \neq 0$ and 0 and ∞ if $b = 0$]

2. Solved sum. Determine the image of the rectangle with vertices $O(0, 0)$, $A(2, 0)$, $B(2, 1)$, $C(0, 1)$ under the transformation $w = (1 + i) z + (1 - 2i)$ **(1978)**

Now $$u + iv = (1 + i)(x + iy) + (1 - 2i)$$

$$u = x - y + 1, v = x + y - 2$$

Equation of the image of the line OA, $y = 0$: It is the eliminant of x, y from

$$y = 0,$$
$$u = x - y + 1, \qquad \therefore \quad u - v = 3.$$
$$v = x + y - 2,$$

Equation of the image of the line CB, $y = 1$: It is the eliminant of x, y from

$$y = 1,$$
$$u = x - y + 1, \qquad \therefore \quad u - v = 1$$
$$v = x + y - 2.$$

Equation of the image of the line OC, $x = 0$: It is the eliminant of x, y from

$$x = 0,$$
$$u = x - y + 1, \qquad \therefore \quad u + v = -1.$$
$$v = x + y - 2.$$

Equation of the image of the line AB, $x = 2$: It is the eliminant of x, y from

$$x = 2,$$
$$u = x - y + 1, \qquad \therefore \quad u + v = 3.$$
$$v = x + y - 2.$$

So the image is the rectangle whose sides are the lines $u - v = 3$, $u - v = 1$, $u + v = -1$, $u + v = 3$.

3. Find the image of the rectangle bounded by the lines $x = 0$, $y = 0$, $x = 2$, $y = 1$ under the transformation $w = z + (1 - 2i)$.

[**Ans.** Rectangle bounded by $u = 1$, $v = -2$, $u = 3$, $v = -1$]

4. Show that the transformations given below maps the domains given in column 2 onto the domains given in column 3.

Transformation	*Domain in z plane*	*Domain in w plane*
(*i*) $w = iz$	$x > 0$	$v > 0$
(*ii*) $w = iz + 1$	$x > 0$	$v > 0$
(*iii*) $w = iz + i$	$x > 0$	$v > 1$
(*iv*) $w = (1 + i) z$	$y > 0$	$v > u$
(*v*) $w = iz + 1$	$x > 0, 0 < y < 2$	$-1 < u < 1. v > 0$
(*vi*) $w = (1 - i) z$	$y > 1$	$u < 0$

Fixed points

5. Solved sum. Find the fixed points of the following mappings :

(*i*) $w = \dfrac{2zi+5}{z-4i}$ (*ii*) $w = \dfrac{z-1-i}{z+2}$

(*i*) The fixed points are given by

$$z = \frac{2zi+5}{z-4i} \quad or \quad z^2 - 4iz = 2zi + 5$$

i.e.,
$$z^2 - 6iz - 5 = 0 \text{ or } z = \frac{6i \pm \sqrt{-36+20}}{2}$$

or
$$z = 5i, i.$$

(*ii*)
$$z = \frac{z-1-i}{z+2} \quad \text{or} \quad z^2 + 2z - z + 1 + i = 0$$

i.e., $z^2 + z + (1+i) = 0$ or $z = \dfrac{-1 \pm \sqrt{1-4(1+i)}}{2}$

or
$$z = \frac{-1 \pm \sqrt{-3-4i}}{2} = \frac{-1 \pm \sqrt{3+4i}}{2}$$

But $\sqrt{3+4i} = \sqrt{5}(\cos\theta + i\sin\theta)^{1/2} = \sqrt{5}\,(\cos \theta/2 + i \sin \theta/2)$, where $\theta = \tan^{-1} 4/3$. So the fixed points are

$$\frac{-1 \pm i\sqrt{5}(\cos \theta/2 + i \sin \theta/2)}{2} \quad \text{or} \quad \frac{-1 \pm \sqrt{5}(i\cos \theta/2 - \sin \theta/2)}{2},$$

where $\theta = \tan^{-1} 4/3$.

6. Find the fixed points under the following transformations :

(*i*) $w = \dfrac{2z-5}{z+4}$ (*ii*) $w = \dfrac{z-2}{z+3}$

(*iii*) $w = \dfrac{z-1}{z+1}$ (*iv*) $w = \dfrac{4z-4}{z}$

(*v*) $w = \dfrac{6z-9}{z}$ (*vi*) $w = \dfrac{1}{z-2i}$

(*vii*) $w = \dfrac{5z+4}{z+5}$

[**Hint:** (*i*) $z = (2z-5)/(z+4)$ gives $z^2 + 2z + 5 = 0$; $z = -1 \pm 2i$
(*ii*) $1 \pm i$ (*iii*) $z = \pm i$ (*iv*) $z = 2$
(*v*) $z = 3$ (*vi*) $z = i$ (*viii*) $z = \pm 2$].

7. Show that $w = (z-1)/(z+1)$ maps the imaginary axis in the z plane onto the circle $|w| = 1$. What portion of the z plane corresponds to the interior of the circle $|w| = 1$? **(1977)**

[**Hint:** $z = i\lambda$, $-\infty < \lambda < \infty$, gives $w = |i\lambda - 1| / |i\lambda + 1| = 1$; $z = 1$ goes to $w = 0$. So right half plane goes onto $|w| \le 1$].

8. Show that $w = (z - i) / (z + i)$ maps the real axis in the z plane onto the circle $|w| = 1$. What portion of z plane corresponds to the interior of $|w| = 1$? **(1980)**

[**Hint:** $z = \lambda, -\infty < \lambda < \infty$, gives $|w| = 1$; $z = i$ goes to $w = 0$. So the upper half plane goes onto $|w| \le 1$].

9. Show that both the transformations,

$$w = \frac{1+z}{1-z}, \qquad w = \frac{z+1}{z-1}$$

map the left half plane **Re** $z \le 0$ onto $|w| \le 1$. **(1979)**

[**Hint:** $|1 + z| = |1 - z|$ gives $x = 0$]

10. Show the both the transformations,

$$w = \frac{z+i}{z-i}, \qquad w = \frac{i+z}{i-z}$$

map the lower half plane **Im** $z \le 0$ onto $|w| \le 1$.

[**Hint:** $|z + i| = |z - i|$ gives $y = 0$]

11. Show that $w = i(1 - z)/(1 + z)$ maps the circular domain $|z| \le 1$ onto the upper half plane **Im** $w \ge 0$. **(1982)**

[**Hint:** $u = \dfrac{2y}{(1+x)^2 + y^2}$, $v = \dfrac{x^2 + y^2 - 1}{(1+x)^2 + y^2}$. Now $x^2 + y^2 = 1$ gives $v = 0$. $z = 0$ goes to $w = i$]

12. Solved sum. Prove that $w = \dfrac{1+iz}{i+z}$ maps the line segment joining -1 and 1 onto a semicircle in the w plane.

✍ The given line segment is

$$z = \lambda, \quad -1 \le \lambda \le 1, \tag{1}$$

where λ is real. The point $z = \lambda$ goes to the point w given by

$$w = \frac{1+i\lambda}{i+\lambda}$$

whose modulus is

$$|w| = \frac{|1+i\lambda|}{|i+\lambda|} = \frac{\sqrt{1+\lambda^2}}{\sqrt{1+\lambda^2}} = 1.$$

So the line segment (1) goes into the points on the unit circle

$$|w| = 1.$$

Now, the images of $z = -1, 0, 1$ which are on the line segment are

$$\frac{1+i(-1)}{i+(-1)}, \ \frac{1}{i}, \ \frac{1+i}{i+1} \quad \text{or} \quad -1, \ -i, \ 1.$$

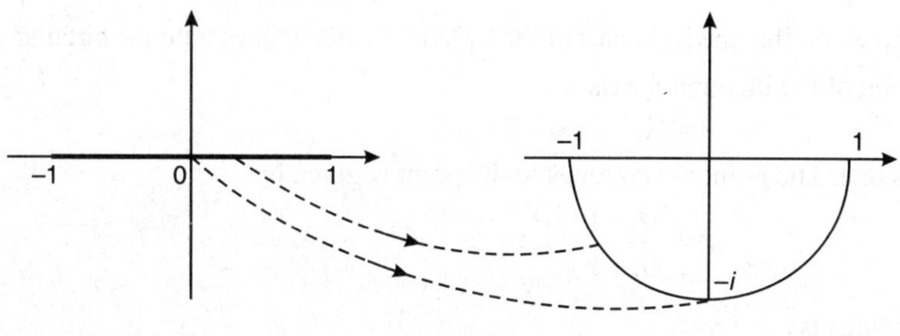

Fig. 7.10.1

This shows that the line segment goes onto the semicircle below the real axis because 0 goes to $-i$.

13. Solved sum. Show that $w = \dfrac{z-i}{z+i}$ maps the real axis in the z plane onto $|w| = 1$ in the w plane. Show also that the upper half of the z plane, **Im** $z \geq 0$, goes onto the circular disc $|w| \leq 1$.

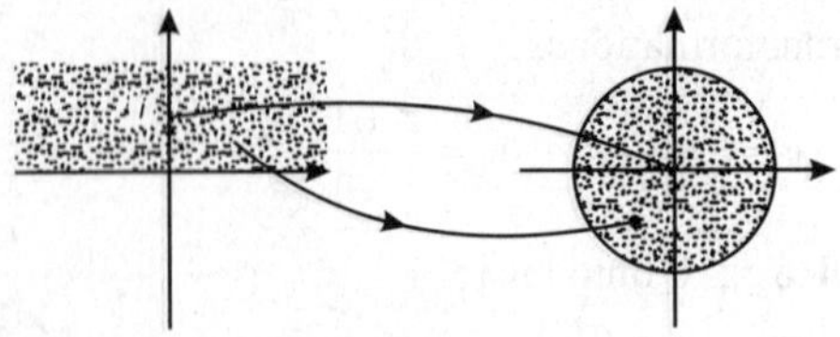

Fig. 7.10.2

The equation of the real axis is,

$$z = \lambda, \quad -\infty < \lambda < \infty,$$

where λ is real. The point $z = \lambda$ goes to the point w given by

$$w = \frac{\lambda - i}{\lambda + i}$$

whose modulus is

$$|w| = \frac{|\lambda - i|}{|\lambda + i|} = \frac{\sqrt{\lambda^2+1}}{\sqrt{\lambda^2+1}} = 1.$$

So the real axis in the z plane goes onto the circle

$$|w| = 1$$

in the w plane. But $z = i$, a point in the upper half plane, goes to

$$w = \frac{i-i}{i+i} = 0$$

which is a point inside the circle $|w| = 1$. This shows that the upper half plane **Im** $z \geq 0$ goes onto the circular disc $|w| \leq 1$.

Remark. There are many forms of transformations which can be rewritten as

$$w = \pm\frac{z-i}{z+i} \quad \text{or} \quad w = \pm\, i\,\frac{z-i}{z+i}.$$

All these transformations transform the upper half plane **Im** $z \geq 0$ onto the circular disc $|w| \leq 1$ and transform the lower half plane **Im** $z \leq 0$ onto the annular region $|w| \geq 1$.

Of course, the way in which the z points go to the w points varies with the transformations.

14. Solved sum. Show that $w = \dfrac{z-1}{z+1}$ maps the imaginary axis in the z plane onto $|w| = 1$ in the w plane. Show also that the right half of the z plane, Re $z \geq 0$, goes onto the circular disc $|w| \leq 1$.

The equation of the imaginary axis is

$$z = i\lambda, \quad -\infty < \lambda < \infty,$$

where λ is real. The point $z = ix\lambda$ goes to the point w given by

$$w = \frac{i\lambda - 1}{i\lambda + 1}$$

whose modulus is

$$|w| = \frac{|i\lambda - 1|}{|i\lambda + 1|} = \frac{\sqrt{\lambda^2 + 1}}{\sqrt{\lambda^2 + 1}} = 1.$$

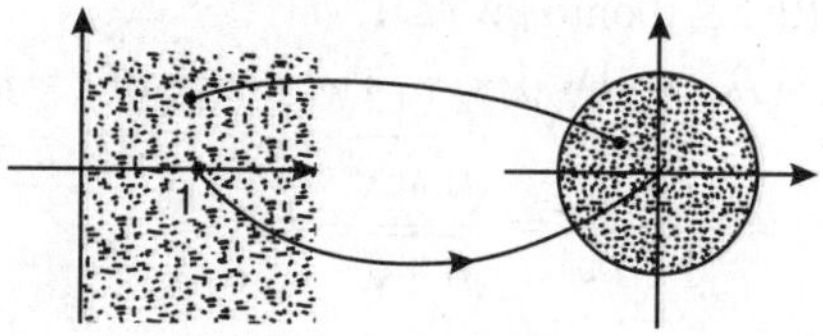

Fig. 7.10.3

So the imaginary axis in the z plane goes onto the circle

$$|w| = 1$$

in the w plane. But $z = 1$, a point in the right half plane, goes to

$$w = \frac{1-1}{1+1} = 0$$

which is a point inside the circle $|w| = 1$. This shows that the right half plane $\text{Re}\, z \geq 0$ goes onto the circular disc $|w| \leq 1$.

Remark. There are many forms of transformations which can be rewritten as

$$w = \pm\frac{z-1}{z+1} \quad \text{or} \quad w = \pm\frac{z-1}{z+1}.$$

All these transformations transform the right half plane **Re** $z \geq 0$ onto the circular disc $|w| \leq 1$ and transforms the left half plane $\text{Re}\, z \leq 0$ onto the annular region $|w| \geq 1$.

15. Solved sum. Show that, under the mapping

$$w = \frac{i - z}{i + z},$$

the image of the region $x^2 + y^2 < 1$ is the entire half of the w plane to the right of the imaginary axis.

✍ The given region $x^2 + y^2 < 1$ is the circular region $|z| < 1$. From (1) we shall obtain z.

$$w(i + z) = i - z \quad \text{or} \quad z(w + 1) = -iw + i$$

$$\therefore \qquad z = \frac{-iw + i}{w + 1} = -i\frac{w-1}{w+1}.$$

Thus the image of $|z| < 1$ is

$$\left|-i\frac{w-1}{w+1}\right| < 1 \quad \text{or} \quad |-i|\frac{|w-1|}{|w+1|} < 1$$

or $$|w - 1| < |w + 1|$$

Taking $w = X + iY$, we get

$$|(X + iY) - 1| < |(X + iY) + 1|$$

or $$|(X - 1) + iY| < |(X + 1) + iY|.$$

$$\therefore \qquad (X - 1)^2 + Y^2 < (X + 1)^2 + Y^2$$

$$\therefore \qquad -4X < 0 \quad \text{or} \quad X > 0.$$

Thus the image range is the entire w plane on the right of the imaginary axis.

16. Show that the transformation

$$w = \frac{1+z}{1-z}$$

maps the left half plane Re $z \leq 0$ onto $|w| \leq 1$.

[**Hint:** Imaginary axis: $z = i\lambda, \quad -\infty < \lambda < \infty$. The image of $z = i\lambda$ is w with

$$|w| = \frac{|1+i\lambda|}{|1-i\lambda|} = \frac{\sqrt{1+\lambda^2}}{\sqrt{1+\lambda^2}} = 1.$$

Now $z = -1$ goes to $w = 0$]

17. Show that the transformation

$$w = \frac{z+i}{z-i}$$

maps the lower half plane **Im** $z \leq 0$ onto $|w| \leq 1$.

[**Hint:** Real axis: $z = \lambda, -\infty < \lambda < \infty$. The image of $z = \lambda$ is w whose modulus is

$$|w| = \frac{|\lambda+i|}{|\lambda-i|} = \frac{\sqrt{\lambda^2+1}}{\sqrt{\lambda^2+1}} = 1.$$

Now $z = -i$ goes to $w = 0$]

18. Show that the transformation

$$w = \frac{z-i}{1-iz}$$

maps the circular disc $|z| \leq 1$ onto the lower half of the w plane.

[**Hint:** $z = i\dfrac{1-iw}{1+iw}$. Hence the image of $|z| \leq 1$ is

$$\left| i\frac{1-iw}{1+iw} \right| \leq 1 \quad \text{or} \quad |i| \frac{|1-iw|}{|1+iw|} \leq 1.$$

Hence $\quad |1-iw| \leq |1+iw|$. If $w = X + iY$, then

$$|(1+Y) - iX| \leq |(1-Y) + iX| \quad \text{or} \quad Y \leq 0]$$

19. Show that

$$w = i\frac{1-z}{1+z}$$

maps the circular domain $|z| \leq 1$ onto the upper half plane Im $w \geq 0$.

[**Hint:** $z = i\dfrac{1+iw}{w+i}$. Hence the image of $|z| \leq 1$ is

$$\left| i\frac{1+iw}{w+i} \right| \leq 1 \quad \text{or} \quad |i| \frac{|1+iw|}{|w+i|} \leq 1.$$

Hence $|1 + iw| \le |w + i|$. If $w = X + iY$, then
$|(1 - Y) + iX| \le |X + i(1 + Y)|$ or $Y \ge 0]$

20. Solved sum. Find the bilinear transformation that maps the points ∞, i, 0 onto 0, i, ∞ respectively.

- **Method 1.** Now the image of 0 is ∞ and the image of ∞ is 0. Thus the transformation is $w = 1/z$ or more generally $w = a/z$. Since i goes to i itself

$$i = \frac{a}{i} \quad \text{or} \quad a = -1.$$

Thus the required transformation is $w = -1/z$.

- **Method 2.** Let the transformation be

$$\frac{(w - w_1)(w_2 - w_3)}{(w_1 - w_2)(w_3 - w)} = \frac{(z - z_1)(z_2 - z_3)}{(z_1 - z_2)(z_3 - z)}.$$

Since ∞ goes to 0, setting $z_1 = \infty$, $w_1 = 0$,

$$\frac{(w)(w_2 - w_3)}{(-w_2)(w_3 - w)} = \frac{(-z_1)(z_2 - z_3)}{(z_1)(z_3 - z)}$$

or

$$\frac{w(w_2 - w_3)}{(-w_2)(w_3 - w)} = -\frac{z_2 - z_3}{z_3 - z}.$$

Since 0 goes to ∞, setting $z_3 = 0$, $w_3 = \infty$,

$$\frac{w(-w_3)}{-w_2(w_3)} = -\frac{z_2}{-z} \quad \text{or} \quad \frac{w}{w_2} = \frac{z_2}{z}.$$

Since i goes to i, setting $z_2 = i$, $w_2 = i$,

$$\frac{w}{i} = \frac{i}{z} \quad \text{or} \quad w = -\frac{1}{z}.$$

This is the required bilinear transformation.

Remark. Substitution of ∞, 1, 0 for z_1, z_2, z_3 etc. can also be done in one stroke as given in the next example.

21. Solved sum. Find the bilinear transformation which maps the points 1, i, –1 onto the points 0, 1, ∞. Show that the transformation maps the interior of the unit circle of the z plane onto the upper half of the w plane. **(1981)**

The required transformation is

$$\frac{(w - 0)(1 - \infty)}{(0 - 1)(\infty - w)} = \frac{(z - 1)(i + 1)}{(1 - i)(-1 - z)}.$$

Since $1 - \infty = -\infty$, $\infty - w = \infty$, we get the mapping as

$$\frac{w - 0}{0 - 1}(-1) = -\frac{z - 1}{z + 1}\,\frac{1 + i}{1 - i}$$

or

$$w = -\frac{z - 1}{z + 1}i = \frac{-iz + i}{z + 1} \quad \text{since} \quad \frac{1 + i}{1 - i} = i.$$

Next, to find the image of $|z| < 1$, we shall get z.

$$w(z+1) = -iz + i \quad \text{or} \quad z(w+i) = -w + i$$

$$\therefore \qquad z = \frac{-w+i}{w+i} = -\frac{w-i}{w+i}$$

Thus the region $|z| < 1$ goes onto the region

$$\left|\frac{w-i}{w+i}\right| < 1.$$

Taking w as $X + iY$, we get the image region as

$$|(X+iY) - i| < |(X+iY) + i|$$

or $\qquad |X + i(Y-1)| < |X + i(Y+1)|$

$\therefore \qquad X^2 + (Y-1)^2 < X^2 + (Y+1)^2$

$\therefore \qquad -4Y < 0 \quad \text{or} \quad Y > 0.$

22. Solved sum. Find the bilinear transformation which maps the points $z = -2i, i, \infty$ onto the points $w = 0, -3, \frac{1}{3}$ respectively. Find the image of $|z| < 1$.

✍ The required transformation is

$$\frac{(w-0)(-3-\frac{1}{3})}{(0+3)(\frac{1}{3}-w)} = \frac{(z+2i)(i-\infty)}{(-2i-i)(\infty - z)}$$

Since $i - \infty = -\infty, \infty - z = \infty$, we get the mapping as

$$\frac{w(-\frac{10}{3})}{1-3w} = \frac{z+2i}{-3i}(-1) \quad \text{or} \quad \frac{-10w}{3-9w} = \frac{z+2i}{3i}$$

or $\qquad \dfrac{-10w}{1-3w} = \dfrac{z+2i}{i} \quad \text{or} \quad -10iw = z + 2i - 3wz - 6iw$

or $\qquad w(3z-4i) = z + 2i \quad \text{or} \quad w = \dfrac{z+2i}{3z-4i}$

Next, to find the image of $|z| < 1$, we shall get z. Then

$$w(3z-4i) = z + 2i$$

or $\qquad z = -2i\dfrac{1+2w}{1-3w}.$

Thus the region $|z| < 1$ goes onto the region

$$\left|2i\frac{1+2w}{1-3w}\right| < 1 \quad \text{or} \quad 2\frac{|1+2w|}{|1-3w|} < 1.$$

Taking w as $X + iY$, we get the region as

$$2|1 + 2(X+iY)| < |1 - 3(X+iY)|$$

or $\qquad 2|(1+2X) + i2Y| < |(1-3X) - i3Y|$

$\therefore \qquad 4[(1+2X)^2 + 4Y^2] < (1-3X)^2 + 9Y^2$

$\therefore \qquad 7X^2 + 7Y^2 + 22X + 3 < 0.$

23. Find the bilinear transformations which map the following z points into w points :

(*i*) $z = 0, -i, -1;\ w = i, 1, 0$ **(1978)**

(*ii*) $z = 2, i, -2;\ w = 1, i, -1$

(*iii*) $z = -i, 0, i$; $w = -1, i, 1$ **(1980)**

(*iv*) $z = 2, 1, 0$; $w = 1, 0, i$

(*v*) $z = 0, \infty$; $w = 0, \infty$

(*vi*) $z = 0$ (only fixed point) ; $w = 0$ (only fixed point)

[**Hint:** (*i*) Set $z_1 = 0, z_2 = -i, z_3 = -1, w_1 = i, w_2 = 1, w_3 = 0$ in (7.1.2) ;

$$\frac{w-i}{w(1-i)} = \frac{z(-i+1)}{(z+1)(-i)}; \quad w = \frac{-i(z+1)}{z-1}$$

(*ii*) $w = \dfrac{3z+2i}{iz+6}$ (*iii*) $w = (-i)\,\dfrac{z-1}{z+1}$ (*iv*) $\dfrac{2i(z-1)}{(1+i)z-2}$

(*v*) $0 = \dfrac{a(0)+b}{c(0)+d}$ gives $b = 0$; $\infty = \dfrac{a+(b/\infty)}{c+(d/\infty)}$ gives $c = 0$; $w = (a/d)\,z$ or $w = a'z$

(*vi*) In (7.1.4) set $p = 0$; $w = z/(kz+1)$].

24. Show that $|(z-1)/(z+1)| = \frac{1}{2}$ represents a circle. Find its centre and radius. Show that $z = 1$ and $z = -1$ are inverse points with respect to the circle. **(1981)**

[**Hint:** $z = x + iy$ gives $x^2 + y^2 - (10/3)x + 1 = 0$; Centre C : (5/3, 0); Radius r : 4/3 ; If (1, 0), (– 1, 0) are P, Q ; $CP = 2/3$, $CQ = 8/3$; $CP \cdot CQ = r^2$].

7.4. Transformation $w = z^2$

25. Solved sum. Show that the transformation $w = z^2$ transforms the families of lines $x = h$ and $y = k$ into confocal parabolas, having $w = 0$ as the common focus. **(1975, 78, 81)**

✍ If $z = x + iy$ and $w = u + iv$, then

$$u = x^2 - y^2, \quad v = 2xy.$$

(*i*) *Image of* $x = h$: To obtain the image of $x = h$ (Fig. 7.9.1), we have to eliminate x, y from

$$u = x^2 - y^2, \quad v = 2xy, \quad x = h.$$

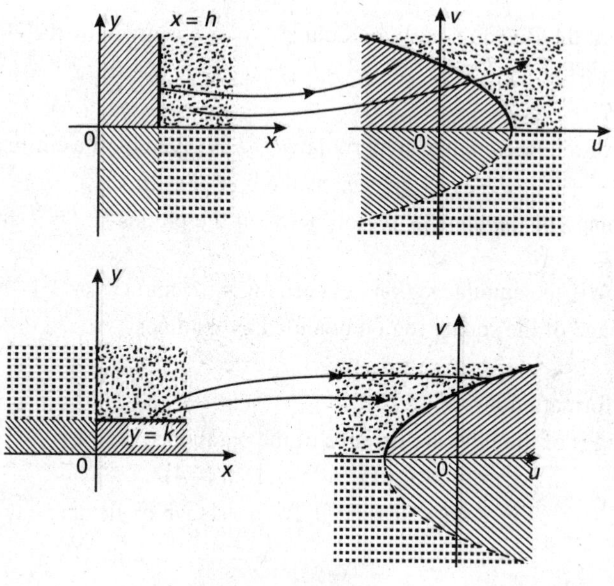

Fig. 7.10.4

First eliminating x from the first and the second,

$$u = h^2 - y^2, \quad v = 2hy.$$

Now, eliminating y from these two,

$$u = h^2 - (v/2h)^2 \quad \text{or} \quad v^2 = -4\,h^2\,(u - h^2).$$

As h varies, this represents a family of parabolas having vertices at $w = h^2$ i.e., $(h^2, 0)$ and foci at $w = 0$ i.e., $(0, 0)$.

(*ii*) *Image of* $y = k$: Eliminating x, y from

$$u = x^2 - y^2, \quad v = 2xy, \quad y = k,$$

we get the equation of the image of the family $y = k$ as

$$v^2 = 4k^2\,(u + k^2)$$

As k varies, this represents a family of parabolas with vertices at $w = -k^2$ i.e., $(-h^2, 0)$ and foci at $w = 0$ i.e., $(0, 0)$.

Conclusion. The image parabolas in the above two transformations have the same point (0, 0) as their foci. So the images of the families $x = h$ and $y = k$ are confocal parabolas foci at $w = 0$ i.e., $(0, 0)$.

26. Obtain the images of the lines $x = 0$ (Imaginary axis) and $y = 0$ (Real axis) and the rectangular hyperbolas $x^2 - y^2 = h$ and $xy = h$ under the transformation $w = z^2$.

[**Hint:** (*i*) The image of $x = 0$ is given by the eliminant of x and y from

$$u = x^2 - y^2, \quad v = 2xy, \quad x = 0.$$

Substituting $x = 0$ in the first two, we get

$$u = -y^2, \quad v = 0.$$

$v = 0$ is the real axis. But $u = -y^2$ and hence u is negative. So the image is the left half of the real axis.

(*ii*) The image of $y = 0$ is given by the eliminant of x, y from

$$u = x^2 - y^2, \quad v = 2xy, \quad y = 0.$$

Substituting $y = 0$ in the first two, we get

$$u = x^2, \quad v = 0$$

$v = 0$ is the real axis. But $u = x^2$ and hence u is positive. So the image is the right half of the real axis.

(*iii*) The image of the Rectangular hyperbola $x^2 - y^2 = h$ is given by the eliminant of x, y from

$$u = x^2 - y^2, \quad v = 2xy, \quad x^2 - y^2 = h$$

as the line $u = h$.

(*iv*) The image of the Rectangular hyperbola $xy = h$ is given by the eliminant of x, y from

$$u = x^2 - y^2, \quad v = 2xy, \quad xy = h \quad \text{as the line} \quad v = 2h.]$$

27. State the image of the annular region between the circles $|\,z\,| = 2$ and $|\,z\,| = 3$ under the transformation $w = z^2$.

[**Ans.** The image is the annular region between $|\,w\,| = 2^2$ and $|\,w\,| = 3^2$]

28. Find the image of the square region bounded by the lines

$$x = 1, \quad x = 2, \quad y = 1, \quad y = 2$$

under the transformation $w = z^2$

[**Hint:** The images of the lines are the arcs of the parabolas

$$v^2 = -4\,(u - 1), \quad v^2 = -16\,(u - 4),$$

$$v^2 = 4\,(u + 1), \quad v^2 = 16\,(u + 4) \text{ which lie above the real axis]}$$

29. Solved sum. Plot the image, under the transformation $w = z^2$, of the triangular region bounded by $x = 1, y = 1, x + y = 1$.

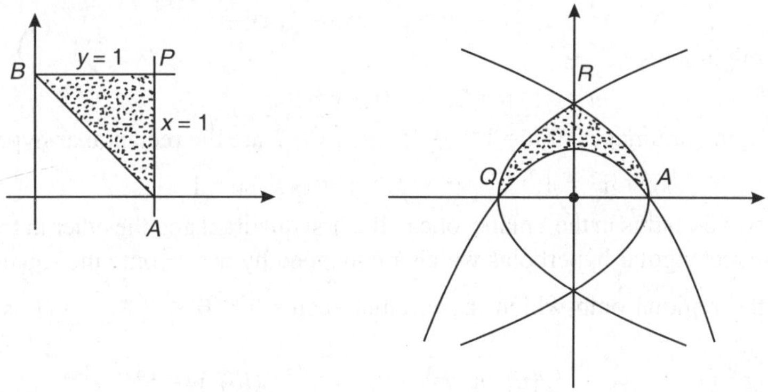

Fig. 7.10.5

The vertices of the triangle A, B, P are

$$z = 1, \quad i, \quad 1 + i$$

Their image A, Q, R, under the transformation are

$$w = 1^2, \; i^2, \; (1 + i)^2 \text{ or } 1, \; -1, \; 2i$$

(*i*) The image of the line AP, $x = 1$, is a parabola, opening on the left, since

$$\left.\begin{aligned} u &= x^2 - y^2 \\ v &= 2xy \\ x &= 1 \end{aligned}\right\} \Rightarrow \left.\begin{aligned} u &= 1 - y^2 \\ v &= 2y \end{aligned}\right\} \Rightarrow u = 1 - \left(\frac{v}{2}\right)^2$$

or $$v^2 = -4(u - 1). \tag{1}$$

(*ii*) The image of the line BP, $y = 1$, is a parabola, opening on the right, since

$$\left.\begin{aligned} u &= x^2 - y^2 \\ v &= 2xy \\ y &= 1 \end{aligned}\right\} \Rightarrow \left.\begin{aligned} u &= x^2 - 1 \\ v &= 2x \end{aligned}\right\} \Rightarrow u = \left(\frac{v}{2}\right)^2 - 1$$

or $$v^2 = 4(u + 1) \tag{2}$$

(*iii*) The equation of the line AB is

$$x + y = 1 \text{ or } y = 1 - x$$

So, substituting in $u = x^2 - y^2$, $v = 2xy$, we have

$$u = x^2 - (1 - x)^2 = 2x - 1, \quad v = 2x(1 - x)$$

From the first equation, we get

$$u = 2x - 1 \text{ or } x = \tfrac{1}{2}(u + 1)$$

Substituting in the second equation,

$$v = (u + 1)\tfrac{1}{2}(1 - u) \text{ or } v = -\tfrac{1}{2}(u^2 - 1)$$

or $$u^2 = -2(v - \tfrac{1}{2}) \tag{3}$$

It is a parabola opening below. Thus the required image is the region bounded by the parabolas (1), (2), (3).

30. Solved sum. Find the preimage in the z plane of the square domain in the w plane bounded by the lines $u = 1, u = 2, v = 1, v = 2$ under the transformation $w = z^2$.

Now $u = x^2 - y^2, \; v = 2xy$

Hence the transformation $w = z^2$ maps the rectangular hyperbolas

$$x^2 - y^2 = 1, \; x^2 - y^2 = 2, \; 2xy = 1, xy = 1$$

into the straight lines

$$u = 1, \; u = 2, \; v = 1, \; v = 2$$

Conversely, the preimages of $u = 1, u = 2, v = 1, v = 2$ are the rectangular hyperbolas

$$x^2 - y^2 = 1, \; x^2 - y^2 = 2, \; xy = \tfrac{1}{2}, \; xy = 1$$

There are two domains in the z plane, one in the first quadrant and the other in the third quadrant, enclosed by the rectangular hyperbolas which are mapped by $w = z^2$ onto the square in the w plane.

31. Find the regions onto which the circular sector $0 < \theta < \frac{1}{4}\pi$, $r < 1$ is mapped by the transformations

(*i*) $w = z^2$ (*ii*) $w = z^3$ (*iii*) $w = z^4$

[**Ans.** (*i*) $0 < \theta < \frac{1}{2}\pi, r < 1$ (*ii*) $0 < \theta < \frac{3}{4}\pi, r < 1$ (*iii*) $0 < \theta < \pi, r < 1$]

32. Show that the transformation $w = z^2$ maps the circle $|z - \alpha| = \rho$, where α is real, onto a limacon.

The equation of the circle is

$$z = \alpha + \rho\, e^{i\theta}, \quad 0 \le \theta < 2\pi$$

(Fig. 7.10.6) The equation of the image is

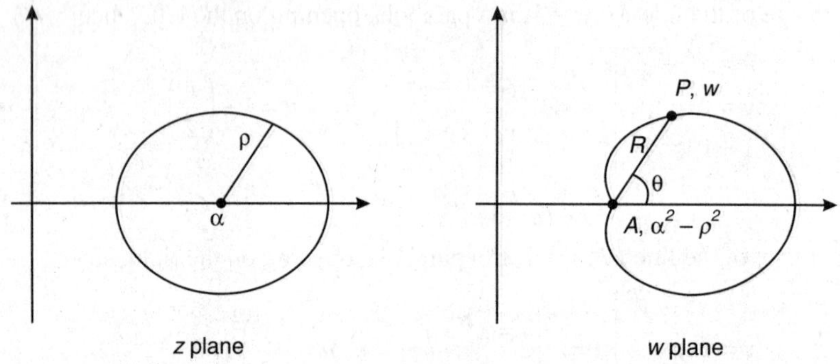

Fig. 7.10.6

$$w = (\alpha + \rho\, e^{i\theta})^2 = \alpha^2 + \rho^2 e^{2i\theta} + 2\alpha\rho\, e^{i\theta}$$

i.e., $$w = (\alpha^2 - \rho^2) + [\rho^2 + \rho^2 e^{2i\theta} + 2\alpha\rho e^{i\theta}]$$

i.e., $$w = (\alpha^2 - \rho^2) + e^{i\theta}\,[\rho^2\,(e^{-i\theta} + e^{i\theta}) + 2\alpha\,\rho]$$

i.e., $$w = (\alpha^2 - \rho^2) + (2\rho^2 \cos\theta + 2\alpha\rho)\, e^{i\theta}, \quad 0 \le \theta < 2\pi$$

Also, denoting w on the image by P and $\alpha^2 - \rho^2$ by A, the equation of the image is

$$w = \overline{OA} + \overline{AP} = (\alpha^2 - \rho^2) + AP\, e^{i\theta}, \quad 0 \le \theta < 2\pi$$

$\therefore$ $$AP = 2\rho^2 \cos\theta + 2\alpha\rho$$

i.e., $$R = 2\rho\,(\alpha + \rho\cos\theta), \quad (AP = R)$$

which is the polar equation of the image when A is chosen as the pole and R, θ as the polar coordinates. This equation represents a limacon.

33. Solved sum. Show that the transformation $w = z^{1/2}$ maps the upper half of the inside of the parabola $y^2 = 4c^2 (x + c^2)$ onto the infinite strip bounded by $0 \le u < \infty$, $0 \le v \le c$, where $w = u + iv$.

(1976)

✍ By the inverse transformation $z = w^2$ the upper half w plane goes onto the entire z plane once (Fig. 7.9.3). The image of the line $v = c$ is obtained by eliminating u and v from

$$x = u^2 - v^2,\ y = 2uv,\ v = c,$$

i.e., elilminating u from $x = u^2 - c^2$, $y = 2uc$ as the parabola

$$y^2 = 4c^2 (x + c^2)$$

In the limiting case when $c = 0$, $v = 0$ goes onto the right semi-real axis. So by the transformation $z = w^2$ the infinite strip

$$0 \le v \le c$$

goes onto the interior of the parabola $y^2 = c^2 (x + c^2)$. But under the transformation

$$w = z^{1/2}$$

the upper half of the interior of the parabola $y^2 = 4c^2 (x + c^2)$

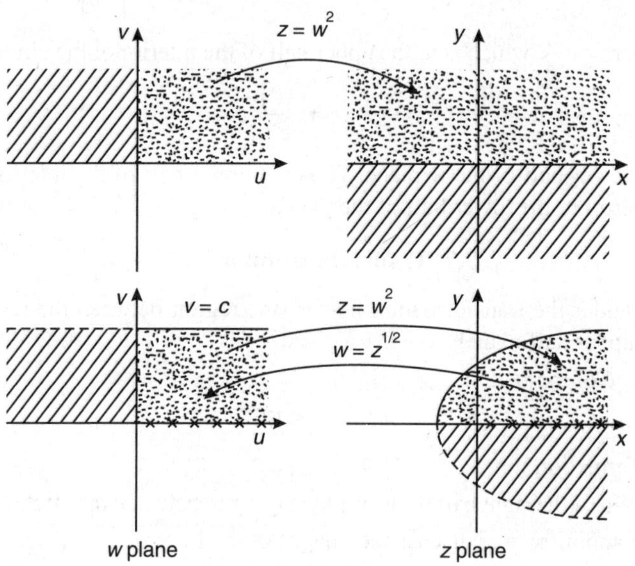

Fig. 7.10.7

goes onto the infinite strip $0 \le u < \infty$, $0 \le v \le c$ because the arguments of the points in the upper half of the interior of the parabola vary from 0 to π and the arguments of the points in the infinite strip $0 \le u < \infty$, $0 \le v \le c$ vary from 0 to $\frac{1}{2}\pi$.

7.6. Transformation $w = z + \dfrac{1}{z}$

34. What are the fixed points of the mapping $w = z + 1/z$.

[**Hint:** $z + 1/z = z$. So ∞ is the only fixed point]

35. Show that the transformation $w = z + k^2/z$ maps the circle $r = a$ onto an ellipse if $k \ne a$.

[Equation of the ellipse $\dfrac{u^2}{(a + k^2/a)^2} + \dfrac{v^2}{(a - k^2/a)^2} = 1$]

36. Discuss the transformation $w = z + k^3/z$, where k is positive.

37. Solved sum. Show that, under the mapping $w = z + 1/z$, the upper half of the interior of the unit circle $|z| = 1$ except $z = 0$ goes onto the lower half of the w plane.

✍ We obtained that, under the transformation

$$w = z = 1/z,$$

the circle $|z| = a$ goes onto the ellipse

$$\frac{u^2}{(a + 1/a)^2} + \frac{v^2}{(a - 1/a)^2} = 1 \quad (1)$$

and the circle $|z| = 1$ goes on to the line segment joining 2 and – 2. As a decreases from 1 to 0.

$$(a + 1/a)^2 \to \infty \quad (a - 1/a)^2 \to \infty$$

So the ellipse magnifies itself and covers the entire w plane. So the interior of $|z| = 1$ goes onto the entire w plane.

We shall consider $z = i/2$ which is in the upper half of the interior of the circle $|z| = 1$. Its image is

$$w = \frac{i}{2} + \frac{2}{i} = \frac{i}{2} - 2i = -\frac{3i}{2}$$

which is in the lower half of the w plane. Thus the upper half of the interior of the unit circle $|z| = 1$ except 0, goes onto the lower half of the plane.

7.7 Transformation $w = e^z$

38. Show that, under the transformation $w = e^z$, the region between the real axis and the line $y = \pi$ goes onto the upper half plane.

39. Plot the image of the region

$$-1 < x < 1, \quad -\pi/2 < y < \pi/2$$

under the transformation $w = e^z$.

[**Hint:** The portion of the right half of the w plane lying in between the circles $|w| = e^{-1}$, $|w| = e$.]

40. Under the mapping $w = e^z$ discuss the images of the lines

$$y = 0, \quad y = \pi/2, \quad y = \pi, \quad y = 3\pi/2, \quad y = 2\pi, \quad -\infty < x < \infty$$

in the w plane. What is the image of the region between $y = 0$ and $y = 2\pi$?

41. Discuss the transformation $w = e^z$ and obtain the image of the rectangle.

$$x = c_1, \quad x = c_2, \quad y = k_1, \quad y = k_2.$$

7.9. Conformal Mapping

42. Show that under a conformal mapping orthogonal curves go onto orthogonal curves.

43. Show that under the transformation $w = z^2$ the constant of magnification and angle of rotation at $z = 2 + i$ are $2\sqrt{5}$ and $\tan^{-1} \frac{1}{2}$.

[**Hint:** $f'(z) = 2z$; At $z = 2 + i$, $|2z| = 2\sqrt{5}$, $\arg(2z) = \tan^{-1} \frac{1}{2}$]

44. Solved sum. Find the critical points of the following mappings

(*i*) $w = z^4 - 1$

(*ii*) $w^2 = (z - \alpha)(z - \beta)$

(i) $f'(z) = 0$ gives the critical points as

$$4z^3 = 0 \quad \text{or} \quad z = 0.$$

(ii) Differentiating $w^2 = (z - \alpha)(z - \beta)$ w.r.t. z,

$$2w\frac{dw}{dz} = 2z - (\alpha + \beta)$$

$$\therefore \quad \frac{dw}{dz} = \frac{z - \frac{1}{2}(\alpha + \beta)}{\sqrt{(z - \alpha)(z - \beta)}}$$

Thus the critical points are given by

$$z - \tfrac{1}{2}(\alpha + \beta) = 0 \quad \text{or} \quad z = \tfrac{1}{2}(\alpha + \beta).$$

45. Find the critical points of the mapping $w = 3z - z^3$. **[Ans.** $z \pm 1$]

46. Find the critical points of the mapping

$$w = z + \frac{1}{z}$$ **[Ans.** $z = \pm 1$]

47. Find the points at which the mapping

$$w = \sin z$$

is not conformal.

[**Hint:** $\cos z$ vanishes at $z = \pm (2n + 1)\, \pi/2$, $n = 0, 1, 2, ..., \infty$].

48. Find the points where $f(z) = \cos z$ is not conformal.

[**Hint:** $\sin z$ vanishes at $z = \pm n\pi$, $n = 0, 1, 2, ..., \infty$].

49. Find the locus of points at which the coefficient of magnification of lengths is 1 under the mapping $w = 3z - z^3$.

[**Hint:** $|f'(z)| = 1$ gives $(x^2 - y^2 - 1)^2 + 4x^2y^2 = \frac{1}{9}$]

50. Find the locus of points at which the infinitesimal lengths are rotated through an angle 45° under the mapping $w = 3z - z^3$.

[**Hint:** $\arg f'(z) = 45°$ or $\tan^{-1} \dfrac{2xy}{x^2 - y^2 - 1} = 45°$.

Therefore $\dfrac{2xy}{x^2 - y^2 - 1} = 1$ or $x^2 - 2xy - y^2 = 1$].

51. Show that the bilinear mapping

$$w = \frac{az + b}{cz + d}, \quad \begin{vmatrix} a & b \\ c & d \end{vmatrix} \neq 0$$

is conformal everywhere in the finite plane except at $z = -\dfrac{d}{c}$.

[**Hint:** $f'(z) = \dfrac{ad - bc}{(cz + d)^2} \neq 0$].

52. Which part of the plane is shrunk and which part is stretched under the following transformations?

(i) $w = z^2$ (ii) $w = z^2 + 2z$ (iii) $w = 1/z$

Ans. (i) $|z| < 1/2$ is shrunk, $|z| = 1/2$ is neither shrunk nor stretched, $|z| = 1/2$ is stretched (ii) $|z + 1| < 1/2$ is shrunk, etc. (iii) $|z| < 1$ is stretched, etc.

CHAPTER 8

COMPLEX INTEGRATION

8.1. SIMPLE RECTIFIABLE ORIENTED CURVES

The integral of a complex function is defined as the limit of a certain sum in the same manner as an integral of a real function along a line is defined. So, to define complex integration, we need to define a curve in the complex plane along which the integration has to be done.

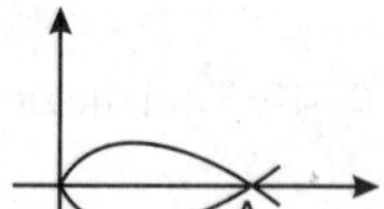

Multiple point: If a curve intersects itself at a point then, the point is said to be a multiple point of the curve. In the figure, A is a multiple point of the curve. A is called a point of self intersection.

Simple curve: If a continuous curve which does not have a point of self intersection is called a simple curve. Simple curves are called *Jordan curves.*

Simple arc: Suppose $z(t)$ is a continuous complex function defined in a closed interval $[h, k]$. Then the set of complex points

$$C : \{z(t), h \leq t \leq k\}$$

is called a simple arc (Jordan arc) if, for any two different values of t, namley, t_1 and t_2, $z(t_1) \neq z(t_2)$ (Fig. 8.1.1 (a)).

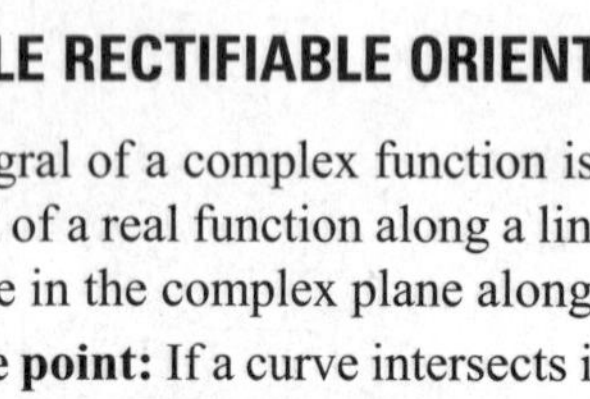

Fig. 8.1.1 (a)

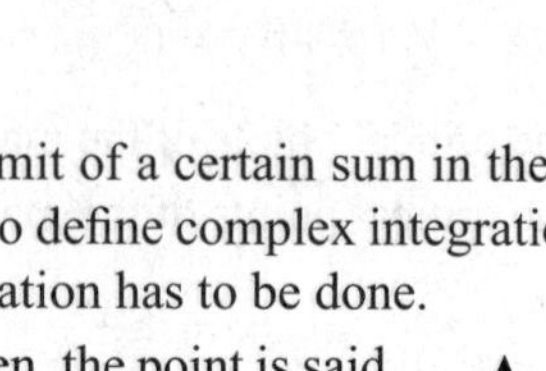

Fig. 8.1.1 (b)

Simple closed curve: The above simple arc is called a simple closed curve (Jordan curve) if $z(h) = z(k)$ and, for any other two different t_1 and t_2, $z(t_1) \neq z(t_2)$ (Fig. 8.1.1 (b)).

Examples. (*i*) The equation

$$z = \cos t + i \sin t, \quad 0 \leq t \leq \pi$$

represents a simple arc. This arc is a semicircle above the real axis.

(*ii*) The equation $z = \cos t + i \sin t$, $0 \leq t \leq 2\pi$, represents a closed simple curve. It is the unit circle.

Orientation of a simple closed curve: Let z_0 be a point inside a simple closed curve C. Then C is said to be *positively-oriented* if arg $\{z(t) - z_0\}$ increases by 2π when t increases from h to k, i.e., if the traversal of C is in that sense for which the region enclosed by C is always on the left ; otherwise, C is said to be *negatively-oriented.*

Note: If the region lies to the left of a person when he travels along C, then the curve C is called a positively-oriented simple closed curve.

For example, the unit circles,

$$z = \cos t + i \sin t, \qquad 0 \le t \le 2\pi$$
$$z = \cos(-t) + i \sin(-t), \qquad 0 \le t \le 2\pi$$

are positively-oriented and negatively-oriented respectively (Fig. 8.1.2). If one of these circles is denoted by Γ, then the other is denoted by $-\Gamma$.

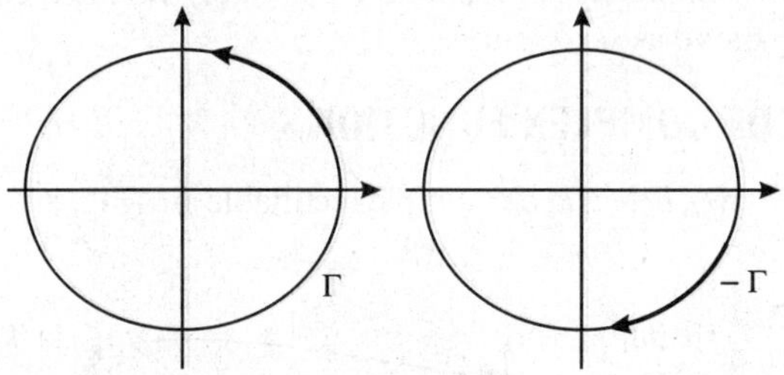

Fig. 8.1.2

Rectifiability of a simple arc. Consider a simple arc C

$$z = z(t), \quad h \le t \le k.$$

Suppose the closed interval $[h, k]$ is divided into n subintervals $[t_0, t_1], [t_1, t_2], ..., [t_{n-1}, t_n]$, where

$$h = t_0 < t_1 < t_2 < ... < t_{n-1} < t_n = k$$

(Fig. 8.1.3). Denote this *division of partition* by $\wp$. The maximum of the lengths of the subintervals is called the *norm* of $\wp$ and is denoted by $|\wp|$. Now

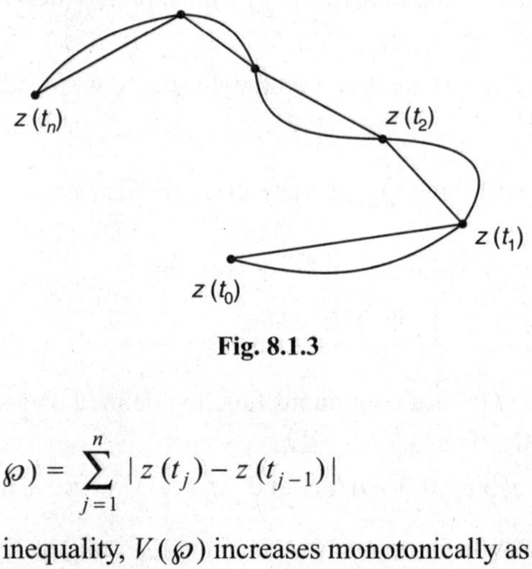

Fig. 8.1.3

the positive number

$$V(\wp) = \sum_{j=1}^{n} |z(t_j) - z(t_{j-1})|$$

varies with $\wp$. By triangle inequality, $V(\wp)$ increases monotonically as the subintervals are further divided into smaller subintervals. If the suprimum (*lub*) of the set $V(\wp)$, namely

$$\sup_{\wp} \sum_{j=1}^{n} |z(t_j) - z(t_{j-1})|$$

is finite, then C is said to be rectifiable. This finite number is denoted by

$$\lim_{|\wp| \to 0} \sum_{j=1}^{n} |z(t_j) - z(t_{j-1})| \tag{8.1.1}$$

This number is the length of the arc C and is denoted by $l(C)$.

The length of the arc C is the integral $\int_C |dz| = \int_C |dx + id_y|$.

From real analysis, if $z(t) = x(t) + iy(t)$, where $x(t)$ and $y(t)$ are of bounded variation in $[h, k]$, then the length of the arc C is

$$l(C) = \int_h^k \sqrt{[x'(t)]^2 + [y'(t)]^2}\, dt. \tag{8.1.2}$$

Scr and Scro curves: We henceforth use the abbreviation 'scr curve' for simple closed rectifiable curve and 'scro curve' for simple closed rectifiable positively-oriented curve. We read scr curve as *es.s ē .är* curve and scro curve as *skrō* curve.

8.2. INTEGRATION OF COMPLEX FUNCTIONS

Definition. Let $C : z = z(t), h \le t \le k$, be a simple rectifiable arc (Fig. 8.2.1). Let $f(z)$ be a continuous function defined on C.

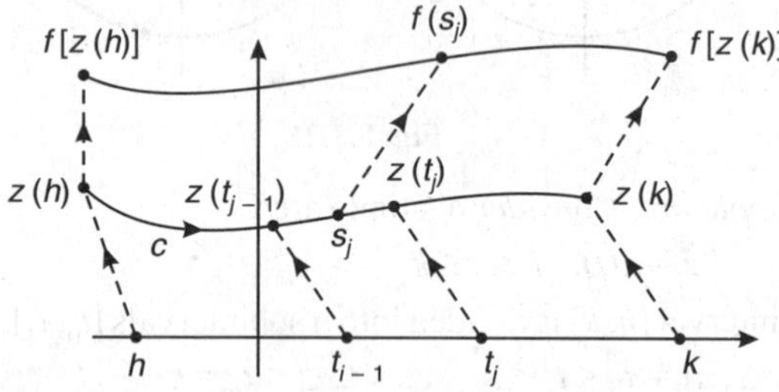

Fig. 8.2.1

If $\wp$ is a partition of the closed interval $[h, k]$ with subintervals

$$[t_0, t_1], [t_1, t_2], \dots [t_{n-1}, t_n],$$

where $t_0 = h$, $t_n = k$ and if s_j is any point on the jth subarc with end points $z(t_{j-1})$ and $z(t_j)$, then the existence of the limit,

$$\lim_{|\wp| \to 0} \sum_{j=1}^{n} [z(t_j) - z(t_{j-1})]\, f(s_j) \quad \text{as } |\wp| \to 0$$

is called the integral of $f(z)$ along C and is denoted by

$$\int_C f(z)\, dz. \tag{8.2.1}$$

Theorem 8.1. Suppose $f(z)$ is a continuous function defined on a simple rectifiable arc

$$C : z = z(t), \quad h \le t \le k.$$

If $\quad z(t) = x(t) + iy(t)$ and $f(z) = u(x, y) + iv(x, y)$, then

$$\int_C f(z)\, dz = \int_C (u\, dx - v\, dy) + i \int_C (u\, dy - v\, dx). \tag{8.2.2}$$

Proof: Suppose $\wp$ is a partition of the closed interval $[h, k]$ and s_j is a point on the subarc with endpoints $z(t_{j-1})$ and $z(t_j)$, then

$$\begin{aligned}[z(t_j) - z(t_{j-1})]\, f(s_j) &= [\{x(t_j) + iy(t_j)\} - \{x(t_{j-1}) + iy(t_{j-1})\}] \times (u + iv) \\ &= \{x(t_j) - x(t_{j-1})\}\, u - \{y(t_j) - y(t_{j-1})\}\, v + i\, [\{y(t_j) - y(t_{j-1})\}\, u \\ &\quad + \{x(t_j) - x(t_{j-1})\}\, v]\end{aligned}$$

$$\therefore \quad \int_C f(z)\, dz = \lim_{|\wp| \to 0} \sum_{j=1}^{n} [z(t_j) - z(t_{j-1})\, f(s_j)$$

$$= \int_C u\,dx - \int_C v\,dy + i\left[\int_C u\,dy + \int_C v\,dx\right]$$

$$= \int_C (u\,dx - v\,dy) + i\int_C (u\,dy + v\,dx)$$

wherein the real line integrals exist since C is a simple rectifiable arc and $u(x, y)$ and $v(x, y)$ are continuous functions on C.

Remark. It can be shown that, if C is smooth, that is, if $z(t)$ has a continuous derivative $z'(t)$ at every point in $[h, k]$, then

$$\int_C f(z)\,dz = \int_h^k f\{z(t)\}\,z'(t)\,dt. \tag{8.2.3}$$

This result is valid even if C is piece-wise smooth. A curve is said to be piece-wise smooth if C is not smooth at a finite number of points on it.

Complex integral in parameters: Suppose the parametric equations of C are

$$x = g(t), \quad y = h(t)$$

and t_1, t_2 denote the end points of C. Then

$$\int_C f(z)\,dz = \int_{t_1}^{t_2} f[z(t)]\frac{dz}{dt}\,dt.$$

Contour integral: Integrals along scr curves are called contour integrals while the paths are called contours.

Theorem 8.2. If $l(C)$ is the length of the simple rectifiable arc C and if max $|f(z)|$ on C is M, then

$$\left|\int_C f(z)\,dz\right| \le M\,l(C).$$

Proof: From the definition of complex integral (8.2.1),

$$\left|\int_C f(z)\,dz\right| = \left|\lim_{|\wp|\to 0}\sum_{j=1}^{n}\{z(t_j) - z(t_{j-1})\}\,f(s_j)\right|$$

$$\le \lim_{|\wp|\to 0}\sum_{j=1}^{n}|z(t_j) - z(t_{j-1})|\,|(f(s_j)|$$

$$\le \lim_{|\wp|\to 0}\sum_{j=1}^{n}|z(t_j) - z(t_{j-1})|\,M$$

$$= M\left[\lim_{|\wp|\to 0}\sum_{j=1}^{n}|z(t_j) - z(t_{j-1})|\right]$$

$$= Ml(C) \text{ by (8.1.1).} \tag{8.2.4}$$

The following results are the properties of the complex integrals :

(*i*) $\int_C [f_1(z) + f_2(z)]\,dz = \int_C f_1(z)\,dz + \int_C f_2(z)\,dz$

(*ii*) $\int_C cf(z)\,dz = c\int_C f(z)\,dz$, ($c$: complex constant)

(iii) $\int_{C_1+C_2} f(z)\,dz = \int_{C_1} f(z)\,dz + \int_{C_2} f(z)\,dz$ ($C_1 + C_2$: Curve C_1 followed by curve C_2)

(iv) $\int_C f(z)\,dz = -\int_{-C} f(z)\,dz$

8.3. SIMPLE INTEGRALS USING DEFINITION

First we integrate $f(z) = 1$ along C whose end points are

$$a = z(t_0), \quad b = z(t_n)$$

$$\int_C 1\,dz = \lim_{|\wp|\to 0} \sum_{j=1}^{n} [z(t_j) - z(t_{j-1})](1)$$

$$= \lim_{|\wp|\to 0} [\{z(t_1) - z(t_0)\} + \{z(t_2) - z(t_1)\} + ... + \{z(t_n) - z(t_{n-1})\}]$$

$$= \lim_{|\wp|\to 0} [z(t_n) - z(t_0)] = \lim_{|\wp|\to 0} (b-a) = b - a \tag{8.3.1}$$

This value depends only on the initial and final points of the arc C. That is, for any arc joining a and b, the value of the integral is $b - a$. If C is a closed curve the value is zero.

Next, we integrate $f(z) = z$ along C. The integral can be written either as

$$\lim_{|\wp|\to 0} \sum_{j=1}^{n} [z(t_j) - z(t_{j-1})]\, z(t_j)$$

or as

$$\lim_{|\wp|\to 0} \sum_{j=1}^{n} [z(t_j) - z(t_{j-1})]\, z(t_{j-1}).$$

Adding these two and dividing by two,

$$\int_C z\,dz = \frac{1}{2} \lim_{|\wp|\to 0} \sum_{j=1}^{n} [\{z(t_j)\}^2 - \{z(t_{j-1})\}^2]$$

$$= \frac{1}{2} \lim_{|\wp|\to 0} [\{z(t_n)\}^2 - \{z(t_0)\}^2]$$

$$= \frac{1}{2} \lim_{|\wp|\to 0} (b^2 - a^2) - \frac{1}{2}(b^2 - a^2) \tag{8.3.2}$$

which also is independent of C but depends on the end points a and b. If C is a closed curve, then the integral vanishes.

Remark. Later we will prove that if $f(z)$ is analytic in a simply connected domain, then the values of the integrals of $f(z)$ along all paths in the region joining two fixed points are the same.

Theorem 8.3. $\int_C \frac{1}{z-a}\,dz = 2\pi i$, (indepenent of r and a),

where C is the positively-oriented circle whose radius is r and centre is $z = a$.

Proof: The equation of C is

$$C : z = a + re^{it}, \quad 0 \le t \le 2\pi$$

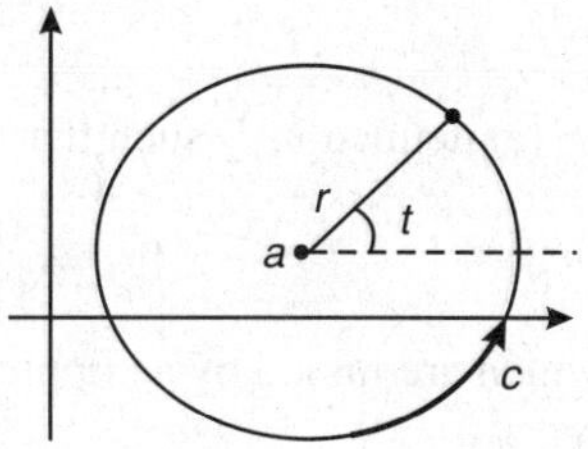

Fig. 8.3.1

(Fig. 8.3.1). Using (8.2.3),

$$\int_C \frac{1}{z-a}\,dz = \int_0^{2\pi} \frac{1}{z-a}\frac{dz}{dt}\,dt = \int_0^{2\pi} \frac{1}{(a+re^{it})-a}\frac{d}{dt}(a+re^{it})\,dt$$

$$= \int_0^{2\pi} i\,dt = 2\pi i \qquad (8.3.3)$$

Theorem 8.4. If C is a simple arc of length L and if $f(z)$ is a continuous function defined on C and if on C, $\max|f(z)| \le M$, then

$$\left|\int_C f(z)\,dz\right| \le ML.$$

Proof:
$$\int_C f(z)\,dz = \lim_{n\to\infty}\sum_{k=1}^{n} f(\lambda_k)\,\Delta z_k,$$

where $|\Delta z_k| \to 0$. Taking modulus on both sides,

$$\left|\int_C f(z)\,dz\right| = \left|\lim_{n\to\infty}\sum_{k=1}^{n} f(\lambda_k)\Delta z_k\right|$$

$$= \lim_{n\to\infty}\left|\sum_{k=1}^{n} f(\lambda_k)\,\Delta z_k\right|$$

$$\le \lim_{n\to\infty}\sum_{k=1}^{n} |f(\lambda_k)\,\Delta z_k| \text{ since } |z_1+z_2| \le |z_1|+|z_2|$$

$$= \lim_{n\to\infty}\sum_{k=1}^{n} f(\lambda_k)|\,|\Delta z_k| \text{ since } |z_1 z_2| = |z_1||z_2|$$

$$\le \lim_{n\to\infty}\sum_{k=1}^{n} M|\Delta z_k|$$

$$= M\left[\lim_{n\to\infty}\sum_{k=1}^{n}|\Delta z_k|\right] = ML$$

8.4. DEFINITE INTEGRALS

Suppose there exists a function $F(z)$ defined in D such that

$$\frac{d}{dz} F(z) = f(z).$$

Let a and b be two points in D which are joined by a simple rectifiable arc

$$C : z = z(t), \quad h \le t \le k,$$

where $z'(t)$ is continuous in $[h, k]$ (Fig. 8.4.1). Then

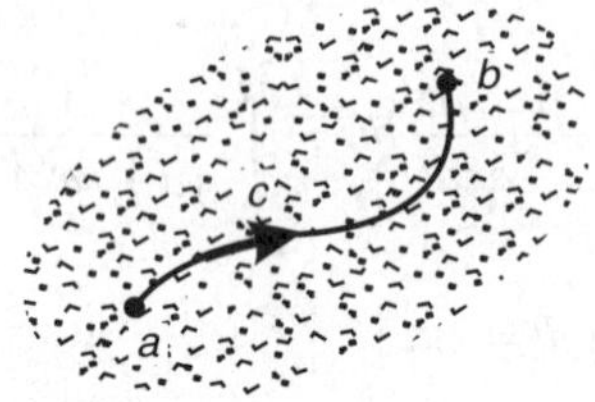

Fig. 8.4.1

$$\int_C f(z)\, dz = \int_C F'(z)\, dz = \int_h^k F'\{z(t)\} \frac{dz}{dt}\, dt \text{ by } (8.2.3)$$

$$= \int_h^k \frac{d}{dt} F\{z(t)\}\, dt = F\{z(t)\}\Big|_{t=h}^{t=k}$$

$$= F(b) - F(a). \tag{8.4.1}$$

This integral is said to be a definite integral. This does not depend on C_i but depends on the end points $z = a$ and $z = b$. It is written as

$$\int_a^b f(z)\, dz.$$

Illustrtion 1. The derivative of $z^{n+1}/(n+1)$, $n \neq -1$ is z^n. So

$$\int_C z^n\, dz = \int_a^b z^n\, dz = \frac{z^{n+1}}{n+1}\Bigg|_a^b = \frac{b^{n+1} - a^{n+1}}{n+1}.$$

Illustration 2. If a and b are any two finite points in the plane,

$$\int_a^b e^z\, dz = e^b - e^a, \quad \int_a^b \cos z\, dz = \sin b - \sin a, \quad \int_a^b \sin z\, dz = -(\cos b - \cos a).$$

8.5. INTERIOR AND EXTERIOR OF A CLOSED CURVE

A simple closed curve C divides the rest of the complex plane into two disjoint regions of which one is bounded and the other is unbounded. The bounded region is called the *interior of* C and is denoted by C_i and the other is called the *exterior of* C and is denoted by C_e (Fig. 8.5.1). Now C is the set of boundary points of C_i and also the set of boundary points of C_e. The union of C_i and C is a closed region which is denoted by C^*.

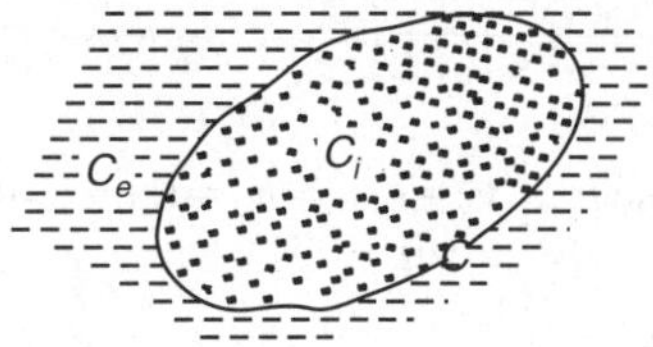

Fig. 8.5.1

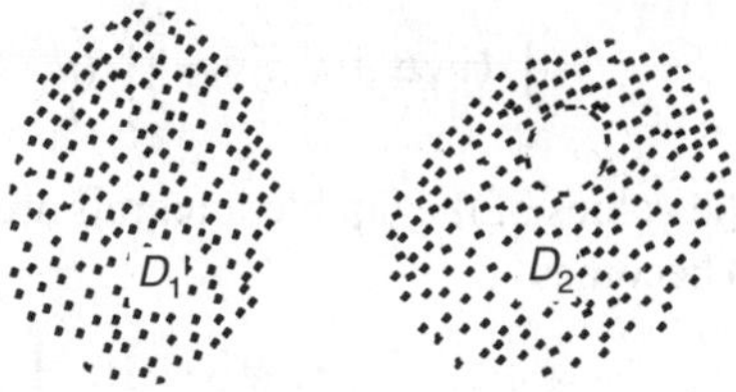

Fig. 8.6.1

8.6. SIMPLY-CONNECTED REGION

A region D is said to be simply-connected if, for every closed curve C in D, C_i is contained in D.

Intuitively speaking, simply-connectedness of a region is equivalent to the absence of holes in it or the situation in which every closed curve in D shrink to a point all the while being in D itself. In Fig. 8.6.1 D_1 is simply-connected but D_2 is not simply-connected. The exterior C_e of any closed curve C is not simply-connected.

Multiply-connected region. A region which is not simply-connected is called multiply-connected. In Fig. 8.6.1 D_2 is multiply-connected.

Annular region. A region which has only one hole is an annular region. For example,

$$|z-a|<r, \quad r<|z-a|<R$$

are annular regions. In the figure D_2 is an annular region.

8.7. CAUNCHY'S FUNDAMENTAL THEOREM

It was Cauchy who enunciated an important *fundamental theorem* connected with vanishing of integrals along scr curves. The theorem is named differently as

(*i*) Cauchy's theorem

(*ii*) Cauchy's fundamental theorem

(*iii*) Cauchy's integral theorem.

The weaker version of this theorem is as follows:

Theorem 8.5. Suppose

(*i*) D is a simply-connected region

(*ii*) $f(z)$ is analytic in D and $f'(z)$ is continuous in D

(*iii*) C is an scr curve in D.

Then

$$\int_C f(z)\, dz = 0.$$

Proof: Let $f(z) = u(x, y) + iv(x, y)$. Then, from (8.2.2.),

$$\int_C f(z)\, dz = \int_C (u\, dx - v\, dy) + i\int_C (v\, dx + u\, dy) \tag{8.7.1}$$

From the hypothesis, u, v, u_x, u_y, v_x, v_y are all continuous in C^*. The Green's theorem in two dimensions is that, if P, Q, Q_x, P_y are functions in x and y, continuous in C^*, then

$$\int_C P\, dx + Q\, dy = \iint_{C^*} (Q_x - P_y)\, dx\, dy.$$

So $$\int_C (u\, dx - v\, dy) = \iint_{C^*} (-v_x - u_y)\, dx\, dy$$

and $$\int_C (v\,dx + u\,dy) = \iint_{C^*} (u_x - v_y)\,dx\,dy\,.$$

But the values of the double integrals are zero because $u_y + v_x = 0, u_x - v_y = 0$ by C-R equations. So (8.7.1) becomes

$$\int_C f(z)\,dz = 0$$

It was Goursat who showed first that continuity of $f'(z)$ was not necessary and just the existence of $f'(z)$ would do for the integral to vanish. He first proved a lemma which involved partitioning of C^* into a finite number of closed regions, squares and truncated squares, such that these regions satisfy a certain property. Then he used this lemma to prove Cauchy's theorem. The lemma is called Goursat's lemma.

A further refinement of the proof was later furnished by Pringsheim which is not stated here.

Goursat's lemma. Suppose

(*i*) D is a simply-connected region,

(*ii*) $f(z)$ is analytic in D and

(*iii*) C is an scr curve in D.

Then C^* can be partitioned into a finite number of closed subregions, full or truncated squares, such that each subregion, satisfies the property P that, for any $\in > 0$, the region has an a with

$$\left|\frac{f(z) - f(a)}{z - a} - f'(a)\right| < \in \quad \text{for all } z \text{ in that region.}$$

Proof: Suppose D cannot be partitioned into a finite number of closed subregions such that each subregion satisfies P. Take a square A, enclosing C^*, with its sides parallel to the real and imaginary axis (Fig. 8.7.1). Let the projection of this square on the real axis be the closed interval $[h, k]$. Then the side of the square is $k - h$. Divide A into four smaller squares. Then, C^* in general, gets partitioned

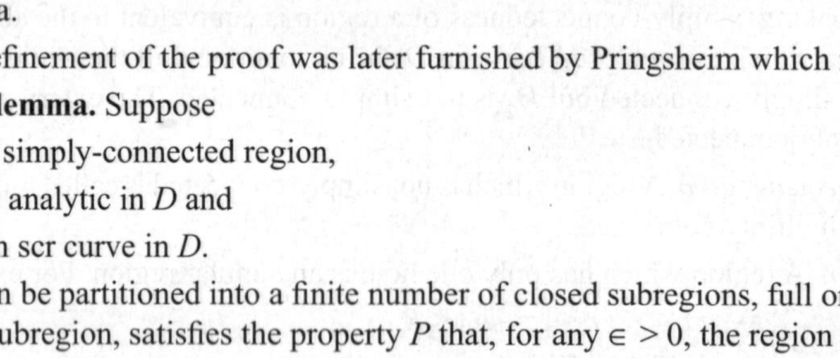

Fig. 8.7.1

into four closed subregions and at least one of them cannot satisfy the property P. Let this particular subregion be C_1^* and let it be contained in the square A_1. If the projection of A_1 on the real axis is $[h_1, k_1]$, then the side of A_1 is

$$k_1 - h_1 = \frac{k - h}{2}.$$

Next, again, divide A_1 into four squares. Then C_1^* gets partitioned into 4 closed subregions and at least one of them does not satisfy the property P. Let this subregion be C_2^* and the corresponding square be A_2. Continuing this process repeatedly, we get that the closed subregion, say C_n^*, cannot satisfy the property P, the side of respective square A_n being

$$k_n - h_n = \frac{k-h}{2^n}.$$

The real sequence $(k_n - h_n)$ converges to zero, that is,

$$\lim_{n\to\infty} (k_n - h_n) = 0$$

This implies that

$$\lim_{n\to\infty} k_n = \lim_{n\to\infty} h_n.$$

Let this limit be α. Similarly let the projection of A_n on the imaginary axis be the interval $[l_n, m_n]$. Then

$$m_n - l_n = \frac{k-h}{2^n}$$

Let $$\lim_{n\to\infty} l_n = \lim_{n\to\infty} m_n = \beta.$$

Now $a = \alpha + i\beta$ is a point which is common to all the closed regions C^*, C_1^* C_2^*, ... and the squares $A, A_1, A_2, ...$

Since $f(z)$ is analytic in D, $f'(a)$ exists and so, for the given $\in$, there exists a δ such that

$$\left| \frac{f(z) - f(a)}{z-a} - f'(a) \right| < \in$$

for all z in the disc

$$\Gamma : |z-a| < \delta.$$

That is, the disc Γ stisfies the property P (Fig. 8.7.2).

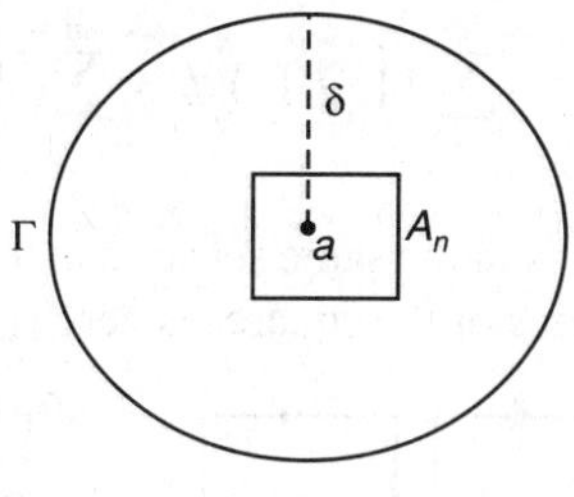

Fig. 8.7.2

The closed region C_n^* and the square A_n will be contained in the disc Γ as soon as n is large enough so that the diagonal d of A_n is less than δ, i.e., as soon as n is large enough so that

$$\frac{\sqrt{2}\,(k-h)}{2^n} < \delta.$$

$\therefore$ C_n^* satisfies P which is a contradiction. So the original assumption is wrong. This implies the result as stated.

Theorem 8.6. Cauchy's theorem and proof by Goursat. Suppose

(*i*) D is a simply-connected region,

(*ii*) $f(z)$ is analytic in D and

(*iii*) C is an scr curve in D.

Then

$$\int_C f(z)\, dz = 0.$$

Alternative statement. If $f(z)$ is analytic on and in the interior of an scr curve C, then the integral of $f(z)$ along C is zero.

Proof : Choose arbitrarily a small positive $\in$. Then, from Goursat's lemma, it is possible to partition C^* into a finite number of closed regions, squares or truncated squares, such that each region contains an a so that

$$\left|\frac{f(z)-f(a)}{z-a}-f'(a)\right|<\in \quad \text{for all } z \text{ in that region.}$$

Let the boundaries of the full and truncated square regions be

$$S_1, S_2, \ldots, S_n; \quad T_1, T_2, \ldots, T_m$$

respectively. Let their sides be

$$\lambda_1, \lambda_2, \ldots, \lambda_n; \quad \mu_1, \mu_2, \ldots, \mu_m.$$

Let the particular points a's in those regions be

$$a_1, a_2, \ldots, a_n; \quad b, b_2, \ldots, b_m.$$

Then, for all z on S_j,

$$\left|\frac{f(z)-f(a_j)}{z-a_j}-f'(a_j)\right|<\in.$$

Denoting the complex number

$$\frac{f(z)-f(a_j)}{z-a_j}-f'(a_j) \text{ by } P_j, \quad |P_j|<\in \tag{8.7.2}$$

for all z on S_j. Now

$$\int_C f(z)\,dz=\sum_{j=1}^{n}\int_{S_j} f(z)\,dz+\sum_{j=1}^{m}\int_{T_j} f(z)\,dz,$$

where integration along S_j and T_j is in the same sense as in the orientation of C, because the sum of the integrals along the common sides of the squares is zero (Fig. 8.7.3)

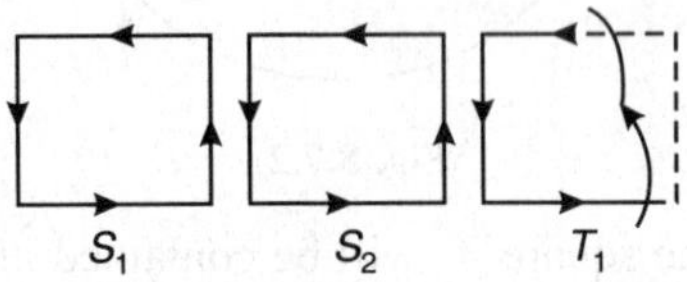

Fig. 8.7.3

$$\left|\int_C f(z)\,dz\right| = \left|\sum_{j=1}^{n}\int_{S_j} f(z)\,dz+\sum_{j=1}^{m}\int_{T_j} f(z)\,dz\right|$$

$$\leq \sum_{j=1}^{n}\left|\int_{S_j} f(z)\,dz\right|+\sum_{j=1}^{m}\left|\int_{T_j} f(z)\,dz\right|$$

But, from (8.7.2),

$$\int_{S_j} f(z)\,dz = \int_{S_j}[f(a_j)+f'(a_j)(z-a_j)+P_j(z-a_j)]\,dz$$

$$= f(a_j)\int_{S_j} 1\,dz+f'(a_j)\int_{S_j} z\,dz - f'(a_j)\,a_j\int_{S_j} 1\,dz+\int_{S_j} P_j(z-a_i)\,dz$$

$$= 0 + 0 + 0 + \int_{S_j} P_j\,(z - a_j)\,dz \text{ by (8.3.1), (8.3.2).}$$

$$\therefore \quad \left| \int_{S_j} f(z)\,dz \right| \le [\text{Max} \mid P_j\,(z - a_j) \mid \text{on } S_j\,]\; l\,(S_j)$$

$$< \in \text{(diagonal of } S_j) \, (4\lambda_j) \text{ since } | P_j | < \in \text{ and } | z - a_j | < \text{diagonal of } S_j$$

$$= \in (\sqrt{2}\,\lambda_j)\,(4\,\lambda_j) = 4\sqrt{2}\,(\lambda_j)^2 \in .$$

Similarly, using the result obtained from (8.7.2) by replacing a_j, P_j, S_j respectively by b_j, Q_j, T_j,

$$\left| \int_{T_j} f(z)\,dz \right| \le [\text{Max} \mid Q_j\,(z - b_j) \mid \text{on } T_j]\; l\,(T_j).$$

But $l\,(T_j) < 4\mu_j + l\,(\gamma_j)$ where γ_j is the arc of C intercepted by T_j (Fig. 8.7.4) so that $\Sigma\,\gamma_j = C$ and $l\,(\gamma_j) < l\,(C)$.

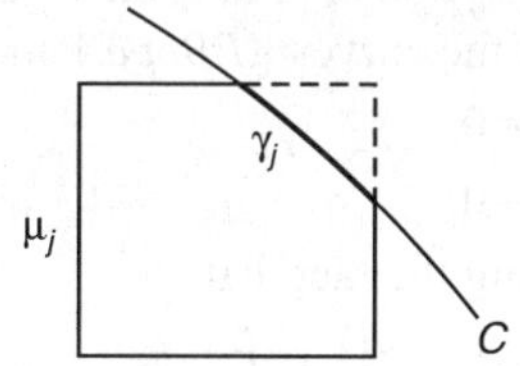

Fig. 8.7.4

$$\therefore \quad \left| \int_{T_j} f(z)\,dz \right| < \in \text{(diagonal of } T_j)\,[4\,\mu_j + l\,(C)]$$

$$= \in (\sqrt{2}\,\mu_j)\,[4\mu_j + l\,(C)]$$

$$\therefore \quad \left| \int_C f(z)\,dz \right| < \left[4\sqrt{2} \left(\sum_{j=1}^{n} \lambda_{j^2} + \sum_{j=1}^{m} \mu_{j^2} \right) + \sqrt{2} \left(\sum_{j=1}^{m} \mu_j \right) l\,(C) \right] \in$$

$$= [\,4\sqrt{2} \text{ (sum of the areas of the squares)}$$

$$+ \sqrt{2} \text{ (sum of the perimeters of the } T\text{-squares) } l\,(C)] \in$$

$$< [4\,\sqrt{2} \text{ (Area of } A) + \sqrt{2} \text{ (Perimeter of } A) \text{ (length of } C)] \in$$

$$= [4\,\sqrt{2}\,(k - h)^2 + 4\,\sqrt{2}\,(k - h)\,l\,(C)] \in\; = M \in,$$ say where M is finite and fixed but $\in$ is arbitrary. This implies that

$$\int_C f(z)\,dz = 0.$$

Extension to Cauchy's fundamental theorem. Cauchy's fundamental theorem pertains to a simply-connected region. This theorem can easily be extended to a multiply-connected region.

Theorem 8.7. Suppose

(*i*) D is an annulus region,

(*ii*) $f(z)$ is analytic in D and

(*iii*) C_0, C_1 are scro curves such that

1. C_1 lies in the interior of C_0,

2. the annulus region between C_0 and C_1 does not contain a hole while the interior of C_1 contains holes (Fig. 8.7.5.).

Then $$\int_{C_0} f(z)\,dz = \int_{C_1} f(z)\,dz.$$

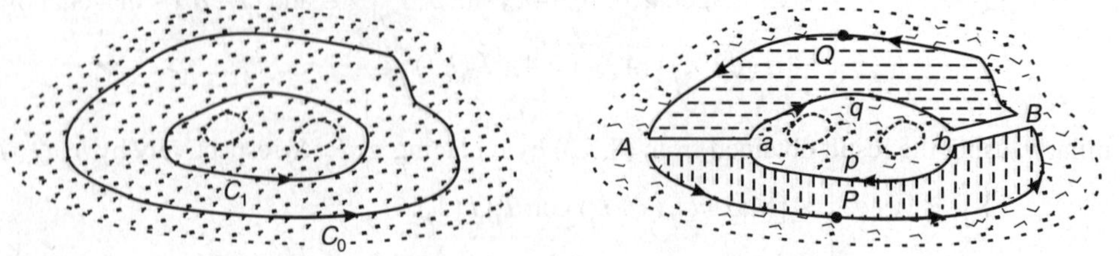

Fig. 8.7.5

Proof: Take points A, P, B, Q on C_0 and a, p, b, q on C_1 and join Aa and Bb as in the figure. $f(z)$ is analytic on and in the interior of the curves $APBbpaA$ and $BQAaqbB$. So

Integral along $APBbpaA = 0$,

Integral along $BQAaqbB = 0$.

Adding these two integrals and using the fact that

$$\int_{Aa} + \int_{aA} = 0 \quad \int_{Bb} + \int_{bB} = 0,$$

Integral along $APBQA$ + integral along $aqbpa = 0$

i.e., Integral C_0 + Integral along $(-C_1) = 0$.

$$\therefore \quad \int_{C_0} f(z)\,dz = \int_{C_1} f(z)\,dz.$$

Corollary. Suppose $C_0, C_1, C_2, \ldots C_n$ are scr curves of same orientation lying in a multiply-connected region D such that all of $C_1, C_2, \ldots, C_n$ lie in the interior of C_0 as in Fig. 8.7.6. If $f(z)$ is a function analytic in D, then dividing the region between C_0 and the other curves into two simply-connected regions,

$$\int_{C_0} = \int_{C_1} + \int_{C_2} + \ldots + \int_{C_n}.$$

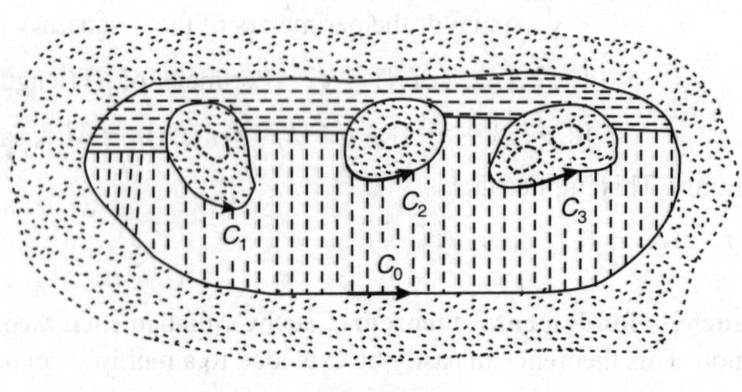

Fig. 8.7.6

8.8. INTEGRAL ALONG AN ARC JOINING TWO POINTS

Theorem 8.8. Suppose $f(z)$ is an analytic function in a simply-connected region D and $A(a)$ and $B(b)$ are two points in D.

Let C be any arbitrarily chosen simple rectifiable arc in D oriented from A to B (Fig. 8.8.1). Then the integral

$$\int_C f(z)\, dz$$

does not depend on C but depends on a and b.

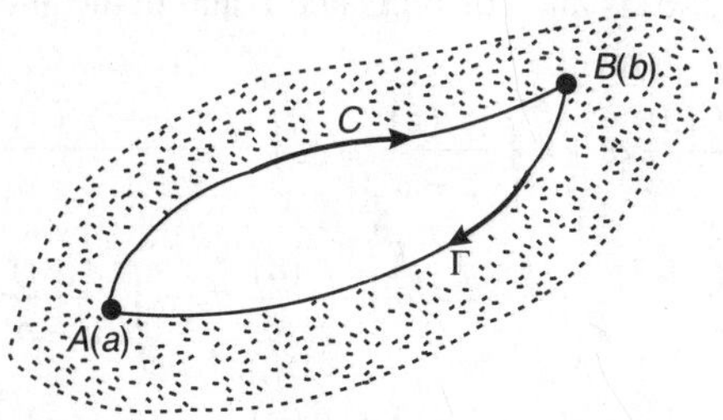

Fig. 8.8.1

Proof: Let Γ be a fixed simple rectifiable arc oriented from B to A. Then $C+\Gamma$ is a scr curve. So, by Cauchy's fundamental theorem,

$$\int_{C+\Gamma} f(z)\, dz = 0$$

i.e., $$\int_C f(z)\, dz + \int_\Gamma f(z)\, dz = 0$$

or $$\int_C f(z)\, dz - \int_{-\Gamma} f(z)\, dz = 0$$

i.e., $$\int_C f(z)\, dz = \int_{-\Gamma} f(z)\, dz.$$

Here the integral on the right is a constant. So the integral on the left is independent of C and depends on the points $A(a)$ and $B(b)$ as already seen in section 8.4. This path-independent integral can be written as

$$\int_a^b f(z)\, dz.$$

Remark. When A is fixed and B (z) moves in D, the integral

$$\int_a^z f(z)\, dz \tag{8.8.1}$$

defines a single-valued function of z in D.

8.9. CAUCHY'S INTEGRAL FORMULA AND FORMULAS FOR DERIVATIVES

Theorem 8.9. Cauchy's integral formula. Suppose

(*i*) D is a simply-connected region,

(*ii*) $f(z)$ is analytic in D and

(*iii*) C is a scro curve in D.

Then for any a in C_i

$$f(a) = \frac{1}{2\pi i} \int_C \frac{f(z)}{z-a}\, dz .$$

Proof: The function $\dfrac{f(z)}{z-a}$ is analytic in D except at $z = a$.

In C_i take a positively oriented circle Γ,

$$|z - a| = \rho.$$

Now the function $f(z)/z - a)$ is analytic on C and Γ and in the annulus region between C and Γ. So, by theorem 8.6,

$$\begin{aligned}
\int_C \frac{f(z)}{z-a}\, dz &= \int_\Gamma \frac{f(z)}{z-a}\, dz = \int_\Gamma \frac{f(z)-f(a)+f(a)}{z-a}\, dz \\
&= \int_\Gamma \frac{f(z)-f(a)}{z-a}\, dz + \int_\Gamma \frac{f(a)}{z-a}\, dz \\
&= \int_\Gamma \frac{f(z)-f(a)}{z-a}\, dz + f(a)\int_\Gamma \frac{1}{z-a}\, dz \\
&= \int_\Gamma \frac{f(z)-f(a)}{z-a}\, dz + f(a)(2\pi i) \text{ by (8.3.3)}
\end{aligned}$$

or $$\int_C \frac{f(z)}{z-a}\, dz - (2\pi i)\, f(a) = \int_\Gamma \frac{f(z)-f(a)}{z-a}\, dz$$

$$\therefore \quad \left| \int_C \frac{f(z)}{z-a}\, dz - (2\pi i)\, f(a) \right| = \left| \int_\Gamma \frac{f(z)-f(a)}{z-a}\, dz \right|$$

$$\leq \left(\max \left| \frac{f(z)-f(a)}{z-a} \right| \text{ on } \Gamma \right) l(\Gamma) \tag{8.9.1}$$

Now $f(z)$ is analytic in D. Therefore its derivative exists at all points in D and hence it exists at $z = a$ also. So

$$\lim_{z \to a} \frac{f(z)-f(a)}{z-a} = f'(a).$$

Denoting $\dfrac{f(z)-f(a)}{z-a} - f'(a)$ by $\phi(z, a)$ and, taking limit as z tends to a, we get

$$\lim_{z \to a} \phi(z, a) = 0$$

Hence, given an $\in > 0$, we can find a δ such that

$$|\phi(z, a) - 0| < \in \quad \text{or} \quad |\phi(z, a)| < \in$$

for all z in the disc $|z - a| < \delta$. Therefore, choosing ρ to be less than δ, for all z on and inside the circle $\Gamma : |z - a| = \rho$, $|\phi(z, a)| < \in$ and for this z

$$\begin{aligned}
\left| \frac{f(z)-f(a)}{z-a} \right| &= |f'(a) + \phi(z, a)| \leq |f'(a)| + |\phi(z, a)| \\
&< |f'(a)| + \in = M, \text{ say,}
\end{aligned}$$

where M is finite because $f'(a)$ and $\in$ are finite. Thus on Γ

$$\max\left|\frac{f(z)-f(a)}{z-a}\right| < M.$$

Therefor (8.9.1) becomes

$$\left|\int_C \frac{f(z)}{z-a}\,dz-(2\pi i)\,f(a)\right| < M(2\pi\rho).$$

Taking limit as ρ tends to zero,

$$\left|\int_C \frac{f(z)}{z-a}\,dz-(2\pi i)\,f(a)\right| = 0.$$

$$\therefore \qquad \int_C \frac{f(z)}{z-a}\,dz-(2\pi i)\,f(a) = 0$$

or

$$f(a) = \frac{1}{2\pi i}\int_C \frac{f(z)}{z-a}\,dz.$$

Illustration. We shall rewrite the Cauchy's integral formula as

$$\int_C f(z)\frac{1}{z-a}\,dz = 2\pi i\,f(a).$$

In this, $f(z)$ is analytic on and inside C and $z = a$ is a point inside C.

Remark. There are problems which can be solved by Cauchy's integral formula and also by Residue thseorem 8.10.

An immediate application of theorem 8.9 is theorem 8.10.

Theorem 8.10. Cauchy's integral formula for first dierivatie. Suppose

(*i*) D is a simply-connected region,

(*ii*) $f(z)$ is analytic in D and

(*iii*) C is a scro curve in D.

Then for any a in C_i

$$f'(a) = \frac{1!}{2\pi i}\int_C \frac{f(z)}{(z-a)^2}\,dz.$$

Proof: Now a belongs to C_i and the integral

$$\frac{1}{2\pi i}\int_C \frac{f(z)}{(z-a)^2}\,dz$$

is a function of a defined in D. Denote this function by $\phi(a)$. Take another point $a + b$ in C_i. Now the Cauchy's integrals for $f(z)$ corresponding to a and $a + b$ are

$$f(a) = \frac{1}{2\pi i}\int_C \frac{f(z)}{z-a}\,dz, \quad f(a+b) = \frac{1}{2\pi i}\int_C \frac{f(z)}{z-a-b}\,dz$$

$$\therefore \qquad \left|\frac{f(a+b)-f(a)}{b}-\phi(a)\right|$$

$$= \left|\frac{1}{2\pi ib}\int_C \left[\frac{f(z)}{z-a-b}-\frac{f(z)}{z-a}-\frac{bf(z)}{(z-a)^2}\right]dz\right|$$

$$= \frac{1}{2\pi}\left|\int_C \frac{(z-a)^2-(z-a-b)(z-a)-b(z-a-b)}{b(z-a-b)(z-a)^2} f(z)\,dz\right|$$

$$= \frac{1}{2\pi}\left|\int_C \frac{bf(z)}{(z-a-b)(z-a)^2}\,dz\right|$$

$$\leq \frac{|b|}{2\pi}\operatorname*{Max}_{\text{on C}} \frac{|f(z)|}{|z-a-b||z-a|^2}\, l(C).$$

Take any simple closed curve C_1 in C_i encircling a and $a+b$. Let δ be the minimum distance between the points of the curves C_1, and C (Fig. 8.9.1).

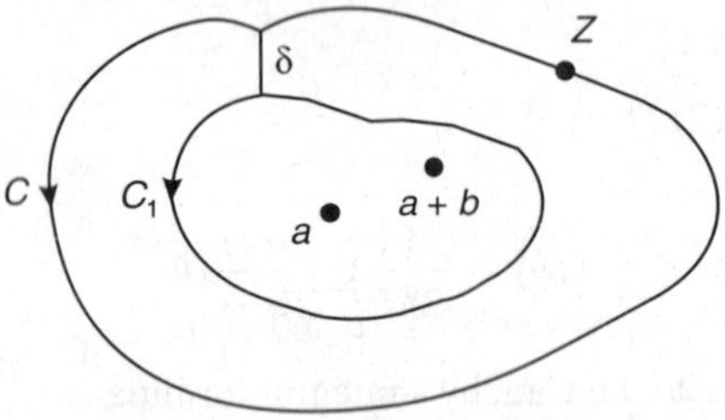

Fig. 8.9.1

If z is any point on C, then

$$|z-a| > \delta, \quad |z-(a+b)| > \delta$$

$$\therefore \qquad \frac{1}{|z-a|} < \frac{1}{\delta}, \quad \frac{1}{|z-a-b|} < \frac{1}{\delta}$$

So, for any a and $a+b$ in the interior of C_1.

$$\left|\frac{f(a+b)-f(a)}{b} - \phi(a)\right| < \frac{1}{2\pi}\frac{|b|M}{\delta^3}\, l(C),$$

where M is max $|f(z)|$ on C. Taking limit as $b \to 0$,

$$|f'(a)-\phi(a)| = 0 \quad \text{or} \quad f'(a)-\phi(a) = 0.$$

$$\therefore \qquad f'(a) = \frac{1}{2\pi i}\int_C \frac{f(z)}{(z-a)^2}\,dz.$$

Theorem 8.11. Integral formula for *n*th derivative. Suppose

(*i*) D is a simply-connected region,

(*ii*) $f(z)$ is analytic in D and

(*iii*) C is a scro curve in D.

Then for any a in C_i

$$f^n(a) = \frac{n!}{2\pi i}\int_C \frac{f(z)}{(z-a)^{n+1}}\,dz.$$

Proof: We use the following lemma in the proof of the theorem.

LEMMA. If n is an integer,

$$\frac{b^{-n}-a^{-n}}{b-a} + na^{-n-1} = (b-a)(1\cdot a^{-2}b^{-n} + 2a^{-3}\,b^{-n+1} + \ldots + na^{-n-1}\,b^{-1})$$

Now $\quad \dfrac{b^{-n}-z^{-n}}{b-z} = -\left[\left(\dfrac{1}{b}\right)^n\left(\dfrac{1}{z}\right)^1 + \left(\dfrac{1}{b}\right)^{n-1}\left(\dfrac{1}{z}\right)^2 + \ldots + \left(\dfrac{1}{b}\right)^1\left(\dfrac{1}{z}\right)^n\right]$

whose truth can be seen by summing the series on the right. Differentiating both sides of this result with respect to z and then multiplying both sides by $b-z$ and then replacing z by a, we get the result of the lemma.

To prove the theorem we use the method of induction. Suppose the derivatives of $f(z)$ of orders upto m exist in D and, for an a in C_i,

$$f^{(m)}(a) = \frac{m!}{2\pi i}\int_C \frac{f(z)}{(z-a)^{m+1}}\,dz \tag{8.9.2}$$

Then, choosing another point $a+b$ in C_i, we get

$$\frac{f^{(m)}(a+b)-f^{(m)}(a)}{b} = \frac{m!}{2\pi i}\int_C \frac{f(z)}{b}\left[\frac{1}{(z-a-b)^{m+1}} - \frac{1}{(z-a)^{m+1}}\right]dz$$

$$= \frac{m!}{2\pi i}\int_C f(z)\,\frac{(z-a-b)^{-(m+1)}-(z-a)^{-(m+1)}}{-[(z-a-b)-(z-a)]}\,dz$$

Now a is in C_i. So the integral

$$\frac{(m+1)!}{2\pi i}\int_C \frac{f(z)}{(z-a)^{m+2}}\,dz$$

defines a function of a in C_i. Let this function be $g(a)$. Then

$$\frac{f^{(m)}(a+b)-f^{(m)}(a)}{b} - g(a) = \frac{m!}{2\pi i}\int_C f(z)\left[\frac{(z-a-b)^{-(m+1)}-(z-a)^{-(m+1)}}{-[(z-a-b)-(z-a)]}\right.$$

$$\left. -(m+1)(z-a)^{-(m+2)}\right] dz.$$

$$= \frac{m!}{2\pi i}\int_C f(z)\,b\,[1(z-a)^{-2}(z-a-b)^{-m-1} + 2(z-a)^{-3}(z-a-b)^{-m} + \ldots$$

$+ (m+1)(z-a)^{-m-2}(z-a-b)^{-1}]\,dz$ by taking $n = m+1$ in the lemma.

Taking C_1 and δ as in the previous theorem, we get here, for $a+b$ in the iterior of C_1,

$$\left|\frac{f^{(m)}(a+b)-f^{(m)}(a)}{b} - g(a)\right| < \frac{m!\,M(f;C)\,|b|}{2\pi\,\delta^{m+3}}[1+2+3+\ldots+(m+1)]\,l(C),$$

where $M(f;C)$ is the maximum of $|f(z)|$ on C. Now, taking limit as $b \to 0$ in any manner, we achieve

$$f^{(m+1)}(a) - g(a) = 0$$

i.e., $$f^{(m+1)}(a) = \frac{(m+1)!}{2\pi i}\int \frac{f(z)}{(z-a)^{m+2}}\,dz.$$

But by the previous theorem the result (8.9.2) is true when $m = 1$. So the result is true for all positive integral values of m.

Corollary. Now it is evident that, if $f^{(m)}(z)$ exists, then $f^{(m+1)}(z)$ also exists. Thus

Theorem 8.11A. A function analytic in D has derivatives of all orders in D. (8.9.3)

Theorem 8.11B. The derivative of an analytic function is also analytic. (8.9.4)

Extensions of Cauchy's formulas. In the same manner in which the theorem 8.6 was extended to the theorem 8.7, the theorems 8.9, 8.10, 8.11 can be extended to the results

$$f(a) = \frac{1}{2\pi i}\int_{C_0}\frac{f(z)}{z-a}\,dz - \frac{1}{2\pi i}\int_{C_1}\frac{f(z)}{z-a}\,dz$$

$$f'(a) = \frac{1}{2\pi i}\int_{C_0}\frac{f(z)}{(z-a)^2}\,dz - \frac{1}{2\pi i}\int_{C_1}\frac{f(z)}{(z-a)^2}\,dz$$

$$f^{(n)}(a) = \frac{n!}{2\pi i}\int_{C_0}\frac{f(z)}{(z-a)^{n+1}}\,dz - \frac{n!}{2\pi i}\int_{C_1}\frac{f(z)}{(z-a)^{n+1}}\,dz$$

From the analyticity of a function, Cauchy's fundamental theorem establishes vanishing of the integral of an analytic function along an scr curve. But the following Morera's theorem establishes analyticity from the vanishing of the integral and continuity of the function. As such, Morera's theorem can be considered as a converse of Cauchy's fundamental theorem.

Theorem 8.12. Morera's theorem. Suppose

(*i*) D is a region,

(*ii*) $f(z)$ is continuous in D,

(*iii*) C is an scr curve in D and

(*iv*) $\int_C f(z)\,dz = 0$ for all C's.

Then $f(z)$ is analytic in D.

Proof: Let z_0 be a fixed point in D. Since the integral of $f(z)$ along every scr curve is zero, the integral

$$\int_{z_0}^{z} f(z)\,dz$$

is path-independent and defines a single-valued function in D (See (8.8.1)). Denote this function by $F(z)$ and take any two points a, $a + b$ in D. Then

$$\begin{aligned}
F(a+b) - F(a) &= \int_{z_0}^{a+b} f(z)\,dz - \int_{z_0}^{a} f(z)\,dz \\
&= \int_{a}^{a+b} f(z)\,dz = \int_{a}^{a+b} [f(z) - f(a) + f(a)]\,dz \\
&= \int_{a}^{a+b} [f(z) - f(a)]\,dz + f(a)\int_{a}^{a+b} 1\,dz \\
&= \int_{a}^{a+b} [f(z) - f(a)]\,dz + f(a)\,[(a+b) - a]
\end{aligned}$$

$$\therefore \quad \frac{F(a+b) - F(a)}{b} - f(a) = \frac{1}{b}\int_{a}^{a+b} [f(z) - f(a)]\,dz\,.$$

Now $f(z)$ is continuous in D. So, for any given $\in > 0$, we can find a δ such that

$$|f(z) - f(a)| < \in \text{ for all } z \text{ in } |z - a| < \delta$$

Fig. (8.9.2). Restrict $a + b$ to lie in the disc $|z - a| < \delta$ and denote the segment joining a and $a + b$ by L. Then

$$\left|\frac{F(a+b) - F(a)}{b} - f(a)\right| = \frac{1}{|b|}\left|\int_L [f(z) - f(a)]\,dz\right|$$

$$\leq \frac{1}{|b|} \underset{\text{on } L}{\text{Max}} \left| [f(z) - f(a)] \right| l(L) < \frac{1}{|b|} \in |b| = \in$$

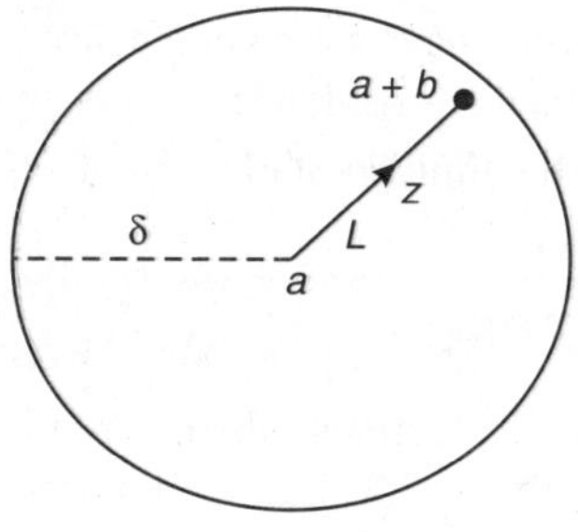

Fig. 8.9.2

Taking limit as $b \to 0$, $|F'(a) - f(a)| < \in$. But $\in$ is arbitrary.

$\therefore$ $\qquad |F'(a) - f(a)| = 0$ or $F'(a) = f(a)$.

Thus the derivative of $F(z)$ exists throughout D. So $F(z)$ is analytic in D. But, from the corollary to theorem 8.10 (Theorem 8.10.B), the derivative of an analytic function is analytic. Therefore $f(z)$ which is the derivative of $F(z)$, is analytic in D.

Corollary. If C is an arc in D, joining a fixed point z_0 and the arbitrary point z, then

$$\frac{d}{dz} \int_C f(z)\, dz = f(z).$$

8.10. ZEROS OF A FUNCTION

ZEROS OF AN ANALYTIC FUNCTION

Given the quadratic equation with real coefficients,

$$x^2 - 3x + 2 = 0,$$

we say that 1 and 2 are its roots. But, given the complex function,

$$f(z) = z^2 - 3z + 2,$$

we say that $z = 1$ and $z = 2$ are the zeros of $f(z)$.

DEFINITION. If a function $f(z)$ is such that

$$f(z) = (z - a)^k \phi(z),$$

where k is a positive integer and $\phi(a) \neq 0$, then a is said to be a zero of $f(z)$ of *order k* or of *multiplicity k*. For example, the function

$$(z - i)^2 (z + 1)^3 e^z$$

has a zero i of order 2 and a zero -1 of order 3.

Remark. If $f(z) = (z - a)^m [a_0 + a_1 (z - a) + a_2 (z - a)^2 + \ldots]$, $a_0 \neq 0$, then $z = a$ is a zero of $f(z)$ of order m.

Theorem 8.13. Zeros of an analytic function are isolated.

Proof: Suppose the analytic function $f(z)$ has a zero $z = a$ of order k. Then

$$f(z) = (z - a)^k \phi(z)$$

where $\phi(z)$ is such that the constant $\phi(a) \neq 0$. Since $f(z)$ and $(x - a)^k$ are analytic at $z = a$, $\phi(z)$ is analytic at $z = a$. Hence $\phi(z)$ is continuous at $z = a$. So, for a given $\in = \frac{1}{2}|\phi(a)|$, there exists a δ such that

$$|\phi(z) - \phi(a)| < \tfrac{1}{2}|\phi(a)| \quad \text{for all } z \text{ in } |z - a| < \delta.$$

Suppose $\phi(z)$ has a zero in this disc. Let it be $z = b$. Then $\phi(b) = 0$. Consequently

$$|0 - \phi(a)| < \tfrac{1}{2}|\phi(a)| \quad \text{or} \quad |\phi(a)| < \tfrac{1}{2}|\phi(a)|$$

which is a contradiction. So $\phi(z)$ cannot have a zero in the disc $|z - a| < \delta$. So also $f(z)$ cannot have a zero in this disc, of course, other than $z = a$. Hence $z = a$ is an isolated zero of $f(z)$.

Remark. We will see later that the limit point of zeros of an analytic function is a singular point of the function.

8.11. RELATED INTEGRAL THEOREM

In this section we prove a set of theorems which are immediate consequences of Cauchy's theorem. These theorems are

(*i*) Cauchy's inequality : Theorem 8.14

(*ii*) Liouville's theorem : Theorem 8.15

(*iii*) Fundamental theorem of Algebra : Theorem 8.16

(*iv*) Maximum modulus theorem : Theorem 8.17

(*v*) Gauss' mean value theorem : Theorem 8.18

(*vi*) Theorem on the mean of the values of a harmonic function on a circle : Theorem 8.19

Theorem 8.14. Cauchy' inequality. Suppose

(*i*) D is a simply-connected region,

(*ii*) $f(z)$ is analytic in D and

(*iii*) Γ is any circle $|z - a| = r$ in D.

Then
$$|f^n(a)| \le \frac{n!\,M(f; r, a)}{r^n},$$

where $M(f; r, a)$ is the maximum of $|f(z)|$ on Γ.

Proof: From Cauchy's formula for n^{th} derivative.

$$f^{(n)}(a) = \pm\frac{n!}{2\pi i}\int_\Gamma \frac{f(z)}{(z-a)^{n+1}}\,dz$$

$$\therefore \qquad |f^{(n)}(a)| = \left|\frac{n!}{2\pi i}\int \frac{f(z)}{(z-a)^{n+1}}\,dz\right|$$

$$= \frac{n!}{2\pi}\left|\int_\Gamma \frac{f(z)}{(z-a)^{n+1}}\,dz\right|$$

$$\le \frac{n!}{2\pi}\left[\underset{\text{on } \Gamma}{\text{Max}} \frac{|f(z)|}{|z-a|^{n+1}}\right] l(\Gamma)$$

$$= \frac{n!}{2\pi}\,\frac{M(f; r, a)}{r^{n+1}}(2\pi r) = \frac{n!\,M(f; r, a)}{r^n}.$$

Corollary. If the power series $a_0 + a_1 z + a_2 z^2 + \ldots$ converges to $f(z)$ and R is the radius of convergence, then

$$|a_n| \le \frac{M(f; r, 0)}{r^n}, \quad r < R. \qquad \textbf{(1979)}$$

Proof: Differentiate the series term-by-term n times and set $z = 0$. This gives

$$f^{(n)}0) = n!\, a_n.$$

Now $f(z)$ is analytic in $|z| < R$. Take a circle $|z| = r < R$. Then

$$|n!\, a_n| \le \frac{n!\, M(f; r, 0)}{r^n} \quad \text{or} \quad |a_n| \le \frac{M(f; r, 0)}{r^n}.$$

Theorem 8.15. Liouville's theorem. If an entire function is bounded in the finite plane, then it is a constant.

Proof: Entire functions are functions which are analytic in the finite plane. Let $f(z)$ be an entire function bounded in the finite plane. Then there exists an M such that $|f(z)| \le M$ for all finite z. Now, setting $n = 1$ in the Cauchy's inequality, for any finite a,

$$|f'(a)| \le \frac{M(f; r, a)}{r}$$

$$\therefore \qquad |f'(a)| \le \frac{M}{r}.$$

This inequality is true for any large r. So $f'(a)$ is necessarily zero. Since a is arbitrary, $f'(z) \equiv 0$. Hence $f(z)$ is a constant.

Corollary. Every function analytic in the extended plane is a constant.

Proof: The function is analytic everywhere in the extended plane. So its value is not ∞ anywhere in the extended plane. Thus the function is bounded in the extended plane and, in particular, in the finite plane. So, by Liouville's theorem, the function is a constant in the finite plane. Now the continuity at $z = \infty$ requires the value of the function at $z = \infty$ also to be the same constant. So the result as stated.

Remark. There is no function other than a constant, which is analytic throughout the extended plane.

Theorem 8.16. Fundamental theorem of Algebra. Every polynomial in z of degree equal to or greater than 1, has at least one zero.

Proof: Let $$P(z) = c_0 + c_1 z + c_2 z^2 + \ldots + c_n z^n$$

be the given polynomial and suppose it has no zero. Then the function

$$f(z) = \frac{1}{P(z)}$$

is an entire function. Further $f(z)$ is bounded in the finite plane because, for no value of z, $P(z) = 0$. Hence, by Liouville's theorem, $f(z)$ is a constant. This means that $P(z)$ is constant which is a contradiction. So the assumption that $P(z)$ has no zero, does not hold. Therefore $P(z)$ has at least one zero.

Corollary. If a is one zero, then $P(a) = 0$ and

$$P(z) - P(a) = (c_0 + c_1 z + \ldots c_n z^n) - (c_0 + c_1 a + \ldots + c_n a^n)$$

i.e., $$P(z) = c_1(z - a) + c_2(z^2 - a^2) + \ldots + c_n(z^n - a^n)$$
$$= (z - a)\, Q(z),$$

where $Q(z)$ is a polynomial of degree $n - 1$. Therefore $Q(z)$ has at least one zero. Continuing this argument we get ultimately that $P(z)$ has exactly n zeros.

Remark. The fundamental theorem of algebra will be proved again later as an application of Rouche theorem.

Maximum absolute value of an analytic function. Next theorem reveals where the maximum of the absolute value of an analytic function occurs. If $f(z)$ is analytic inside and on an scr curve C, then the maximum always occurs at a boundary point.

Theorem 8.17. Maximum modulus theorem. Suppose

(*i*) D is a simply-connected region,

(*ii*) $f(z)$ is analytic in D and

(*iii*) C is any scr curve in D.

Then the maximum of $|f(z)|$ in the whole of C_i and C occurs only on C unless $f(z)$ is a constant.

Proof: Let k be any positive integer. Then $[f(z)]^k$ is analytic in D. So, for any a in C_i, from Cauchy's integral theorem,

$$[f(a)]^k = \frac{1}{2\pi i} \int_C \frac{[f(z)]^k}{z-a}\, dz$$

$$\therefore \qquad |[f(a)]^k| = \left| \frac{1}{2\pi i} \int_C \frac{[f(z)]^k}{z-a}\, dz \right|$$

$$= \left| \frac{1}{2\pi i} \right| \left| \int_C \frac{[f(z)]^k}{z-a}\, dz \right|$$

$$\leq \frac{1}{2\pi} \underset{\text{on } C}{\text{Max}} \left| \frac{[f(z)]^k}{z-a} \right| l(C)$$

$$= \frac{1}{2\pi} \frac{[M(f;C)]^k}{d} l(C),$$

where $M(f;C)$ is the maximum of $|f(z)|$ on C and d is the minimum distance of $z = a$ from the points on C. Hence, taking k^{th} root,

$$|f(a)| \leq M(f;C) \left[\frac{l(C)}{2\pi d} \right]^{1/k}$$

Taking limit as $k \to \infty$,

$$|f(a)| \leq M(f;C).$$

Since a is arbitrary in C_i, the equality is possible only when $f(z)$ is constant. If $f(z)$ is not a constant, then for all z in the interior of C, $|f(z)|$ is less than the maximum of $|f(z)|$ on C. Thus, in C_i and C, the maximum of $|f(z)|$ occurs on C.

Remark. The maximum of $|f(z)|$, for all z in C_i occurs at some point on the boundary C.

Theorem 8.18. Gauss' mean value theorem. Suppose

(*i*) D is a simply-connected region,

(*ii*) $f(z)$ is analytic in D,

(*iii*) Γ is any positively oriented circle in D.

Then the mean of the values of the function on Γ is the value of the function at its centre.

Proof: The equation of the circle Γ is

$$z = a + re^{it}, \quad 0 \leq t \leq 2\pi$$

where r is the radius and a, the centre (Fig. 8.11.1). By Cauchy's integral formula,

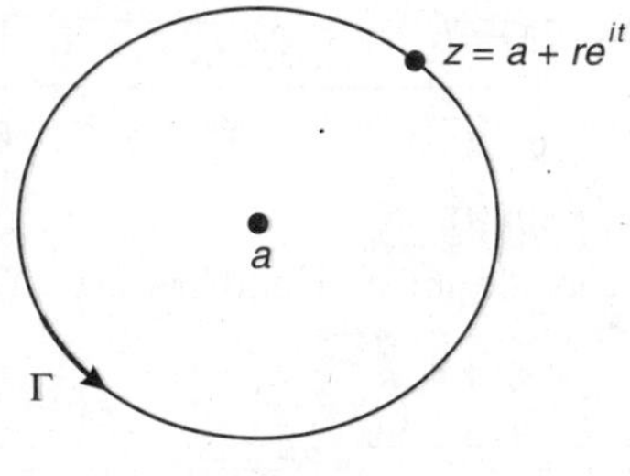

Fig. 8.11.1

$$f(a) = \frac{1}{2\pi i}\int_{\Gamma} \frac{f(z)}{z-a}\,dz$$

$$= \frac{1}{2\pi i}\int_0^{2\pi} \frac{f(a+re^{it})}{(a+re^{it})-a}\frac{d}{dt}(a+re^{it})\,dt$$

$$= \frac{1}{2\pi i}\int_0^{2\pi} \frac{f(a+re^{it})}{re^{it}}(i\,re^{it})\,dt = \frac{1}{2\pi}\int_0^{2\pi} f(a+re^{it})\,dt$$

which is the mean value of the values of $f(z)$ on Γ.

Mean value of the values of a harmonic function on a circle.

Theorem 8.19. If $u(x, y)$ is a function harmonic in a simply-connected region D, then the mean value of $u(x, y)$ taken along a circle in D is always equal to its value at the centre.

Proof: Let the harmonic conjugate of $u(x, y)$ be $v(x, y)$. Then $f(z) = u + iv$ is analytic in D and by the previous theorem the mean value of $f(z)$ on Γ, $|z-a| = r$, in D, is

$$f(a) = \frac{1}{2\pi}\int_0^{2\pi} f(a+re^{it})\,dt.$$

If $a = \alpha + i\beta$, then $f(a) = u(\alpha, \beta) + iv(\alpha, \beta)$ and $f(a + re^{it}) = u(\alpha + r\cos t, \beta + r\sin t) + iv(\alpha + r\cos t, \beta + r\sin t)$

$$\therefore \quad u(\alpha, \beta) + iv(\alpha, \beta) = \frac{1}{2\pi}\int_0^{2\pi} u(\alpha + r\cos t, \beta + r\sin t)\,dt$$

$$+ \frac{i}{2\pi}\int_0^{2\pi} v(\alpha + r\cos t, \beta + r\sin t)\,dt$$

Equating the real parts,

$$u(\alpha, \beta) = \frac{1}{2\pi}\int_0^{2\pi} u(\alpha + r\cos t, \beta + r\sin t)\,dt$$

which is the mean value of the values of $u(x, y)$ on Γ.

Corollary. A nonconstant function $u(x, y)$ harmonic in D has neither a maximum nor a minimum at any point of D. Suppose it has a maximum at (α, β). But $u(\alpha, \beta)$ is the mean of the values of $u(x, y)$ on the circle Γ. It means that some of the values of $u(x, y)$ are greater than $u(\alpha, \beta)$ and the rest are less than $u(\alpha, \beta)$ which contradicts the supposition.

Theorem 8.20. (Poisson's integral). Suppose the function $u(x, y)$ is harmonic inside and on a circle $\Gamma : |z| = R$. If $re^{i\theta}$ is a point inside this circle, then

$$u(r, \theta) = \frac{1}{2\pi} \int_0^{2\pi} \frac{u(R, \phi)(R^2 - r^2)}{R^2 + r^2 - 2Rr\cos(\theta - \phi)} d\phi,$$

where $u(r, \theta)$ stands for $u(r\cos\theta, r\sin\theta)$.

Proof: Suppose $f(z)$ is analytic inside and on Γ and has $u(x, y)$ as its real part. Then, by Cauchy's integral formula,

$$f(re^{i\theta}) = \frac{1}{2\pi i} \int \frac{f(z)}{z - re^{i\theta}} dz .$$

Now the point $R^2e^{i\theta}/r$ is inverse to the point $re^{i\theta}$ with respect to Γ and it lies outside Γ. So the function $f(z)/\{z - (R^2/r)e^{i\theta}\}$ is analytic inside and on Γ. Hence, by Cauchy's fundamental theorem,

$$0 = \frac{1}{2\pi i} \int \frac{f(z)}{z - (R^2/r)e^{i\theta}} dz$$

Subtracting the second integral from the first,

$$f(re^{i\theta}) = \frac{1}{2\pi i} \int_\Gamma f(z) \left(\frac{1}{z - re^{i\theta}} - \frac{r}{rz - R^2e^{i\theta}} \right) dz$$

$$= \frac{1}{2\pi i} \int_\Gamma f(z) \frac{-R^2e^{i\theta} + r^2e^{i\theta}}{(z - re^{i\theta})(rz - R^2e^{i\theta})} dz$$

$$= \frac{1}{2\pi i} \int_\Gamma f(z) \frac{-(R^2 - r^2)e^{i\theta}}{(z - re^{i\theta})(re^{-i\theta} - R^2/z)e^{i\theta}z} dz$$

$$= \frac{1}{2\pi i} \int_0 f(\mathrm{Re}^{i\phi}) \frac{-(R^2 - r^2)}{(\mathrm{Re}^{i\phi} - re^{i\theta})(re^{-i\theta} - \mathrm{Re}^{-i\phi})} i d\phi$$

since, on Γ, $z = Re^{i\phi}$ and $dz = iRe^{i\phi}\, d\phi = iz\, d\phi$

$$= \frac{1}{2\pi} \int_0^{2\pi} f(Re^{i\phi}) \frac{R^2 - r^2}{R^2 + r^2 - 2Rr\cos(\theta - \phi)} d\phi .$$

Equating the real parts we get the result as stated.

Corrollary. Setting $u(r, \theta) = 1$.

$$\frac{1}{2\pi} \int_0^{2\pi} \frac{R^2 - r^2}{R^2 + r^2 - 2Rr\cos(\theta - \phi)} d\phi = 1, \quad R > r .$$

8.12. TERM-BY-TERM DIFFERENTIATION AND INTEGRATION OF UNIFORMLY CONVERGENT SERIES

In chapter 6 we saw that a power series can be differentiated term-by-term in the interior of its circle of convergence. Now we consider term-by-term differentiation and integration of a uniformly convergent functional series. The following theorem asserts the possibility of term-by-term integration.

Theorem 8.21. Suppose

(*i*) L is a simple rectifiable curve,

(*ii*) $f_1(z) + f_2(z) + f_3(z) + \ldots$ is uniformly convergent on L, whose sum is $f(z)$ and

(*iii*) $f_n(z)$ is continuous on L ($n = 1, 2, 3, \ldots$).

Then the series $f_1(z) + f_2(z) + f_3(z) + \ldots + \ldots = f(z)$ can be integrated term by term along L.

Proof: Since the limit function of a uniformly convergent series of continuous functions is also continuous, $f(z)$ is continuous on L (Theorem 5.6). So $f(z)$ is integrable along L. Further, since the series is uniformly convergent on L, given any $\in > 0$, we can find an integer $N(\in) > 0$ such that

$|s_n(z) - f(z)| < \in/l$ for all $n > N$ and for all z on L,

where $s_n = f_1(z) + f_2(z) + \ldots + f_n(z)$ and l is the length of L.

Now consider the finite series

$$\left[\int_L f_1(z)\,dz + \int_L f_2(z)\,dz + \int_L f_3(z)\,dz + \ldots + \int_L f_n(z)\right] - \int_L f(z)\,dz$$

$$= \int_L [s_n(z) - f(z)]\,dz$$

$$\therefore \quad \left|\left(\sum_{j=1}^{n} \int_L f_j(z)\,dz\right) - \int_L f(z)\,dz\right| = \left|\int_L [s_n(z) - f(z)]\,dz\right|$$

$$\leq \underset{\text{on } L}{\text{Max}} |s_n(z) - f(z)|\; l(L)$$

$$< \frac{\in}{l} l = \in \text{ for all } n > N(\in).$$

Since $\in$ is arbitrary, this implies that

$$\left|\sum_{j=1}^{\infty} \int_L f_j(z)\,dz - \int_L f(z)\,dz\right| = 0$$

or
$$\sum_{j=1}^{\infty} \int_L f_j(z)\,dz = \int_L f(z)\,dz.$$

Term-by-term differentiation. The following Weierstrass theorem on uniformly convergent series of analytic functions asserts the possibility of term-by-term differentiation of a functional series.

Theorem 8.22. Suppose

(*i*) D is a region,

(*ii*) $f_1(z) + f_2(z) + f_3(z) + \ldots$ is a series uniformly convergent in every compact region in D, whose sum is $f(z)$ and

(*iii*) $f_n(z)$ is analytic in D ($n = 1, 2, 3, \ldots$).

Then (*i*) the sum function $f(z)$ is analytic in D and

(*ii*) $f_1(z) + f_2(z) + \ldots = f(z)$ can be differentiated in D term-by-term any number of times.

Proof: Let a be an arbitrary point in D. Take a simply-connected subregion D' in D, containing a. Let C be an arbitrary scr curve in D'. Because of the analyticity, $f_1(z), f_2(z), \ldots$ are continuous on C and hence $f(z)$ is continuous on C (Theorem 5.6). Since the series

$$f_1(z) + f_2(z) + f_3(z) + \ldots = f(z) \tag{8.12.1}$$

is uniformly convergent on C, integrating along C term by term,

$$\int_C f(z)\,dz = \int_C f_1(z)\,dz + \int_C f_2(z)\,dz + \ldots$$

$$= 0 + 0 + \ldots \text{ by Cauchy's fundamental theroem.}$$

Since C is arbitrary by Morera's theorem $f(z)$ is analytic in D'. But D' is arbitrary. Hence $f(z)$ is analytic in D. Now multiplying (8.12.1) by

$$\frac{1}{2\pi i}\frac{1}{(z-a)^2}$$

and integrating term by term along C,

$$\frac{1}{2\pi i}\int_\Gamma \frac{f_1(z)}{(z-a)^2}dz + \frac{1}{2\pi i}\int_\Gamma \frac{f_2(z)}{(z-a)^2}dz + \ldots = \frac{1}{2\pi i}\int \frac{f(z)}{(z-a)^2}dz$$

i.e.,

$$f_1'(a) + f_2'(a) + f_3'(a) + \ldots = f'(a).$$

Similarly multiplying by $\dfrac{n!}{2\pi i}\dfrac{1}{(z-a)^{j+1}}$ and integrating along C,

$$f_1^{(j)}(a) + f_2^{(j)}(a) + \ldots = f^{(j)}(a).$$

Since a is arbitrary in D this differentiation holds good at all points in D.

8.13. EXAMPLES

8.2 Integration of Complex Functions

1. Solved sum. If C is a simple recifiable arc of length $l(C)$, then show that

$$\int_C |dz| = l(C).$$

✍ Setting $f(z) = 1$ in (8.2.1) and using $|dz| = |z(t_j) - z(t_{j-1})|$,

$$\int_C |dz| = \lim_{|\wp|\to 0} \sum_{j=1}^{n} |z(t_j) - z(t_{j-1})|$$

$$= \lim_{|\wp|\to 0} \sum_{j=1}^{n} \text{(Length of the chord } z_{j-1}`z_j) = l(C).$$

2. Show that

$$I_1 = \int_C x\,dz = \frac{1}{2}\pi i, \quad I_2 = \int_C y\,dz = -\frac{1}{2}\pi$$

if C is the semicircle $|z| = 1, 0 \le \arg z \le \pi$ with initial point $z = 1$.

[**Hint:** Semicircle : $x = \cos t, y = \sin t, 0 \le t \le \pi$,

$$I_1 = \int_0^\pi x(dx + i\,dy) = -\int_0^\pi \cos t \sin t\,dt + i\int_0^\pi \cos^2 t\,dt, \text{ etc.}]$$

3. Show that $\int_C |z|^2\,dz = -1 + i$, where C is the square with vertices $O(0, 0)$, $A(1, 0)$, $B(1, 1)$, $C(0, 1)$.

[**Hint:** $I = \int_C (x^2 + y^2)(dx + i\,dy)$. Lines and integrals :

OA :	$x = t$,	$y = 0$,	$0 \le t \le 1$;	$I_{OA} = 1/3$
AB :	$x = 1$,	$y = t$,	$0 \le t \le 1$;	$I_{AB} = 4i/3$
BC :	$x = 1 - t$,	$y = 1$,	$0 \le t \le 1$;	$I_{BC} = -4/3$

$CO: \ x = 0, \quad y = 1 - t, \quad 0 \le t \le 1; \quad I_{CO} = - i/3]$

4. Show that $\int_{C_1} \bar{z}\, dz = \pi i, \int_{C_2} \bar{z}\, dz = -\pi i$ if C_1 is the upper half of the circle $|z| = 1$ with $z = 1$ as initial point and C_2 is the same semicircle with $z = -1$ as initial point.

[**Hint:** $C_1 : x = \cos t, y = \sin t, 0 \le t \le \pi; C_2 = - C_1$]

5. Show that $\int_C \bar{z}^2\, dz = 0$ or $4\pi i$ according as C is $|z| = 1$ or $|z - 1| = 1$.

[**Hint:** (*i*) $x = \cos t, y = \sin t, 0 \le t \le 2\pi$ (*ii*) $z - 1 = e^{it}$ or $x = 1 + \cos t, y = \sin t, 0 \le t \le 2\pi$]

8.3 Simple Integrals Using Definition

6. Show that if, C is the semicircle $|z| = R$, **Im** $z > 0$ and if $\lim_{R \to \infty} \max |z\, f(z)| = 0$, then

$$\lim_{R \to \infty} \int_C f(z)\, dz = 0. \qquad \textbf{(1981)}$$

[**Hint:** $\left| \int_C f(z)\, dz \right| \le \underset{\text{on } C}{\text{Max}} |f(z)|\, \pi R = \underset{\text{on } C}{\text{Max}} |z\, f(z)|\, \pi$.]

7. Show that the value of $\int \frac{1}{z-2}\, dz$ is $2\pi i$ if it is along the circle $|z - 2| = 1$ or along the circle $|z - 2| = 4$.

[**Hint:** Use Th. 8.3]

8.4. Definite Integrals

8. Solved Sum. Find the integral of $f(z) = x^2$ along the segment of the line $y = 2x$ from (0, 0) to (1, 2).

✍ In $y = 2x$, set $x = t$. Then $y = 2t$. So the parametric equations of the given segment, say C, are

$$x = t, \quad y = 2t, \quad 0 \le t \le 1.$$

Then we have

$$dx = dt, \quad dy = 2dt.$$

$$\therefore \quad \int_C f(z)\, dz = \int_C x^2\, (dx + i\, dy)$$

$$= \int_0^1 t^2 (dt + 2i\, dt) = (1 + 2i) \int_0^1 t^2 dt$$

$$= (1 + 2i) \left[\frac{t^3}{3} \right]_0^1 = (1 + 2i) \left[\frac{1}{3} - 0 \right]$$

$$= \frac{1}{3} (1 + 2i).$$

Note: Here the end points may be given also as $z = 0, z = 1 + 2i$.

9. Solved sum. Find the integrals of $f(z) = z^2$.

(*i*) along the parabolic arc $y = x^2$ from (0, 0) to (1, 1), (*ii*) along the line segment $x = 3y$ from (0, 0) to (3, 1).

✍ (*i*) In $y = x^2$, set $x = t$. Then $y = t^2$. So the parametric equations of the given arc, say C, are

$$x = t,\ \ y = t^2,\ \ 0 \le t \le 1.$$

Then we have

$$dx = dt,\ \ dy = 2t\,dt.$$

$$\therefore \quad \int_C f(z)\,dz = \int_C (x^2 - y^2 + i\,2xy)\,(dx + idy)$$

$$= \int_0^1 (t^2 - t^4 + i2t^3)\,(dt + i2t\,dt)$$

$$= \int_0^1 (t^2 - t^4 + i2t^3)\,(1 + i2t)\,dt$$

$$= \int_0^1 [(t^2 - 5t^4) + 2i\,(2t^3 - t^5)]\,dt$$

$$= \left[\frac{t^3}{3} - t^5\right]_0^1 + 2i\left[\frac{t^4}{2} - \frac{t^6}{6}\right]_0^1$$

$$= \left[\frac{1}{3} - 1 - 0\right] + 2i\left[\frac{1}{2} - \frac{1}{6} - 0\right]$$

$$= \frac{-2}{3} + i\frac{2}{3} = \frac{2}{3}(-1 + i).$$

(*ii*) In $x = 3y$, set $y = t$. Then $x = 3t$. So the parametric equations of the given segment, say C, are

$$y = t,\ \ x = 3t,\ \ 0 \le t \le 1.$$

Then we have

$$dy = dt,\ \ dx = 3\,dt.$$

$$\therefore \quad \int_C f(z)\,dz = \int_C (x^2 - y^2 + i\,2xy)\,(dx + idy)$$

$$= \int_0^1 (9t^2 - t^2 + i6t^2)\,(3dt + idt)$$

$$= \int_0^1 (8 + 6i)\,t^2\,(3 + i)\,dt$$

$$= (18 + 26i)\int_0^1 t^2 dt = (18 + 26i)\left[\frac{t^3}{3}\right]_0^1$$

$$= (18 + 26i)\left[\frac{1}{3} - 0\right] = \frac{18 + 26i}{3}.$$

10. Solved sum. Evaluate $\int_C \frac{z+2}{z}\,dz$, where C is the upper half of the circle $|z| = 2$, from right to left.

The parametric equation of C is

$$z = 2e^{it}, \quad 0 \le t \le \pi.$$

So we have

$$dz = 2ie^{it}\,dt.$$

$$\therefore \quad \int_C f(z)\,dz = \int_C \frac{z+2}{z}\,dz = \int_0^\pi \frac{2e^{it}+2}{2e^{it}}\,2ie^{it}\,dt$$

$$= 2i\int_0^\pi (e^{it}+1)\,dt = 2i\left[\frac{e^{it}}{i}+t\right]_0^\pi$$

$$= 2i\left[\frac{e^{i\pi}}{i}+\pi-\frac{1}{i}-0\right]$$

$$= 2i\left[-\frac{1}{i}+\pi-\frac{1}{i}\right] = -4+2\pi i.$$

11. Solved sum. Show that $\int_C \sin z\,dz = 1-\cosh 1,$ where C is the line segment joining $z = 0$ and $z = i$.

$$\int_C \sin z\,dz = \int_C [\sin(x+iy)]\,(dx+idy).$$

Now C is the line segment on the y axis, joining (0, 0) and (0, 1). So, on C, $x = 0$, $dx = 0$ and y varies from 0 to 1.

$$\therefore \quad \int_C \sin z\,dz = \int_0^1 \sin(0+iy)\,(idy)$$

$$= \int_0^1 (i\sinh y)\,(i\,dy) = -[\cosh y]_0^1$$

$$= -[\cosh 1 - \cosh 0]$$

$$= 1-\cosh 1.$$

12. Solved sum. Evaluate the integrals

(*i*) $\int_{C_1} z^2\,dz$ (*ii*) $\int_{C_2} z^2\,dz$, where

C_1 is the straight line joining the points 0 and $1+i$ **(1981)**

C_2 is the circle $x = r\cos\theta$, $y = r\sin\theta$, from $\theta = \alpha$ to $\theta = \beta$. **(1980)**

(*i*) Equation of the line is $y = x$ or, in parameters, $x = t$, $y = t$, $0 \le t \le 1$. If $u + iv = (x+iy)^2$, then $u = x^2 - y^2$, $v = 2xy$. Using (8.2.2),

$$I_1 = \int_0^1 (t^2-t^2)\,dt - 2t^2\,dt + i\int_0^1 (t^2-t^2)\,dt + 2t^2\,dt$$

$$= -\frac{2}{3}+\frac{2}{3}i = -\frac{2(1-i)}{3}.$$

Alternative method. If $F(z) = z^3/3$, then $F'(z) = z^2$. So, by (8.4.1),

$$I_1 = F(1-i) - F(0) = (1+i)^3/3 = -2(1-i)/3.$$

(*ii*) $$I_2 = F(r \text{ cis } \beta) - F(r \text{ cis } \alpha) = r^3 (\text{cis } 3\beta - \text{cis } 3\alpha)/3.$$

13. If C is the curve $y = x^3 - 3x^2 + 4x - 1$ joining the points $1 + i$ and $2 + 3i$, show that

$$\int_C (12z^2 - 4iz) = -156 + 38\,i.$$ **(1975)**

14. Evaluate the integral $\int_C (z+2)\, e^{iz}\, dz$, where C is the parabolic arc $x^2 = \pi^2 y$ from $(0, 0)$ to $(\pi, 1)$. **(1981)**

[**Hint:** Integrating $(z + 2)\, e^{iz}$ by parts, $F(z) = \{(z + 2)\, e^{iz}/i\} + e^{iz}$; $F(\pi + i) - F(0) = (1/e)(-2 + \pi i + 2i) - 1 + 2i$].

15. Show that $\int_0^1 ze^{2z}\, dz = (e^2 + 1)/4.$ **(1979)**

16. Find $\int z^n\, dz$ along the line joining the points $z = 0$ and $z = 2 + i$. Show that the integral is independent of the path. **(1978)**

[**Hint:** The line is a continuous line; Using (8.4.1), the integral is $[(2 + i)^{n+1} - (0)^{n+1}]/(n + 1)$ which is independent of the path].

8.7. Cauchy's Fundamental Theorem

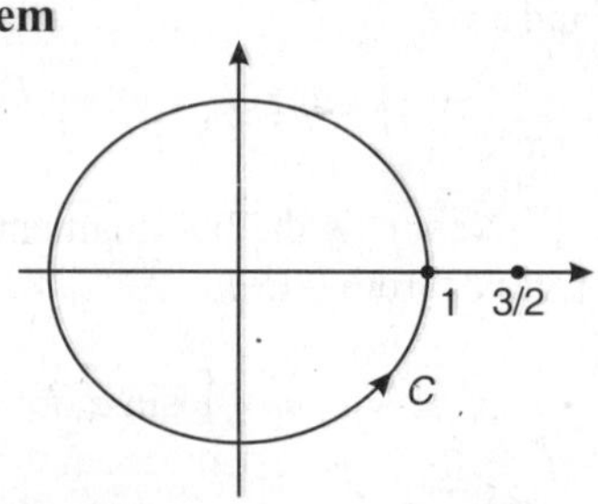

17. Solved sum. Evaluate the following integrals:

(*i*) $\int_C \frac{1}{2z-3}\, dz$; C is $|z| = 1$

(*ii*) $\int_C \frac{1}{2z+3}\, dz$; C is $|z| = 1$

✍ (*i*) $f(z) = \frac{1}{2z-3} = \frac{1}{2} \cdot \frac{1}{z - \frac{3}{2}}$. So $f(\frac{3}{2}) = \infty$.

Hence $\frac{1}{z - \frac{3}{2}}$ is not analytic only at $z = \frac{3}{2}$. But $z = \frac{3}{2}$ lies outside $|z| = 1$. So $f(z)$ is analytic inside and on C. So, by Cauchy's theorem,

$$I = \frac{1}{2} \int_C \frac{1}{z - \frac{3}{2}}\, dz = 0.$$

(*ii*) $f(z) = \frac{1}{2z+3} = \frac{1}{2} \cdot \frac{1}{z + \frac{3}{2}}$. So it is not analytic only at $z = -\frac{3}{2}$ which lies outside C. So $f(z)$ is analytic inside and on C. So, by Cauchy's theorem, the integral is zero.

18. Solved sum. What is the value of

$$\int_C \frac{3z^2 + 7z + 1}{z+1}\, dz$$

where C is $|z| = \frac{1}{2}$?

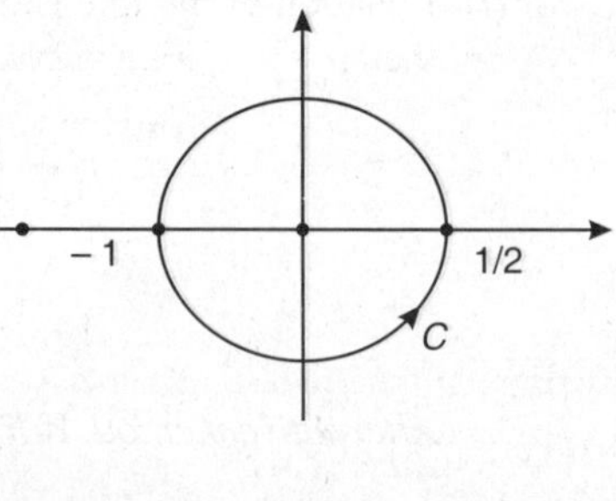

✍ Now $f(z) = \frac{3z^2 + 7z + 1}{z+1}$ is a rational function such that the denominator vanishes at $z = -1$. So $f(-1) = \infty$. Thus $f(z)$ is not analytic at $z = -1$ only. But $z = -1$ lies outside $|z| = \frac{1}{2}$. So $f(z)$ is analytic on and inside C. Hence, by Cauchy's theorem,

$$\int_C \frac{3z^2 + 7z + 1}{z + 1}\, dz = 0.$$

19. State a contour C such that

$$\int \frac{1}{z - 2}\, dz = 0.$$

[**Ans.** $|z| = 1$ which does not contain $z = 2$]

20. Evaluate $\int_C \frac{4z^2 + 4z + 5}{z - 3}\, dz$, where C is the ellipse

$$9x^2 + 4y^2 = 36.$$

[**Hint:** $f(z)$ is not analytic at $z = 3$ i.e., at $(3, 0)$. But $9(3^2) + 4(0^2) - 36 > 0$. So $z = 3$ lies outside the ellipse. Thus the integral is zero]

21. Show that $\int_C (z + 1)\, dz = 0$, where C is the square with vertices $0, 1, 1 + i, i$.

[**Hint:** $f(z)$ is a polynomial. So it is analytic everywhere]

22. If C is the unit circle $|z| = 1$, show that

$$\int_C f(z)\, dz = 0$$

if (*i*) $f(z) = z^2$ (*ii*) $f(z) = ze^z$ (*iii*) $f(z) = ze^{-z}$ (*iv*) $f(z) = e^z \sin z$ (*v*) $f(z) = \frac{1}{z - 2}$, (*vi*) $f(z) = \frac{1}{z^2 - 4}$

(*vii*) $f(z) = \frac{z^2}{z - 3}$ (*viii*) $f(z) = \frac{1}{z^2 + 4}$ (*ix*) $f(z) = \frac{1}{z^2 - 2z + 2}$.

[**Hint:** All $f(z)$'s are analytic inside and on C. So, by Cauchy's fundamental theorem, all the integrals vanish; In (*v*) $f(2) = \infty$; but $z = 2$ lies in C_e; In (*ix*) $f(z) = \infty$ when $z = 1 \pm i$; but $|1 \pm i| > 1$. So the points lie outside C.]

23. Verify Cauchy's theorem for $f(z) = z^2$ along the unit circle. (1978)

[**Hint:** (8.2.3) is $\int_0^{2\pi} (\cos t + i \sin t)^2 (-\sin t + i \cos t)\, dt$].

24. If n is a positive or negative integer and C is the positively oriented circle $|z - a| = \rho$, show that

$$\int_C (z - a)^n\, dz = 0 \quad \text{or} \quad 2\pi i$$

according as $n \neq -1$ or $n = -1$. **(1979)**

[**Hint:** If $n = -1$, $I = 2\pi i$ (theorem 8.3). If $n \neq -1$, use (8.2.3) with $z = a + \rho e^{it}$, $0 \le t \le 2\pi$.]

25. By evaluating $\int e^z\, dz$ along the circle $|z| = 1$, show that

$$\int_0^{2\pi} e^{\cos\theta} \cos(\theta + \sin\theta)\, d\theta = \int_0^{2\pi} e^{\cos\theta} \sin(\theta + \sin\theta)\, d\theta = 0.$$

[**Hint:** By Cauchy's theorem the integral vanishes. Also evaluate it using (8.2.3) and $z = \cos\theta + i\sin\theta, 0 \le \theta \le 2\pi$].

26. Evaluate $\int \frac{2z^2+5}{(z+2)^3(z^2+4)z^2}\,dz$ along the square with vertices at $1+i$, $2+i$, $2+2i$, $1+2i$. **(1981)**

[**Hint:** Integral = 0 since $f(z)$ is not analytic at $z = -2, \pm 2i, 0$ all of which lie outside the square].

Extension to Cauchy's fundamental theorem

27. Solved sum. Evaluate the following integrals:

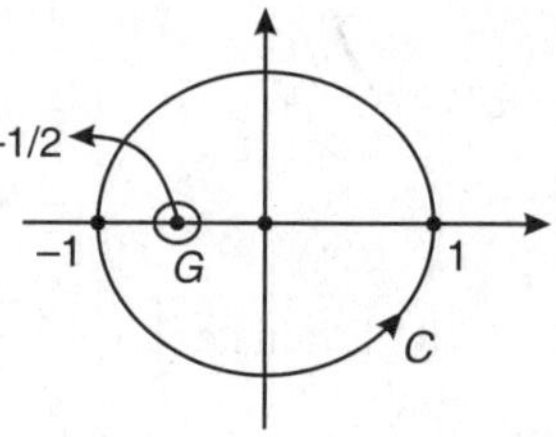

(*i*) $\int_C \frac{1}{2z+1}\,dz$; C is $|z| = 1$

(*ii*) $\int_C \frac{1}{2z+3}\,dz$; C is $|z| = 2$

✍ (*i*) $f(z) = \frac{1}{2z+1} = \frac{1}{2}\frac{1}{z+-\frac{1}{2}}$. So it is not analytic at $z = -1/2$.

But $z = -1/2$ lies inside C. Take a small circle Γ

$$|z - \tfrac{1}{2}| = r$$

inside C. Then, by the extension of Cauchy's theorem, and, by Theorem 5.1. we get

$$\frac{1}{2}\int_C \frac{1}{z+\frac{1}{2}}\,dz = \frac{1}{2}\int_\Gamma \frac{1}{z+\frac{1}{2}}\,dz = \frac{1}{2}\int_\Gamma \frac{1}{z-(-\frac{1}{2})}\,dz = \frac{1}{2}(2\pi i) = \pi i$$

(*ii*) $f(z) = \frac{1}{2z+3} = \frac{1}{2}\cdot\frac{1}{z+3/2}$. So it is not analytic at $-3/2$ which lies inside C. Take a small circle Γ.

$$\left|z - \tfrac{3}{2}\right| = r$$

inside C. Again, by extension of Cauchy's theorem,

$$\frac{1}{2}\int_C \frac{1}{z+\frac{3}{2}}\,dz = \frac{1}{2}\int_\Gamma \frac{1}{z-(-\frac{3}{2})}\,dz = \frac{1}{2}(2\pi i) = \pi i.$$

28. Solved sum. Show that the values of the following integrals are zero, where C is the circle $|z| = 3$;

(*i*) $\int_C \frac{1}{z^2-4}\,dz$ (*ii*) $\int_C \frac{1}{z^2+4}\,dz$.

✍ (*i*) $f(z) = \frac{1}{z^2-4} = \frac{1}{(z-2)(z+2)} = \frac{\frac{1}{4}}{z-2} - \frac{\frac{1}{4}}{z+2}$.

$$\therefore \int_C \frac{1}{z^2-4}\,dz = \frac{1}{4}\int_C \frac{1}{z-2}\,dz - \frac{1}{4}\int_C \frac{1}{z+2}\,dz \qquad (1)$$

Now $\frac{1}{z-2}, \frac{1}{z+2}$ are not analytic respectively at $z = 2, z = -2$ which lie inside the circle C. Let Γ_1, Γ_2 be small circles

$$|z-2| = r_1, |z+2| = r_2$$

inside C. Then, by extension of Cauchy's theorem,

$$\int_C \frac{1}{z-2}\,dz = \int_{\Gamma_1} \frac{1}{z-2}\,dz = 2\pi i$$

$$\int_C \frac{1}{z+2}\,dz = \int_{\Gamma_2} \frac{1}{z+2}\,dz = 2\pi i$$

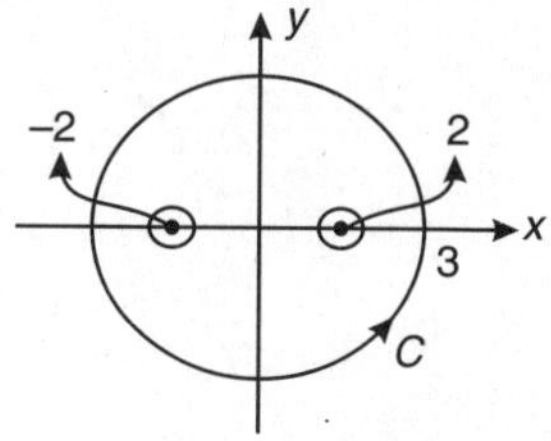

Therefore (1) becomes

$$\int_C \frac{1}{z^2-4}\,dz = \frac{1}{4}(2\pi i) - \frac{1}{4}(2\pi i) = 0\,.$$

(*ii*) $\frac{1}{z^2+4} = \frac{1}{(z-2i)(z+2i)} = \frac{1}{4i}\left(\frac{1}{z-2i} - \frac{1}{z+2i}\right)$ is not analytic at $z = 2i, z = -2i$ which lie on the y axis inside the circle C. Rest is same as in (*i*). The value of the integral is zero.

29. Solved sum. If C is the positively oriented circle $|z-2| = 3$, evaluate

(*i*) $\int_C \frac{1}{z-1}\,dz$ (*ii*) $\int_C \frac{1}{z}\,dz$.

✍ (*i*) At $z = 1$, $1/(z-1)$ is not analytic. Choosing the positively oriented circle $|z-1| = 1$ as C_1 and using theorem 8.6, $\int_C \frac{1}{z-1}\,dz = \int_{C_1} \frac{1}{z-1}\,dz = 2\pi i$ by theorem 8.3.

(*ii*) At $z = 0$, $1/z$ is not analytic. So choosing C_1 as the positively oriented circle $|z| = 1/2$,

$$\int_C \frac{1}{z}\,dz = \int_{C_1} \frac{1}{z}\,dz = 2\pi i\,.$$

30. Evaluate the following integrals if C is a scro curve which does not pass through $z = 1$.

(*i*) $\int_C \frac{1}{z-1}\,dz$ (*ii*) $\int_C \frac{1}{z}\,dz$

[**Hint:** (*i*) Integral = 0 or $2\pi i$ according as $z = 1$ is in C_e or in C_i (*ii*) Integral = 0 or $2\pi i$ according as $z = 0$ is in C_e or in C_i.]

31. Solved sum. Evaluate $\int_C \frac{1}{z^2-1}\,dz$ along C, the positively oriented circle $|z| = 2$.

(1978)

✍ Now $\frac{1}{z^2-1} = \frac{1}{(z-1)(z+1)} = \frac{1}{2}\left(\frac{1}{z-1} - \frac{1}{z+1}\right)$. Choosing two small circles C_1 and C_2 with centres at $z = -1, z = 1$ (Fig. 8.13.1) and using corollary to theorem 8.6,

$$\int_C \frac{1}{z^2-1}\,dz = \frac{1}{2}\int_{C_1}\left(\frac{1}{z-1} - \frac{1}{z+1}\right)dz + \frac{1}{2}\int_{C_2}\left(\frac{1}{z-1} - \frac{1}{z+1}\right)dz$$

$$= \frac{1}{2}(0 - 2\pi i) + \frac{1}{2}(2\pi i - 0) = 0\,.$$

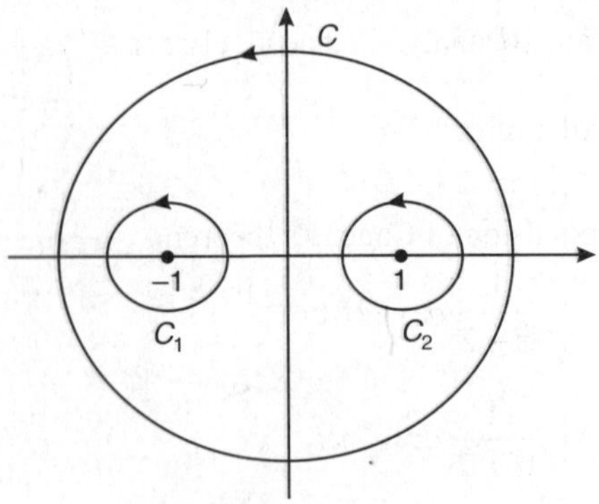

Fig. 8.13.1

32. Show that, if C is the positively oriented circle $|z| = 2$,

(*i*) $\int_C \frac{z}{z^2 - 1}\, dz = 2\pi i$ (*ii*) $\int_C \frac{1}{z^2 + 1}\, dz = 0$

8.9. Cauchy's Integral Formula

33. Solved sum. Evaluate the following contour integrals :

(*i*) $\int_C \frac{e^z}{2z - 1}\, dz$; C is the circle $|z| = 2$

(*ii*) $\int_C \frac{\cos \pi z}{z - 1}\, dz$; C is the circle $|z| = 3/2$

✍ (*i*) $\int_C \frac{e^z}{2z - 1}\, dz = \frac{1}{2}\int_C \frac{e^z}{z - 1/2}\, dz$.

To use the integral formula, we have $f(z) = e^z$ which is analytic on and inside C and $a = 1/2$ is inside C.

$\therefore \int_C \frac{e^z}{2z - 1}\, dz = 1/2\,[2\pi i\, f(1/2)] = \pi i e^{1/2}$.

(*ii*) In $\int_C \frac{\cos \pi z}{z - 1}\, dz$, $f(z) = \cos \pi z$ which is analytic on and inside C and $a = 1$ is inside C. Thus

$$\int_C \frac{\cos \pi z}{z - 1}\, dz = 2\pi i\, f(1) = 2\pi i \cos \pi = -\,2\pi i.$$

34. Solved sum. If C is the circle $|z - 1| = 1$, show that

$$\int_C \frac{3z^2 + z}{z^2 - 1}\, dz = 4\pi i.$$

✍ Now the integrand is

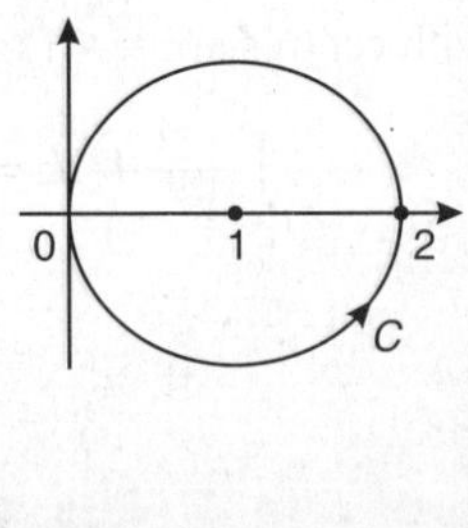

$$\frac{3z^2 + z}{(z + 1)(z - 1)}$$

which is not analytic at

$$z = -1, \quad z = 1$$

of which only $z = 1$ lies inside C. Denoting

$$\frac{3z^2 + z}{z+1} \text{ by } f(z),$$

we see that $f(z)$ is analytic on and inside C with $z = 1$ inside C. Thus, by Cauchy's integral formula,

$$\int_C \frac{3z^2 + z}{z^2 - 1}\, dz = \int_C \frac{3z^2 + z}{z+1} \cdot \frac{1}{z-1}\, dz$$

$$= 2\pi i \left[\frac{3z^2 + z}{z+1}\right]_{z=1} = 2\pi i \left[\frac{4}{2}\right] = 4\pi i\,.$$

35. Solved sum. Using Cauchy's integral formula, evaluate

$$\int_C \frac{z+4}{z^2 + 2z + 5}\, dz,$$

where C is $|z + 1 - i| = 2$.

✍ The points at which the integrand is not analytic are given by

$$z^2 + 2z + 5 = 0 \quad \text{as} \quad z = \frac{-2 \pm \sqrt{4 - 20}}{2}$$

or $$z_1 = -1 + 2i, \quad z_2 = -1 - 2i,$$

The centre of the given circle is $z = -1 + i$. The distances of z_1 and z_2 from this cenre are

$$|(-1 + i) - (-1 + 2i)|, \quad |(-1 + i) - (-1 - 2i)| \quad \text{or} \quad 1, 3.$$

But the radius of the circle is 2. So only z_1 lies inside the circle and z_2 lies outside the circle.

$$\therefore \quad \int_C \frac{z+4}{z^2 + 2z + 5}\, dz = \int_C \frac{z+4}{(z - z_1)(z - z_2)}\, dz$$

$$= \int_C \frac{z+4}{z - z_2} \cdot \frac{1}{z - z_1}\, dz\,.$$

In this, $\dfrac{z+4}{z - z_2}$ is analytic on and inside C and z_1 is inside C. So, by Cauchy's integral formula,

$$\int_C \frac{z+4}{z^2 + 2z + 5}\, dz = 2\pi i \left[\frac{z+4}{z - z_2}\right]_{z = z_1}$$

$$= 2\pi i \frac{z_1 + 4}{z_1 - z_2} = 2\pi i \frac{-1 + 2i + 4}{(-1 + 2i) - (-1 - 2i)}$$

$$= 2\pi i \frac{3 + 2i}{4i} = \frac{\pi(3 + 2i)}{2}\,.$$

36. Solved sum. If C is a simple closed curve enclosing $z = 1$, $z = 2$ and not $z = 3$, show that

(*i*) $\displaystyle\int_C \frac{e^{2e}}{z-1}\, dz = 2\pi i e^2$ (*ii*) $\displaystyle\int_C \frac{e^{2z}}{(z-1)(z-3)}\, dz = -\pi i e^2$

(*iii*) $\int_C \frac{e^{2e}}{(z-1)(z-2)} dz = 2\pi i (e^4 - e^2)$.

(*i*) e^{2z} is analytic everywhere. So it is analytic on and inside C. Now $z = 1$ is a point inside C. Hence, by Cauchy's integral formula,

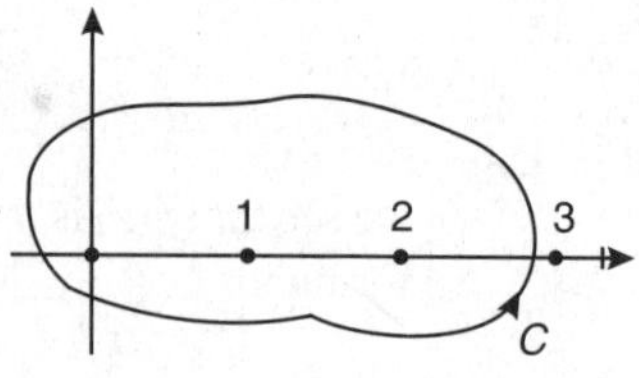

$$\int_C \frac{e^{2z}}{z-1} dz = \int_C e^{2z} \cdot \frac{1}{z-1} dz$$
$$= 2\pi i \, [e^{2z}]_{z=1} = 2\pi i e^2.$$

(*ii*) Now $z = 3$ lies outside C and so

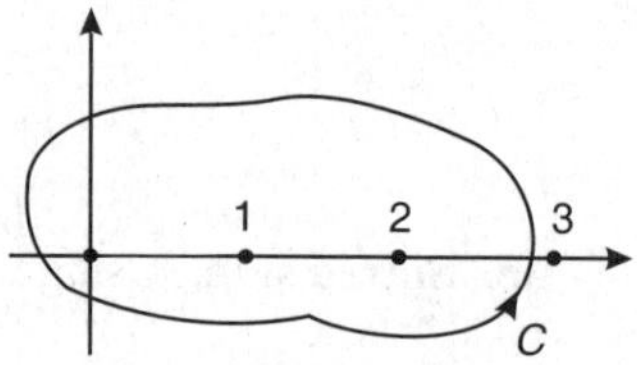

$$\frac{e^{2z}}{z-3}$$

is analytic on and inside C and $z = 1$ is inside C. So, by Cauchy's integral formula,

$$\int_C \frac{e^{2z}}{(z-1)(z-3)} dz = \int_C \frac{e^{2z}}{z-3} \cdot \frac{1}{z-1} dz = 2\pi i \left[\frac{e^{2z}}{z-3}\right]_{z=1} = 2\pi i \left(\frac{e^2}{-2}\right) = -\pi i e^2.$$

(*iii*) In this case we first use extension of Cauchy's theorem. Now the integrand

$$\frac{e^{2z}}{(z-1)(z-2)}$$

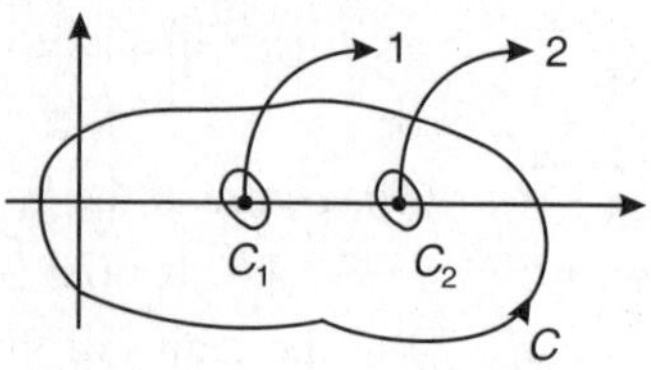

is analytic on and inside C except at $z = 1$ $z = 2$. So, choosing small simple closed curves C_1, C_2 enclosing $z = 1, z = 2$ respectively, by the extension of Cauchy's theorem, we get

$$\int_C \frac{e^{2z}}{(z-1)(z-2)} dz = \int_{C_1} \frac{e^{2z}}{(z-1)(z-2)} dz + \int_{C_2} \frac{e^{2z}}{(z-1)(z-2)} dz.$$

Here $\frac{e^{2z}}{z-2}$ is analytic on and inside C_1, and $z = 1$ is a point inside C_1. Therefore, by Cauchy's integral formula

$$\int_{C_1} \frac{e^{2z}}{(z-1)(z-2)} dz = \int_{C_1} \frac{e^{2z}}{z-2} \cdot \frac{1}{z-1} dz = 2\pi i \left[\frac{e^{2z}}{z-2}\right]_{z=1} = -2\pi i \, e^2.$$

Similarly, for the integral along C_2, we get,

$$\int_{C_2} \frac{e^{2z}}{(z-1)(z-2)} dz = \int_{C_2} \frac{e^{2z}}{z-1} \cdot \frac{1}{z-2} dz = 2\pi i \left[\frac{e^{2z}}{z-1}\right]_{z=2} = 2\pi i \, e^4.$$

$$\therefore \quad \int_C \frac{e^{2z}}{(z-1)(z-2)} dz = -2\pi i e^2 + 2\pi i e^4 = 2\pi i (e^4 - e^2).$$

37. Solved sum. Evaluate $\int_C \frac{4-3z}{z(z-1)(z-2)}\,dz$, where C is the circle $|z| = 3/2$.

✍ Now the integrand is not analytic at

$$z = 0, z = 1, z = 2$$

of which $z = 0$ and $z = 1$ lie inside C. So, by the extension of Cauchy's theorem, choosing small simple closed curves C_1 and C_2 around $z = 0$ and $z = 1$, we get

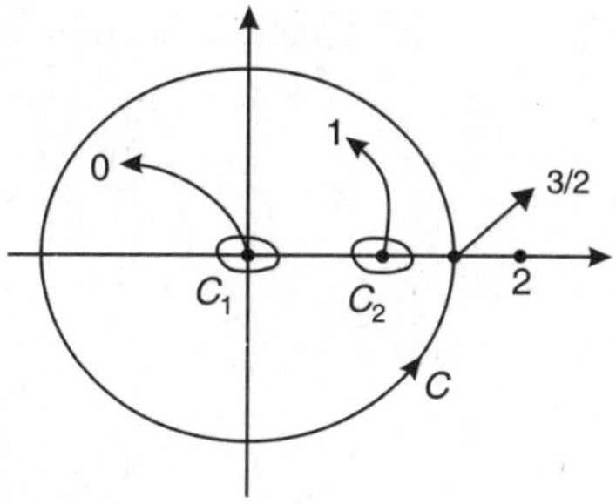

$$\int_C \frac{4-3z}{z(z-1)(z-2)}\,dz =$$

$$\int_{C_1} \frac{4-3z}{z(z-1)(z-2)}\,dz + \int_{C_2} \frac{4-3z}{z(z-1)(z-2)}\,dz.$$

Now, to use Cauchy's integral formula for the integrals on the RHS, we have

(*i*) $\frac{4-3z}{(z-1)(z-2)}$ is analytic on and inside C_1, and $z = 0$ is inside C_1.

(*ii*) $\frac{4-3z}{z(z-2)}$ is analytic on and inside C_2, and $z = 1$ is inside C_2.

$$\therefore \quad \int_{C_1} \frac{4-3z}{z(z-1)(z-2)}\,dz = \int_{C_1} \frac{4-3z}{(z-1)(z-2)} \cdot \frac{1}{(z-0)}\,dz$$

$$= 2\pi i \left[\frac{4-3z}{(z-1)(z-2)}\right]_{z=0}$$

$$= 2\pi i\,[2] = 4\pi i.$$

$$\int_{C_2} \frac{4-3z}{z(z-1)(z-2)}\,dz = \int_{C_2} \frac{4-3z}{z(z-2)} \cdot \frac{1}{z-1}\,dz$$

$$= 2\pi i \left[\frac{4-3z}{z(z-2)}\right]_{z=1}$$

$$= 2\pi i\,[-1] = 2\pi i.$$

$$\therefore \quad \int_C \frac{4-3z}{z(z-1)(z-2)}\,dz = 4\pi i - 2\pi i = 2\pi i.$$

38. Solved sum. Evaluate $\int_C \frac{2z-1}{z^3-z}\,dz$, where C is

(*i*) $|z| = \frac{1}{2}$ (*ii*) $|z| = 2$.

✍ Now the integrand is

$$\frac{2z-1}{z^3-z} = \frac{2z-1}{(z-1)\,z\,(z+1)}$$

which is analytic everywhere, except at

$$z = 1, \quad z = 0, \quad z = -1.$$

(*i*) The integrand is analytic on and inside $C: |z| = \frac{1}{2}$ except at $z = 0$. So by the extension of Cauchy's theorem, choosing a small simple closed curve C_1 around $z = 0$ but lying in C, we get

$$\int_C \frac{2z-1}{z^3 - z}\,dz = \int_{C_1} \frac{2z-1}{z^3 - z}\,dz$$

$$= \int_{C_1} \frac{2z-1}{(z-1)(z+1)} \cdot \frac{1}{(z-0)}\,dz$$

$$= 2\pi i \left[\frac{2z-1}{(z-1)(z+1)}\right]_{z=0} = 2\pi i,$$

by Cauchy's integral formula because $\dfrac{2z-1}{(z-1)(z+1)}$ is analytic on and inside C_1 and $z = 0$ is inside C_1.

(*ii*) The integrand is analytic on and inside

$$C: |z| = 2$$

except at

$$z = 0, \quad 1, \quad -1$$

all of which lie inside C. Choose small simple curves

$$C_1, \quad C_2, \quad C_3$$

enclosing $z = 0$, $z = 1$, $z = -1$ respectively. Then, by the extension of Cauchy's integral theorem,

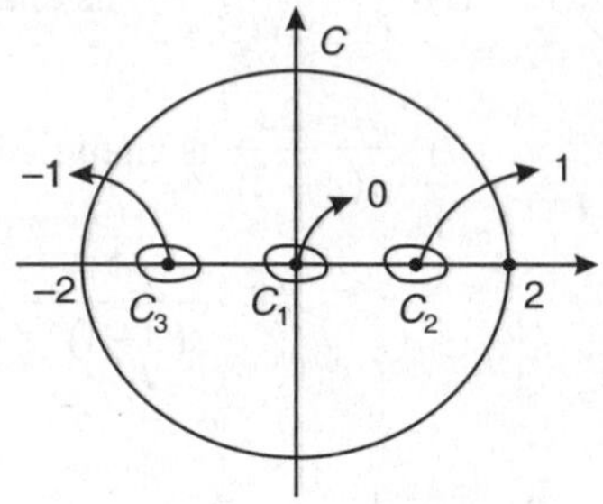

$$\int_C \frac{2z-1}{z^3 - z}\,dz = \int_{C_1} \frac{2z-1}{z^3 - z}\,dz + \int_{C_2} \frac{2z-1}{z^3 - z}\,dz + \int_{C_3} \frac{2z-1}{z^3 - z}\,dz$$

$$= I_1 + I_2 + I_3, \text{ say.}$$

$$I_1 = 2\pi i \text{ as in } (i).$$

Again, using the Cauchy's integral formula,

$$I_2 = \int_{C_2} \frac{2z-1}{z^3 - z}\,dz = \int_{C_2} \frac{2z-1}{z(z+1)} \cdot \frac{1}{z-1}\,dz$$

$$= 2\pi i \left[\frac{2z-1}{z(z+1)}\right]_{z=1} = 2\pi i \left[\frac{1}{2}\right] = \pi i.$$

$$I_3 = \int_{C_3} \frac{2z-1}{z^3 - z}\,dz = \int_{C_3} \frac{2z-1}{z(z-1)} \cdot \frac{1}{z+1}\,dz$$

$$= 2\pi i \left[\frac{2z-1}{z(z-1)}\right]_{z=-1} = 2\pi i \left[-\frac{3}{2}\right] = -3\pi i$$

$$\therefore \quad \int_C \frac{2z-1}{z^3 - z}\,dz = 2\pi i + \pi i - 3\pi i = 0.$$

39. Using Cauchy's integral formula evaluate

$$\int_C \frac{1}{(z^2+1)(z^2-4)}\,dz,$$

where C is $|z| = 3/2$.

[**Hint:** The integrand is analytic on and inside C except at $z = i, -i$. If C_1, C_2 are small simple closed curves around $z = i$, $z = -i$, by Cauchy's integral formula, the value of the integral is

$$2\pi i\left[\frac{1}{(z+i)(z^2-4)}\right]_{z=i} + 2\pi i\left[\frac{1}{(z-i)(z^2-4)}\right]_{z=-i} = 0].$$

40. Evaluate $\int_C \frac{z+4}{z^2+2z+5}\,dz$, where C is the circle

$|z+1+i| = 2.$

[**Hint:** $z^2 + 2z + 5 = 0$ gives $z_1 = -1 + 2i$, $z_2 = -1 - 2i$ of which z_2 lies inside C. The value of the integral is

$$2\pi i\left[\frac{z+4}{z-z_1}\right]_{z=z_2} = 2\pi i\left[\frac{z_2+4}{z_2-z_1}\right] = -\frac{\pi(3-2i)}{2}].$$

41. Evaluate $\int_C \frac{z+1}{z^2+2z+4}\,dz$, where C is the circle

$|z+1+i| = 2.$

[**Hint:** $z^2 + 2z + 4 = 0$ gives $z_1 = -1 + \sqrt{3}\,i$, $z_2 = -1 - \sqrt{3}\,i$ of which z_2 lies inside C. The value of the integral is

$$2\pi i\left[\frac{z+1}{z-z_1}\right]_{z=z_2} = 2\pi i\,\frac{z_2+1}{z_2-z_1} = \pi i].$$

42. Show that $\int_C \frac{\cos \pi z^2}{(z-1)(z-2)}\,dz = 2\pi i$, where C is the circle $|z| = 3/2$.

43. Show that $\int_C \frac{\cos \pi z^2}{(z-1)(z-2)}\,dz = 4\pi i$, where C is the circle $|z| = 3$.

44. Solved sum. If C is a scro curve not passing through $z = a$ and $z = \frac{1}{2}a$, evaluate the integrals

(*i*) $\int_C \frac{z^2}{z-a}\,dz$ (*ii*) $\int_C \frac{e^z}{2z-a}\,dz$.

(*i*) If $z = a$ lies in C_i, then by Cauchy's integral formula

$$a^2 = \frac{1}{2\pi i}\int_C \frac{z^2}{z-a}\,dz \quad \text{or} \quad \int_C \frac{z^2}{z-a}\,dz = 2\pi i\,a^2.$$

If $z = a$ lies in C_e, then $z^2/(z-a)$ is analytic inside and on C and hence, by Cauchy's fundamental theorem, the integral vanishes.

(*ii*) $\int_C \frac{e^z}{2z-a}\,dz = \int_C \frac{\frac{1}{2}e^z}{z-\frac{1}{2}a}\,dz = 2\pi i\,(\frac{1}{2}e^{a/2})$ or 0 according as $z = \frac{1}{2}a$ is in C_i or in C_e.

45. Evaluate $\int_C \frac{e^z}{z-2}\,dz$ if C is (*i*) the positively oriented circle $|z| = 3$ (*ii*) the positively oriented circle $|z| = 1$. **[Ans.** (*i*) $2\pi i\, e^2$ (*ii*) 0]

46. If C is the positively oriented square whose sides lie along the lies $x = \pm 2, y = \pm 2$, evaluate the following integrals : **(1980)**

(*i*) $\int \frac{e^{-z}}{z - \frac{1}{2}\pi i}\,dz$ (*ii*) $\int \frac{\cos z}{z(z^2+8)}\,dz$ (*iii*) $\int \frac{z}{2z+1}\,dz$.

[Hint: (*i*) $I = 2\pi i\, f\left(\frac{1}{2}\pi i\right) = 2\pi$ since $f(z) = e^{-z}$

(*ii*) $I = 2\pi i\, f(0) = \pi i/4$ since $f(z) = (\cos z)/(z^2+8)$

(*iii*) $I = 2\pi i\, f\left(-\frac{1}{2}\right) = -\frac{1}{2}\pi i$ since $f(z) = \frac{1}{2}z$].

47. Solved sum. If C is the positively oriented circle $|z - i| = 2$, show that

$$I = \int_C \frac{e^z}{z^2+4}\,dz = \frac{\pi}{2}e^{2i}$$

✍ Now $\frac{1}{z^2+4} = \frac{1}{(z-2i)(z+2i)} = \frac{1}{4i}\left[\frac{1}{z-2i} - \frac{1}{z+2i}\right]$ and $2i$ lies in C_i and $-2i$ does not (Fig. 8.13.2). Therefore

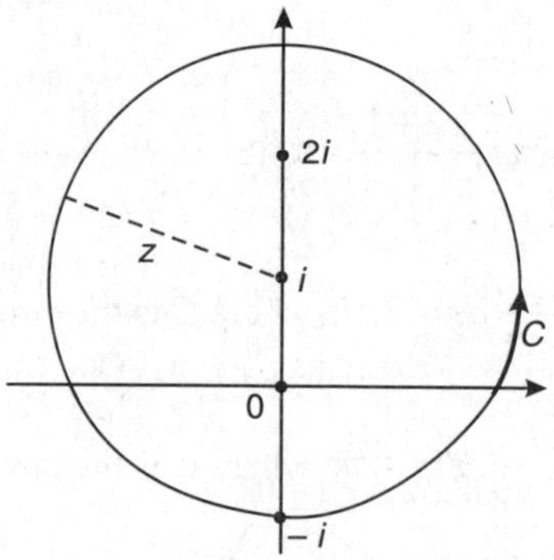

Fig. 8.13.2

$$\int_C \frac{e^z}{z-2i} = 2\pi i\, e^{2i}, \quad \int_C \frac{e^z}{z+2i}\,dz = 0\,.$$

$$\therefore \quad I = \frac{1}{4i}\left\{2\pi i e^{2i} - 0\right\} = \frac{\pi}{2}e^{2i}$$

48. Show that $\frac{1}{2\pi i}\int_C \frac{e^{kz}}{z^2+1}\,dz = \sin k$ if k is real and C is the positively oriented circle $|z| = 2$.

[Hint: $\int_C \frac{e^{kz}}{z^2+1}\,dz = \frac{1}{2i}\int_C \left(\frac{e^{kz}}{z-i} - \frac{e^{kz}}{z+i}\right)dz$]

49. Solved sum. If C is the positively oriented circle $|z| = 3$, show that

$$\int_C \frac{\sin \pi z^2 + \cos \pi z^2}{(z-1)(z-2)}\, dz = 4\pi i .$$

The integrand is analytic everywhere except at $z = 1, z = 2$ both of which lie inside C. So, if C_1, C_2 are two small simple closed curves around $z = 1, z = 2$ respectively, then by the extension of Cauchy's theorem,

$$\int_C \frac{\sin \pi z^2 + \cos \pi z^2}{(z-1)(z-2)}\, dz = \int_{C_1} \frac{\sin \pi z^2 + \cos \pi z^2}{(z-1)(z-2)}\, dz + \int_{C_2} \frac{\sin \pi z^2 + \cos \pi z^2}{(z-1)(z-2)}\, dz$$

$$= I_1 + I_2, \text{ say}$$

Now $\dfrac{\sin \pi z^2 + \cos \pi z^2}{z-2}, \dfrac{\sin \pi z^2 + \cos \pi z^2}{z-1}$ are analytic on and inside C_1, C_2 respectively.

Thus, by Cauchy's integral formula,

$$I_1 = 2\pi i \left[\frac{\sin \pi z^2 + \cos \pi z^2}{z-2}\right]_{z=1}$$

$$= 2\pi i \frac{\sin \pi + \cos \pi}{-1} = 2\pi i,$$

$$I_2 = 2\pi i \left[\frac{\sin \pi z^2 + \cos \pi z^2}{z-1}\right]_{z=2}$$

$$= 2\pi i \left[\frac{\sin 4\pi + \cos 4\pi}{2-1}\right] = 2\pi i .$$

$$\therefore \int_C \frac{\sin \pi z^2 + \cos \pi z^2}{(z-1)(z-2)}\, dz = 2\pi i + 2\pi i = 4\pi i .$$

50. If $f(a) = \displaystyle\int_C \frac{3z^2 + 7z + 1}{z-a}\, dz$, where C is the positively oriented circle $|z| = 2$, find the values of $f(3), f'(1-i), f''(1-i)$. **(1977)**

[**Hint:** Comparing $f(a) = \dfrac{1}{2\pi i}\displaystyle\int_C \frac{2\pi i\,(3z^2 + 7z + 1)}{z-a}\, dz$

with $\qquad f(a) = \dfrac{1}{2\pi i}\displaystyle\int_C \frac{f(z)}{z-a} dz,\ f(z) = 2\pi i\,(3z^2 + 7z + 1)$,

$f'(z) = 2\pi i\,(6z + 7), f''(z) = 2\pi i\,(6)$;

(*i*) Now $z = 3 + 0i$ lies outside C as $|z| = \sqrt{9+0} = 3 > 2$

(*ii*) Now $z = 1 - i$ lies inside C as $|z| = |1 - i| = \sqrt{2} < 2.$ Therefore

(*a*) $f'(1-i) = \pi\,(12 + 26\,i)$, (*b*) $f''(1-i) = 12\pi i$].

51. If $f(a) = \displaystyle\int_C \frac{4z^2 + z + 5}{z-a}\, dz$, where C is the ellipse $(x^2/4) + (y^2/9) = 1$, find the values of $f(i), f'(-1)$. **(1980)**

[**Hint:** If $z = a$ lies in C_i, $f(z) = 2\pi i\,(4z^2 + z + 5)$. When $x = 0, y = 1$, we get $\frac{x^2}{4} + \frac{y^2}{9} - 1$ is – ve. So $z = i$ lies in C_i. So also $z = -1$ lies in C_i.

$$f(i) = 2\pi\,(-1 + i), f'(-1) = -14\pi i]$$

52. If C is a scro curve, using the result that $z \cot z$ is analytic at $z = 0$ if $z \cot z$ is defined to be 1 at $z = 0$, show that

$$f(a) = \frac{1}{2\pi i}\int_C \cot(z-a)\, f(z)\, dz.$$

[**Hint:** In $F(a) = \frac{1}{2\pi i}\int_C \frac{F(z)}{z-a}\,dz$, substitute $F(z) = (z-a)\cot(z-a) f(z)$]

Cauchy's Formula For Derivatives

53. Solved sum. Evaluate $\int_C \frac{1}{(z-a)^m}\,dz$, where C is any simple closed curve which encircles $z = a$.

✍ If $f(z)$ is analytic on and inside C and if $z = a$ is in C_i, then the integral formula for the $(m-1)^{\text{th}}$ derivative is

$$\int_C \frac{f(z)}{(z-a)^m}\,dz = \frac{2\pi i}{(m-1)!} f^{m-1}(a).$$

Now $f(z) = 1$. So its $(m-1)^{\text{th}}$ derivative is zero. So the value of the given integral is 0.

54. Solved sum. If C is the circle $x^2 + y^2 = 4$, find the value of

$$\int_C \frac{e^{-z}}{z^2}\,dz.$$

✍ The function e^{-z} is analytic everywhere (except at ∞). In particular, it is analytic on and inside C and $z = 0$ is a point inside C. Thus, by the integral formula for the first derivative,

$$\int_C \frac{e^{-z}}{z^2}\,dz = \int_C e^{-z}\frac{1}{(z-0)^2}\,dz = 2\pi i\left[\frac{de^{-z}}{dz}\right]_{z=0}$$

$$= 2\pi i\,[-e^{-z}]_{z=0} = -2\pi i.$$

55. Solved sum. Evaluate $\int_C \frac{z+1}{z^3 - 2z^2}\,dz$, where C is the circle $|z| = 1$.

✍ Rewriting the integrand, we have

$$\frac{z+1}{z^3 - 2z^2} = \frac{z+1}{(z-2)\,z^2}.$$

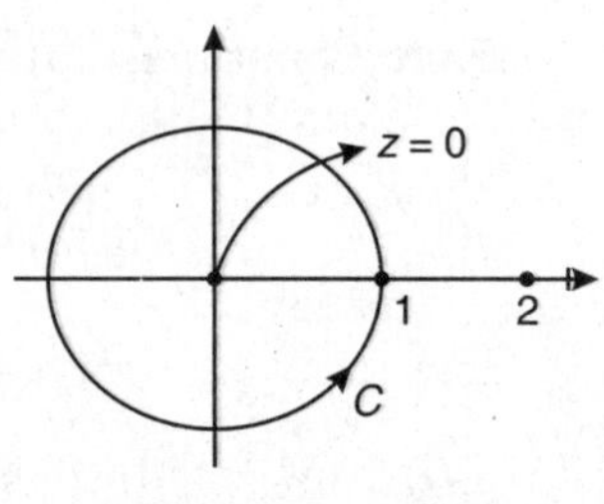

Now we shall denote $\frac{z+1}{z-2}$ by $f(z)$. Then $f(z)$ is analytic everywhere except at $z = 2$ which lies outside C. So, $f(z)$ is analytic on and inside C. But $z = 0$ is a point inside C. So by the integral formula for the first derivative, we get

$$\int_C \frac{z+1}{z-2}\cdot\frac{1}{(z-0)^2}\,dz = 2\pi i\left[\frac{d}{dz}\frac{z+1}{z-2}\right]_{z=0}$$

$$= 2\pi i \left[\frac{(z-2)-(z+1)}{(z-2)^2} \right]_{z=0}$$

$$= 2\pi i \frac{-3}{4} = -\frac{3\pi i}{2}.$$

56. Solved sum. Using Cauchy's integral formula, evaluate

$$\int_C \frac{z+1}{(z^2+2z+4)^2}\, dz,$$

where C is the circle $|z+1+i| = 2$.

✍ The points at which the integrand is not analytic are given by $z^2 + 2z + 4 = 0$ as

$$z_1 = -1+\sqrt{3}\, i, \quad z_2 = -1-\sqrt{3}\, i$$

of which z_2 only lies inside C because

$$|-(1+i) - z_2| = |-1-i+1+\sqrt{3}\, i| = \sqrt{3} - 1 < 2.$$

Thus the integral can be evaluated by using the integral formula for the first derivative as

$$\int_C \frac{z+1}{[(z-z_1)(z-z_2)]^2}\, dz = \int_C \frac{z+1}{(z-z_1)^2} \cdot \frac{1}{(z-z_2)^2}\, dz$$

$$= \frac{2\pi i}{1!} \left[\frac{d}{dz} \frac{z+1}{(z-z_1)^2} \right]_{z=z_2}$$

$$= 2\pi i \left[\frac{(z-z_1)^2 - (z+1)\,2(z-z_1)}{(z-z_1)^4} \right]_{z=z_2}$$

$$= 2\pi i \left[\frac{(z-z_1)(-z-z_1-2)}{(z-z_1)^4} \right]_{z=z_2}$$

$$= 2\pi i \left[\frac{-z_2-z_1-2}{(z_2-z_1)^3} \right] = 2\pi i \frac{0}{(z_2-z_1)^3} = 0.$$

57. Solved sum. If C is the circle $|z| = 3$, evaluate

$$\int_C \frac{e^z}{(z+1)^2\,(z+2)}\, dz.$$

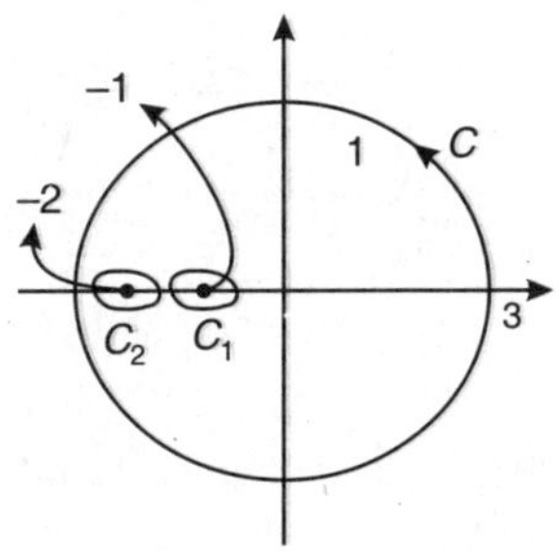

✍ The integrand is analytic everywhere except at

$$z = -1, \quad z = -2$$

both of which lie inside C. So, by the extension of Cauchy's theorem, choosing two small simple closed curves C_1, C_2 encircling $z = -1$, $z = -2$, we get

$$\int_C \frac{e^z}{(z+1)^2\,(z+2)}\, dz = \int_{C_1} \frac{e^z}{(z+1)^2\,(z+2)}\, dz + \int_{C_2} \frac{e^z}{(z+1)^2\,(z+2)}\, dz$$

$$= \int_{C_1} \frac{e^z}{z+2} \frac{1}{(z+1)^2}\, dz + \int_{C_2} \frac{e^z}{(z+1)^2} \frac{1}{z+2}\, dz$$

On the RHS, for the first integral, using the integral formula for the first derivative and, for the second integral, using the integral formula, we get

$$\int_C \frac{e^z}{(z+1)^2(z+2)}\,dz = 2\pi i\left[\frac{d}{dz}\frac{e^z}{z+2}\right]_{z=-1} + 2\pi i\left[\frac{e^z}{(z+1)^2}\right]_{z=-2}$$

$$= 2\pi i\left[e^z\frac{z+1}{(z+2)^2}\right]_{z=-1} + 2\pi i\,\frac{e^{-2}}{1}$$

$$= 0 + 2\pi i\,e^{-2} = 2\pi i e^{-2}.$$

58. Solved sum. Evaluate $\int_C \frac{z^2}{(z-1)^2(z^2-1)}\,dz$, if C is $|z-2| = 2$, using Cauchy's integral formula for a derivative.

✍ Rewriting the integrand, we have

$$\frac{z^2}{(z-1)^3(z+1)}$$

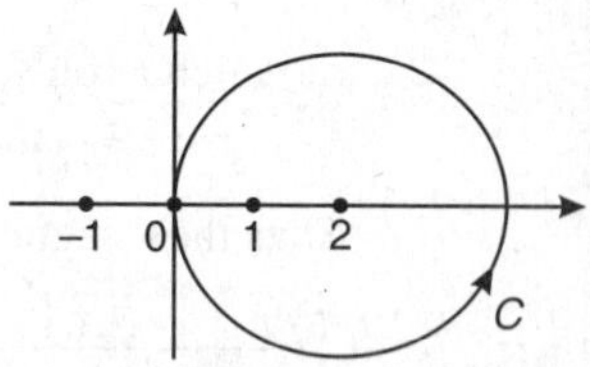

It is evident from the figure that

$$\frac{z^2}{z+1}$$

is analytic on and inside C. But the integrand is not analytic at $z = 1$ which lies inside C. So, by the integral formula for the second derivative, we get

$$\int_C \frac{z^2}{z+1}\frac{1}{(z-1)^3}\,dz = \frac{2\pi i}{2!}\left[\frac{d^2}{dz^2}\frac{z^2}{z+1}\right]_{z=1}$$

$$= \pi i\left[\frac{d}{dz}\frac{(z+1)\,2z - z^2}{(z+1)^2}\right]_{z=1}$$

$$= \pi i\left[\frac{d}{dz}\frac{z^2+2z}{(z+1)^2}\right]_{z=1}$$

$$= \pi i\left[\frac{(z+1)^2(2z+2) - (z^2+2z)\,2(z+1)}{(z+1)^4}\right]_{z=1}$$

$$= \pi i\left[\frac{(z+1)(2z+2) - 2(z^2+2z)}{(z+1)^3}\right]_{z=1}$$

$$= \pi i\left[\frac{8-6}{8}\right] = \frac{\pi i}{4}.$$

59. Solved sum. Evaluate $\int_C \frac{e^z}{z(1-z)^3}\,dz$ in the following cases :

(*i*) When 0 lies in C_i and 1 lies in C_e

(*ii*) When 1 lies in C_i and 0 lies in C_e

(*iii*) When 0 and 1 lie in C_i.

✍ (*i*) By the integral formula, we get

$$\int_C \frac{e^z}{z(1-z)^3}\,dz = \int_C \frac{e^z}{(1-z)^3}\cdot\frac{1}{z-0}\,dz$$

$$= 2\pi i\left[\frac{e^z}{(1-z)^3}\right]_{z=0} = 2\pi i$$

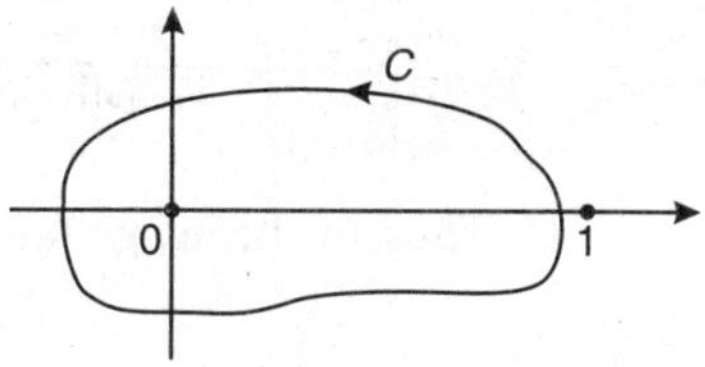

(*ii*) By the integral formula for the second derivative, we get

$$\int_C \frac{e^z}{z}\cdot\frac{1}{(1-z)^3}\,dz = -\int_C \frac{e^z}{z}\cdot\frac{1}{(z-1)^3}\,dz$$

$$= -\frac{2\pi i}{2!}\left[\frac{d^2}{dz^2}\frac{e^z}{z}\right]_{z=1}$$

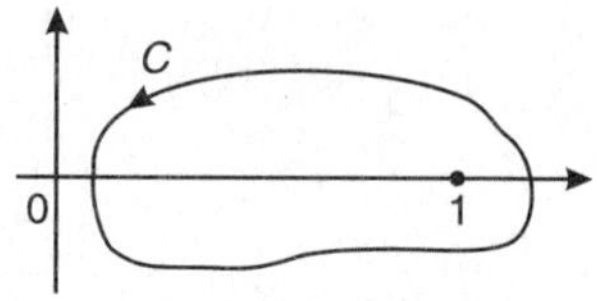

$$= -\pi i\,[e^z(-1)(-2)z^{-3} + e^z\,2(-1)z^{-2} + e^z z^{-1}]_{z=1}$$

by Leibnitz's theorem

$$= -\pi i\,[2e - 2e + e] = -\pi i e.$$

(*iii*) The function $\dfrac{e^z}{z(1-z)^3}$ is analytic everywhere except at $z = 0$ and $z = 1$ which lie inside C. So, by the extension of Cauchy's theorem, choosing two small simple closed curves C_1, C_2 encircling $z = 0$, $z = 1$ respectively, we get

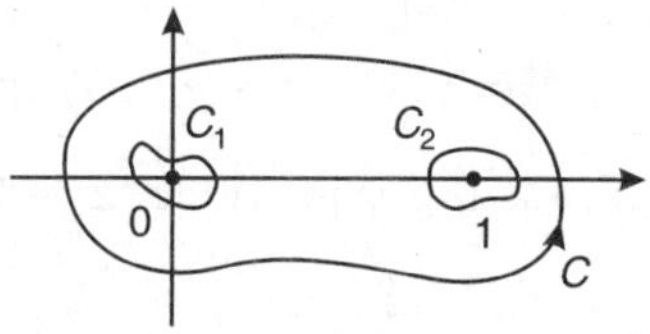

$$\int_C \frac{e^z}{z(1-z)^3}\,dz = \int_{C_1} \frac{e^z}{z(1-z)^3}\,dz + \int_{C_2} \frac{e^z}{z(1-z)^3}\,dz$$

$$= 2\pi i - \pi i e \quad \text{by } (i), (ii)$$

$$= \pi i(2-e).$$

60. Solved sum. If C is the circle $x^2 + y^2 = 4$, evaluate

$$I = \int_C \frac{12z-7}{(z-1)^2(2z+3)}\,dz.$$

✍ The integrand is not analytic at

$$z = 1, z = -3/2$$

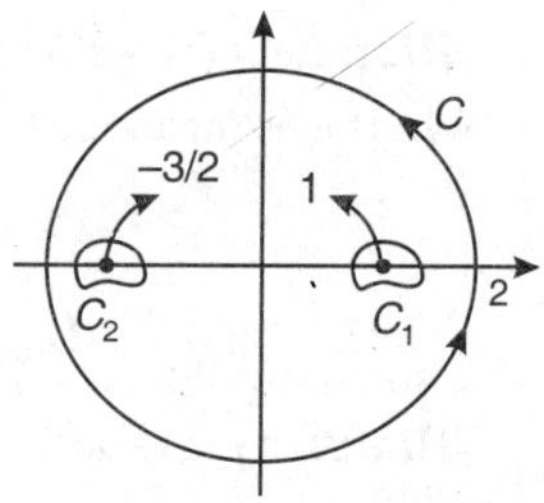

both of which lie inside C. So, by the extension of Cauchy's theorem, choosing two small simple closed curves C_1, C_2 encircling $z = 1$ and $z = -3/2$ respectively, we get

$$I = \int_{C_1} \frac{12z-7}{(z-1)^2(2z+3)}\,dz + \int_{C_2} \frac{12z-7}{(z-1)^2(2z+3)}\,dz.$$

Now we have

(*i*) $\dfrac{12z-7}{2z+3}$ is analytic on and inside C_1 and $z = 1$ lies inside C_1

(ii) $\dfrac{12z-7}{(z-1)^2}$ is analytic on and inside C_2 and $z = -\frac{3}{2}$ lies inside C_2.

Thus, by the integral formula for the first derivative and by the integral formula, we have

$$I = 2\pi i\left[\frac{d}{dz}\frac{12z-7}{2z+3}\right]_{z=1} + 2\pi i\left[\frac{12z-7}{(z-1)^2}\right]_{z=-3/2}$$

$$= 2\pi i\left[\frac{(2z+3)\,12-(12z-7)\,2}{(2z+3)^2}\right]_{z=1} + 2\pi i\left[\frac{12(-\frac{3}{2})-7}{(-\frac{5}{2})^2}\right]$$

$$= 2\pi i\,[2] + 2\pi i\,[-4] = -4\pi i.$$

61. If C is the circle $|z| = 3$, show that

(i) $\displaystyle\int_C \frac{e^z}{(z-2)^2}\,dz = 2\pi i e^2$ (ii) $\displaystyle\int_C \frac{e^z}{(z-2)^3}\,dz = \pi i e^2$

(iii) $\displaystyle\int_C \frac{e^z}{(z-2)^4}\,dz = \frac{\pi i e^2}{3}$

[**Hint:** Let $f(z) = e^z$ (i) $2\pi i f'(2)$ (ii) $\dfrac{2\pi i}{2!}f''(2)$ (iii) $\dfrac{2\pi i}{3!}f'''(2)$]

62. If C is the circle $|z| = 1$, show that

$$\int_C \frac{e^{iz}}{z^3}\,dz = -\pi i.$$

[**Hint:** Let $f(z) = e^{iz}$. Then $I = \dfrac{2\pi i}{2!}f''(0)$].

63. If C is the circle $|z| = 2$, show that

$$\int_C \frac{z^3-2z+1}{(z-i)^2}\,dz = -10\pi i$$

[**Hint:** Let $f(z) = z^3 - 2z + 1$; $I = 2\pi i f'(i)$].

64. If C is the circle $|z| = 2$, show that

$$\int_C \frac{e^{2z}}{(z+1)^4}\,dz = \frac{8\pi i}{3e^2}.$$

[**Hint:** Let $f(z) = e^{2z}$, $I = \dfrac{2\pi i}{3!}f'''(-1)$].

65. If C is the circle $|z| = 1$ show that

$$\int_C \frac{\sin^2 z}{(z-\frac{\pi}{6})^3}\,dz = \pi i.$$

[**Hint:** Let $f(z) = \sin^2 z, f'(z) = \sin 2z$. $I = \dfrac{2\pi i}{2!}f''(\frac{\pi}{6})$].

66. If C is the circle $|z| = 1$, show that

$$\int_C \frac{\sin^6 z}{(z - \pi/6)^3}\, dz = \frac{21\pi i}{16}.$$

[**Hint:** Let $f(z) = s^6$; $f''(z) = 6[5s^4 c^2 - s^6]$. $I = \frac{2\pi i}{2!} f''(\pi/6)$].

67. Using Cauchy's integral formula, evaluate

$$\int_C \frac{2z}{(z-1)(z-3)^2}\, dz,$$

where C is the circle $|z - 4| = 2$.

[**Hint:** Let $z = 3$ lies in C_i and $z = 1$ lies in C_e. $\frac{2z}{z-1}$ is analytic on and inside C. Therefore

$$I = 2\pi i \left[\frac{d}{dz}\frac{2z}{z-1}\right]_{z=3} = -\pi i].$$

68. Show that, if C is the circle $|z - i| = 2$,

$$\int_C \frac{z-1}{(z+1)^2 (z-2)}\, dz = -\frac{2\pi i}{9}.$$

[**Hint:** Let $z = -1$ lies in C_i and $z = 2$ lies in C_e. $\frac{z-1}{z-2}$ is analytic on and inside C. Therefore

$$I = 2\pi i \left[\frac{d}{dz}\frac{z-1}{z-2}\right]_{z=-1}].$$

69. If C is the positively oriented circle $|z| = 2$, show that

$$\int_C \frac{e^z}{(z-1)^2}\, dz = 2\pi ie, \int_C \frac{e^z}{(z-1)^3}\, dz = \pi ie,$$

$$\int_C \frac{e^z}{(z-1)^4}\, dz = \frac{\pi ie}{3}.$$

[**Hint:** Set $f(z) = e^z$ in $f'(1) = \frac{1!}{2\pi i}\int_C \frac{f(z)}{(z-1)^2}\, dz$,

$$f''(1) = \frac{2!}{2\pi i}\int_C \frac{f(z)}{(z-1)^3}\, dz, \quad f'''(1) = \frac{3!}{2\pi i}\int_C \frac{f(z)}{(z-1)^4}\, dz].$$

70. Show that the integral of e^{iz}/z^3 along the positively oriented circle $|z| = 2$ is $-\pi i$. **(1981)**

[**Hint:** In $f''(0) = \frac{2!}{2\pi i}\int_C \frac{f(z)}{(z-0)^3}\, dz$, set $f(z) = e^{iz}$].

71. Show that the integral of $1/(z^2 + 4)^2$ along the positively oriented circle $|z - i| = 2$ is $\pi/16$.

METHOD 1

[**Hint:** $\dfrac{1}{(z^2+4)^2} = \dfrac{-i}{32}\dfrac{1}{z-2i} - \dfrac{1}{16}\dfrac{1}{(z-2i)^2} + \dfrac{i}{32}\dfrac{1}{z+2i} - \dfrac{1}{16}\dfrac{1}{(z+2i)^2}$

The integral is $(-i/32)\,2\pi i + 0 + 0 + 0 = \pi/16$].

METHOD 2

✍ Rewriting the integrand,

$$\frac{1}{(z^2+4)^2} = \frac{1}{[(z+2i)(z-2i)]^2} = \frac{1}{(z+2i)^2(z-2i)^2}$$

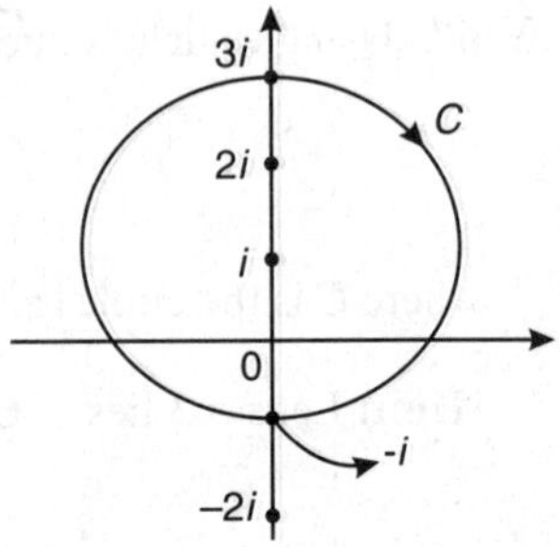

Now we shall denote $\dfrac{1}{(z+2i)^2}$ by $f(z)$. Then $f(z)$ is analytic everywhere except at $z = -2i$ which lies outside C. So $f(z)$ is analytic on and inside C. But $z = 2i$ is a point inside C. Thus, by the integral formula for the first derivative, we get

$$\int_C \frac{1}{(z^2+4)^2}\,dz = \int_C \frac{1}{(z+2i)^2}\frac{1}{(z-2i)^2}\,dz$$

$$= 2\pi i\left[\frac{d}{dz}\frac{1}{(z+2i)^2}\right]_{z=2i}$$

$$= 2\pi i\left[-\frac{2}{(z+2i)^3}\right]_{z=2i} = -4\pi i\,\frac{1}{64\,i^3} = \frac{\pi}{16}.$$

72. If C is the positively oriented circle $|z| = 1$ and k is a real constant, prove that

$$\int_C \frac{e^{kz}}{z}\,dz = 2\pi i.$$

Deduce that $\displaystyle\int_0^{\pi} e^{k\cos\theta}\cos(k\sin\theta)\,d\theta = \pi$.

[**Hint:** By Cauchy's integral formula

$$e^{k.0} = \frac{1}{2\pi i}\int_C \frac{e^{kz}}{z-0}\,dz \quad\text{or}\quad \textbf{Im}\int_C \frac{e^{kz}}{z}\,dz = 2\pi$$

Also, setting $z = e^{i\theta} = \cos\theta + i\sin\theta$, $0 \le \theta \le 2\pi$, in (8.2.3),

$$\int_C \frac{e^{kz}}{z}\,dz = \int_0^{2\pi} e^{k(\cos\theta + i\sin\theta)}\,i\,d\theta$$

$$= \int_0^{2\pi} e^{k\cos\theta}\,[\cos(k\sin\theta) + i\sin(k\sin\theta)]\,i\,d\theta$$

Considering only the imaginary part,

$$\int_0^{2\pi} e^{k\cos\theta}\cos(k\sin\theta)\,d\theta = \int_0^{\pi}(\ldots)d\theta + \int_{\pi}^{2\pi}(\ldots)d\theta = 2\pi .$$

On the right, in the latter, set $\theta = 2\pi - \phi$].

73. Prove that $\int_0^{2\pi} \cos(\cos\theta)\cosh(\sin\theta)\,d\theta = 2\pi$.

[**Hint:** Consider the integral of $(\cos z)/z$ along $|z| = 1$].

74. Solved sum. If C is a scro curve enclosing the origin, show that

$$\left(\frac{a^n}{n!}\right)^2 = \frac{1}{2\pi i}\int_C \frac{a^n e^{az}}{n!\,z^{n+1}}\,dz$$

and deduce that

$$\sum_{n=0}^{\infty}\left(\frac{a^n}{n!}\right)^2 = \frac{1}{2\pi}\int_0^{2\pi} e^{2a\cos\theta}\,d\theta \qquad \textbf{(1974)}$$

✍ Setting $f(z) = e^{az}$ in $f^n(0) = \dfrac{n!}{2\pi i}\displaystyle\int_C \frac{f(z)}{(z-0)^{n+1}}\,dz$,

$$a^n = \frac{n!}{2\pi i}\int_C \frac{e^{az}}{z^{n+1}}\,dz$$

Multiplying both sides by the constant $a^n/(n!)^2$.

$$\left(\frac{a^n}{n!}\right)^2 = \frac{1}{2\pi i}\int_C \frac{a^n e^{az}}{n!\,z^{n+1}}\,dz$$

$$\therefore \quad \sum_{n=0}^{\infty}\left(\frac{a^n}{n!}\right)^2 = \frac{1}{2\pi i}\sum_{n=0}^{\infty}\int_C \frac{a^n e^{az}}{n!\,z^{n+1}}\,dz$$

$$= \frac{1}{2\pi i}\int_C \sum_{n=0}^{\infty} \frac{a^n e^{az}}{n!\,z^{n+1}}\,dz$$

$$= \frac{1}{2\pi i}\int_C \frac{e^{az}}{z}\sum_{n=0}^{\infty}\frac{(a/z)^n}{n!}\,dz = \frac{1}{2\pi i}\int_C \frac{e^{az}}{z}\,e^{a/z}\,dz$$

$$= \frac{1}{2\pi i}\int_C \frac{e^{a(z+1/z)}}{z}\,dz = \frac{1}{2\pi i}\int_{C_1} \frac{e^{a(z+1/z)}}{z}\,dz,$$

where C_1 is the circle $z = e^{i\theta}$, $0 \le \theta \le 2\pi$.

$$= \frac{1}{2\pi i}\int_0^{2\pi} e^{2a\cos\theta}\,i\,d\theta \quad \left(\because\ \frac{1}{z}\frac{dz}{d\theta} = i\right)$$

$$= \frac{1}{2\pi}\int_0^{2\pi} e^{2a\cos\theta}\,d\theta .$$

8.11. Related Integral Theorems

75. Solved sum. Show that, if $f(z)$ is an entire function, and if

$$M(f; r, 0) \le Mr^{\alpha},$$

for fixed positive values of M and α and for a sequence of values of r tending to infinity, then $f(z)$ is a polynomial in z of degree not exceeding α. (Extension to Liouville's theorem). **(1985)**

Now, for any finite z, $f(z)$ can be expanded as a power series in z as

$$\sum_{n=0}^{\infty} a_n z^n .$$

Hence, by Cauchy's inequality, we get, for every finite r,

$$|a_n| \le \frac{M(f; r, 0)}{r^n} \quad \text{or} \quad |a_n| \le \frac{(M)(r^{\alpha})}{r^n}$$

i.e.,
$$|a_n| \le \frac{M}{r^{n-\alpha}} .$$

But $M/r^{n-\alpha}$ tends to zero as $r \to \infty$, if $n > \alpha$. So

$$a_n = 0 \text{ for all } n > \alpha.$$

$$\therefore \quad f(z) = a_0 + a_1 z + a_2 z^2 + \dots + a_{\alpha} z^{\alpha}$$

if α is a positive integer. If α is a positive fraction, then $f(z)$ is a polynomial whose degree is the integral part of α.

CHAPTER 9

TAYLOR'S AND LAURENT'S SERIES

9.1. TAYLOR'S SERIES

In this section, given an analytic function, we find a power series whose sum is the analytic function. Taylor's series is a series of positive powers while Laurent's series is a series of both positive and negative powers.

Given a function $f(z)$ and a point $z = a$, if C is the largest circle with centre at $z = a$ such that $f(z)$ is analytic in C_i, then $f(z)$ can be expanded as a series of non-negative integral powers which is convergent for all z in C_i. This series is called **Taylor's series** for $f(z)$ about $z = a$.

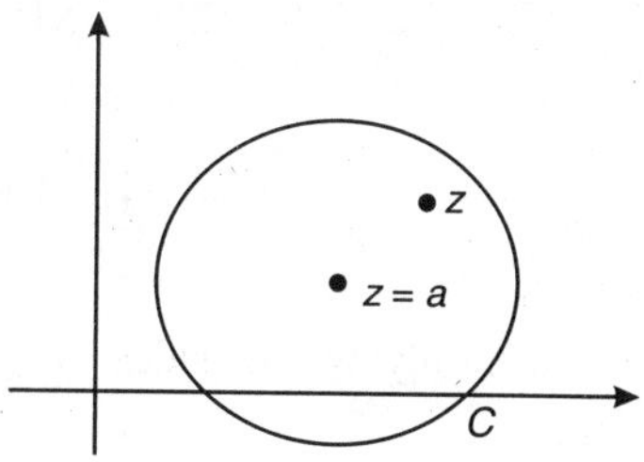

Given a function $f(z)$ and a point $z = a$, if C_1 is the largest circle and C_2 is the smallest circle, both having their centers at $z = a$ such that $f(z)$ is analytic in the annular region between C_1 and C_2, then $f(z)$ can be expressed as a series of positive and negative integral powers of $z - a$ which is convergent for all z in the annular region. This series is called Laurent's series of $f(z)$ about $z = a$.

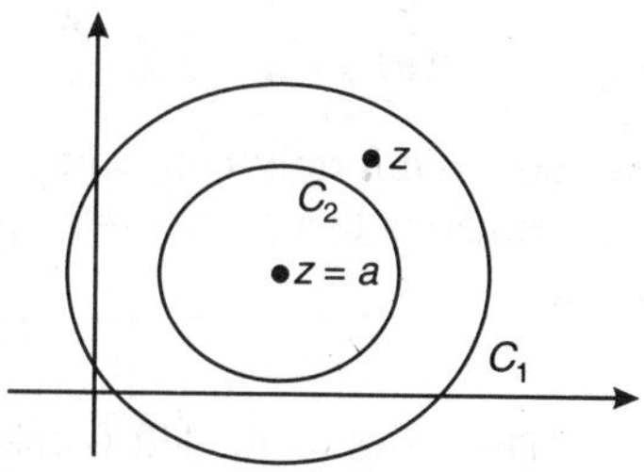

Theorem 9.1. Taylor's series. Suppose

(*i*) D is a simply-connected region.

(*ii*) $f(z)$ is analytic in D and

(*iii*) a is a point in D

(*iv*) Γ is the largest circle whose interior lies in D.

Then, for all z in Γ_i, the power series

$$\sum_{n=0}^{\infty} \frac{f^n(a)}{n!}(z-a)^n$$

is absolutely convergent and its sum is $f(z)$.

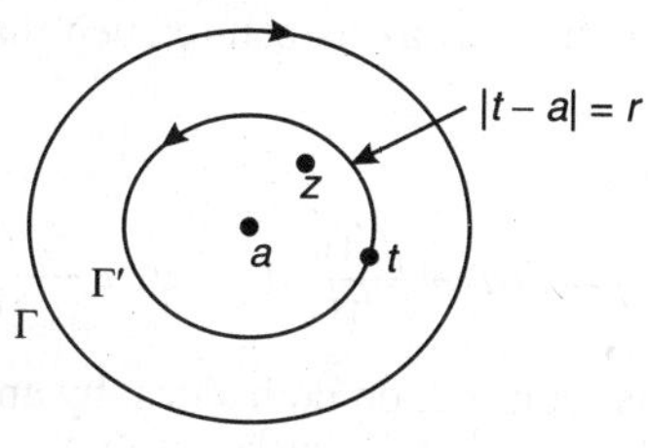

Fig. 9.1.1

Proof: Let Γ' be the positively oriented circle$|\, t - a \,| = r$ lying in Γ' (Fig. 9.1.1), where $z = t$ is any point on the circle and $z = a$ is a constant. Then, from Cauchy's integral formula and formulas for derivatives,

$$\left.\begin{aligned} f(a) &= \frac{1}{2\pi i}\int_{\Gamma'} \frac{f(t)}{t-a}\,dt,\ f'(a) = \frac{1!}{2\pi i}\int_{\Gamma'} \frac{f(t)}{(t-a)^2}\,dt, \\ f''(a) &= \frac{2!}{2\pi i}\int_{\Gamma'} \frac{f(t)}{(t-a)^3}\,dt,\ f^{(n)}(a) = \frac{n!}{2\pi i}\int_{\Gamma'} \frac{f(t)}{(t-a)^{n+1}}\,dt. \end{aligned}\right| \quad (9.1.1)$$

Now, for any z in Γ_i' and for any t on Γ'

$$|z-a| < |t-a| \text{ or } \left|\frac{z-a}{t-a}\right| < 1.$$

So the geometric series

$$1 + \left(\frac{z-a}{t-a}\right) + \left(\frac{z-a}{t-a}\right)^2 + \ldots$$

is aboslutely convergent in Γ_i' and its sum is

$$\frac{1}{1-\dfrac{z-a}{t-a}} = \frac{t-a}{t-a-(z-a)} = \frac{t-a}{t-z}.$$

$$\therefore \qquad 1 + \frac{z-a}{t-a} + \left(\frac{z-a}{t-a}\right)^2 + \ldots = \frac{t-a}{t-z}.$$

Multiplying both sides by $\dfrac{1}{2\pi i}\dfrac{f(t)}{t-a}$,

$$\frac{1}{2\pi i}\frac{f(t)}{t-a} + \frac{1}{2\pi i}\frac{f(t)}{(t-a)^2}(z-a) + \frac{1}{2\pi i}\frac{f(t)}{(t-a)^3}(z-a)^2 + \ldots = \frac{1}{2\pi i}\frac{f(t)}{t-z}$$

which is uniformly convergent on Γ'. Integrating term by term along Γ' and using (9.1.1) and remembering that $z-a$, $(z-a)^2$, etc., are constants

$$f(a) + \frac{f'(a)}{1!}(z-a) + \frac{f''(a)}{2!}(z-a)^2 + \ldots = f(z).$$

The series on the left is absolutely convergent in Γ'. Now Γ', z and r are arbitrary. Hence the series on the left converges to $f(z)$ for all z in circular disc Γ_i. The series is called Taylor's series for $f(z)$ about $z = a$.

Note. Γ is the largest circle with $z = a$ as its centre such that $f(z)$ is analytic in Γ_i but is not analytic somewhere on Γ.

Region of validity. The result

$$f(z) = f(a) + \frac{f'(a)}{1!}(z-a) + \frac{f''(a)}{2!}(z-a)^2 + \ldots$$

is valid for all z in Γ_i, that is, if z is replaced, on both sides, by any complex number in Γ_i, then the result holds good. That is, the series converges to the sum $f(z)$.

If the same $f(z)$ is to be expanded in a series of powers of $z-b$, that is, if $f(z)$ is to be expanded in a series about $z = b$, then we should select the largest circle C' with centre at $z = b$ such that $f(z)$ is analytic in C_i'. In this case we have

$$f(z) = f(b) + \frac{f'(b)}{1!}(z-b) + \frac{f''(b)}{2!}(z-b)^2 + \ldots$$

which is valid for all z in C_i'.

Remark 1. In general, if a function $f(z)$ is analytic in an open circular disc with centre at $z = a$, then for any z in that disc, the series

$$f(a) + \frac{f'(a)}{1!}(z-a) + \frac{f''(a)}{2!}(z-a)^2 + \ldots + \frac{f^{(n)}(a)}{n!}(z-a)^n + \ldots$$

converges to $f(z)$.

Remark 2. Maclaurin's series. Maclaurin's series is the expansion of $f(z)$ about $z = 0$ i.e., in powers of $z - 0$. If a function $f(z)$ is analytic in an open circular disc with centre at $z = 0$, then, for any z in that disc, the series

$$f(0) + \frac{f'(0)}{1!}z + \frac{f''(0)}{2!}z^2 + \frac{f'''(0)}{3!}z^3 + \ldots$$

converges to $f(z)$. This series is called *Maclaurin's series for $f(z)$*.

Series of simple functions. By using Taylor's theorem we get, for all z's except $z = \infty$,

$$e^z = 1 + \frac{z}{1!} + \frac{z^2}{2!} + \frac{z^3}{3!} + \ldots$$

$$\sin z = \frac{z}{1!} - \frac{z^3}{3!} + \frac{z^5}{5!} - \frac{z^7}{7!} + \ldots$$

$$\cos z = 1 - \frac{z^2}{2!} + \frac{z^4}{4!} - \frac{z^6}{6!} + \ldots$$

$$\cosh z = 1 + \frac{z^2}{2!} + \frac{z^4}{4!} + \frac{z^6}{6!} \ldots$$

$$\sinh z = \frac{z}{1!} + \frac{z^3}{3!} + \frac{z^5}{5!} + \frac{z^7}{7!} + \ldots$$

The binomial series is

$$(1+z)^n = 1 + \frac{n}{1}z + \frac{n.(n-1)}{1.2}z^2 + \frac{n.(n-1).(n-2)}{1.2.3}z^3 + \ldots$$

which is valid when $|z| < 1$.

The logarithmic series is

$$\log(1+z) = z - \frac{z^2}{2} + \frac{z^3}{3} - \frac{z^4}{4} + \ldots$$

which is valid when $|z| < 1$, and when $z = +1$ but not when $z = -1$.

STEPS TO FIND A TAYLOR'S SERIES FOR *F (Z)*.

Suppose the expansion needed is about the point $z = a$.Then

1. Find the singular points of $f(z)$.
2. Find the biggest circle C with centre at $z = a$ and passing through the singular point nearest to $z = a$.
3. Expand $f(z)$ as a Taylor's series for all z in C_i.

Theorem 9.2. (Uniqueness Theorem). If $f(z)$ is analytic in an open circular disc with centre at $z = a$, then the Taylor's series about $z = a$ is the only power series in $z - a$ which converges to $f(z)$ in that disc.

Proof: The Taylor's series about $z = a$ for $f(z)$ in the disc is

$$f(a) + \frac{f'(a)}{1!}(z-a) + \frac{f''(a)}{2!}(z-a)^2 + \ldots$$

Suppose there is another series, say

$$a_0 + a_1(z-a) + a_2(z-a)^2 + \ldots \qquad (9.1.2)$$

which converges in that disc to the same function $f(z)$. Then, differentiating (9.1.2.) term by term and successively,

$$\begin{aligned} a_1 + 2a_2(z-a) + 3a_3(z-a)^2 + 4a_4(z-a)^3 \ldots &= f'(z) \\ 2!\, a_2 + 3.2.a_3(z-a) + 4.3.a_4(z-a)^2 + \ldots &= f''(z) \\ 3!\, a_3 + 4.3.2a_4(z-a) + \ldots &= f'''(z) \end{aligned}$$

....

Setting $z = a$ in these series and in (9.1.2),

$$a_0 = f(a), \quad 1!\, a_1 = f'(a), \quad 2!\, a_2 = f''(a), \ \ldots$$

i.e.,
$$a_0 = f(a), \quad a_1 = \frac{f'(a)}{1!}, \quad a_2 = \frac{f''(a)}{2!}, \ \ldots$$

Therefore the series (9.1.2) becomes theTaylor's series itself. Hence Taylor's series is the only power series in $z - a$ which converges to $f(z)$ in the disc.

Different Taylor's series for the same function. If $f(z)$ is analytic in the neighbourhood $N(a)$ and $N(b)$ of $z = a$ and $z = b$, then

$$f(z) = f(a) + \frac{f'(a)}{1!}(z-a) + \frac{f''(a)}{2!}(z-a)^2 + \ldots \text{ for all } z \text{ in } N(a),$$

$$f(z) = f(b) + \frac{f'(b)}{1!}(z-b) + \frac{f''(b)}{2!}(z-b)^2 + \ldots \text{ for all } z \text{ in } N(b).$$

These series are Taylor's series of the same function $f(z)$ about $z = a$ and $z = b$.

Example. Refer Solved sum No. 7, 9.13 Examples. We shall find the Taylor's series for $1/z$ about $z = 1$, $z = 2$ and $z = i$ and the corresponding neighbourhoods (open circular discs) in which they are convergent, that is, valid.

The neighbourhoods of 1, 2, i in which $1/z$ is analytic are

$$|z-1| < 1, \quad |z-2| < 2, \quad |z-i| < 1$$

because $1/z$ is not anlaytic at $z = 0$ and the distances between 0 and 1, 2, i are 1, 2, 1. Therefore

$$\frac{1}{z} = 1 - (z-1) + (z-1)^2 - (z-1)^3 + \ldots \text{ for all } z \text{ in } |z-1| < 1,$$

$$\frac{1}{z} = \frac{1}{2} - \frac{z-2}{2^2} + \frac{(z-2)^2}{2^3} - \frac{(z-2)^3}{2^4} + \ldots \text{ for all } z \text{ in } |z-2| < 2.$$

$$\frac{1}{z} = \frac{1}{i} - \frac{z-i}{i^2} + \frac{(z-i)^2}{i^3} - \frac{(z-i)^3}{i^4} + \ldots \text{ for all } z \text{ in } |z-i| < 1.$$

9.2. ZEROS OF AN ANALYTIC FUNCTION

In section 8.10 we showed that zeros of an analytic function are isolated. The order of a zero can be obtained by expanding the function about the zero.

Example 1. The Taylor's expansion for sin z about $z = 0$ is

$$\sin z = z - \frac{z^3}{3!} + \frac{z^5}{5!} - \ldots = z\left(1 - \frac{z^2}{3!} + \frac{z^4}{5!} - \ldots\right).$$

Therefore $z = 0$ is a zero of sin z of order one.

Example 2. The function $e^z - 1$ has a zero $z = 0$ of order 1. Also

$$e^z - 1 = e^{2\pi i}\,[e^{z-2\pi i} - e^{-2\pi i}]$$

$$= e^{2\pi i}\left[\left\{1 + \frac{z-2\pi i}{1!} + \frac{(z-2\pi i)^2}{2!} + \ldots\right\} - 1\right]$$

$$= (z - 2\pi i)\,[e^{2\pi i} + e^{2\pi i}(z - 2\pi i) + \ldots]$$

has a zero $z = 2\pi i$ of order 1. $-2\pi i, \pm 4\pi i, \pm 6\pi i, \pm 8\pi i, \ldots\ \infty$ are also its zeros. ∞ is the limit point of the set of these zeros.

Next we consider the nature of an analytic function having an infinite sequence of zeros in a region.

Theorem 9.3. Suppose $f(z)$ is analytic in a region D and

$$z_n, \quad n = 1, 2, 3, \ldots,$$

in D are the zeros of $f(z)$, where the sequence $\{z_n\}$ converges to a limit $z = a$ in D. Then $f(z)$ vanishes identically in D.

Proof: The function $f(z)$ can be expanded in Taylor's series about $z = a$ as

$$f(z) = a_0 + a_1(z - a) + a_2(z - a)^2 + \ldots$$

which is true for all z in some disc $|z - a| < R$ in D. Now $f(z)$ is continuous at $z = a$. Therefore, by definition of continuity, $f(a) = 0$ since $z = a$ is the limit point of the zeros of $f(z)$. This implies that $a_0 = 0$ and

$$f(z) = a_1(z - a) + a_2(z - a)^2 + \ldots$$

$$= (z - a)\,[a_1 + a_2(z - a) + \ldots] = (z - a)\,\phi(z), \text{ say.}$$

Since z_n's are zeros of $f(z)$ and $z_n \neq a$, $\phi(z_n) = 0$. But $\phi(z)$ is continuous in D. Hence $\phi(a) = 0$. This implies $a_1 = 0$. Repeating this process indefinitely, we get that

$$a_2 = 0, \quad a_3 = 0, \quad a_4 = 0, \ldots$$

This means that $f(z) \equiv 0$ in this disc.

Let $z = b$ be an arbitrary point in D but not in the disc $|z - a| < R$. Join the points a and b by a rectifiable curve L lying entirely in D. Let the minimum distance of L from the boundary be d. Let L intersect the circle $|z - a| = R$ at a_1. Consider another disc $|z - a_1| < d$. We can find a sequence of points lying in both the discs and converging to a_1. Since $f(z)$ vanishes at these point, $f(z) \equiv 0$ in the second disc also. Continuing in similar manner, we see that a finite number of discs cover L and one of the discs contains z_1. So $f(z)$ vanishes at z_1. Since z_1 is arbitrary, $f(z) \equiv 0$ in the whole of D.

Corollary. Suppose $f(z)$ is not identically zero and analytic in a region D. In any closed bounded region D, $f(z)$ has only a finite number of zeros.

Theorem 9.4. Suppose $f_1(z)$ and $f_2(z)$ are two functions analytic in a region D such that, for all z_n, $f_1(z_n) = f_2(z_n)$, where $\{z_n\}$ is a sequence of points in D coverging to a point in D. Then $f_1(z) \equiv f_2(z)$ in D.

Proof: Let $f(z) = f_1(z) - f_2(z)$. Then, by theorem 9.3., $f_1(z) - f_2(z) \equiv 0$ in D. Hence the theorem.

Remark. This is a uniqueness theorem because it establishes that in a given region only one analytic function can exist such that it assumes specified values at points z_n where the sequence $\{z_n\}$ converges to a point in the region.

9.3. LAURENT'S SERIES

Suppose the radii of convergence of the power series

$$a_0 + a_1 (z-a) + a_2 (z-a)^2 + ...,$$
$$b_0 + b_1 (z-a) + b_2 (z-a)^2 + ...$$

are R_1, R_2 $(R_2 > R_1)$. Then the series are absolutely convergent in

$$|z-a| < R_1, \quad |z-a| < R_2$$

respectively and the series

$$a_0 + \frac{a_1}{z-a} + \frac{a_2}{(z-a)^2} + ...$$

is absolutely convergent in $|1/(z-a)| < R_1$ or in $|z-a| > R_1$, Thus the combined series

$$\left[a_0 + \frac{a_1}{z-a} + \frac{a_2}{(z-a)^2} + ...\right] + [b_0 + b_1 (z-a) + b_2 (z-a)^2 + ...]$$

is absolutely convergent in the annulus $R_1 < |z-a| < R_2$. This type of expansion with positive and negative powers of $z-a$ is called the Laurent's series about $z = a$.

Theorem 9.5. Laurent's series. Suppose

(*i*) D is a multiply-connected region

(*ii*) $f(z)$ is analytic in D and

(*iii*) $\Gamma_1 : |t-a| = R_1$ and $\Gamma_2 : |t-a| = R_2$ are the smallest and biggest circles such that, the annulus region D_1 between Γ_1 and Γ_2 lies in D.

Then, for all z in D_1, the power series

$$\sum_{-\infty}^{\infty} a_n (z-a)^n, \text{ where } a_n = \frac{1}{2\pi i} \int_C \frac{f(t)}{(t-a)^{n+1}} dt$$

and C is any scro curve in D_1 (encircling $z = a$), is absolutely convergent and its sum is $f(z)$.

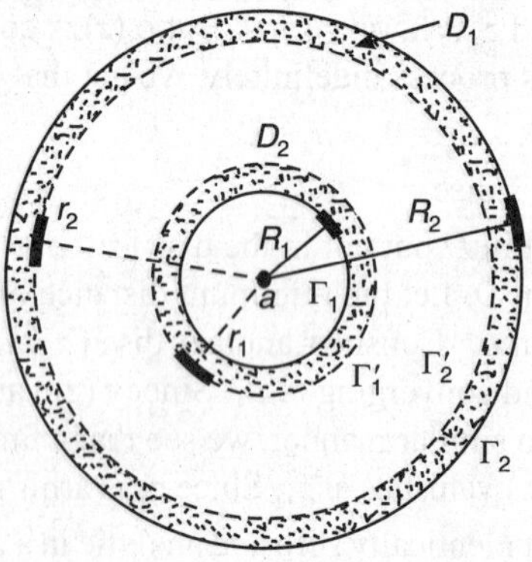

Fig. 9.3.1

Proof : Let $\Gamma_1' : |t-a| = r_1$ and $\Gamma_2' : |t-a| = r_2$ be two positively oriented circles in D_1 of which Γ_1' is smaller (Fig. 9.3.1). Now, by extension to Cauchy's integral formula, for any z in D_2,

$$f(z) = \frac{1}{2\pi i} \int_{\Gamma_2'} \frac{f(t)}{t-z} dt - \frac{1}{2\pi i} \int_{\Gamma_1'} \frac{f(t)}{t-z} dt \tag{9.3.1}$$

Evaluation of the integral along Γ_2'. Now, for any point z in the region D_2 between Γ_1' and Γ_2' and for any t on Γ_2',

$$|z-a|<|t-a| \quad \text{or} \quad \left|\frac{z-a}{t-a}\right|<1.$$

Hence the geometric series

$$1+\frac{z-a}{t-a}+\left(\frac{z-a}{t-a}\right)^2+\left(\frac{z-a}{t-a}\right)^3+\ldots$$

is absolutely convergentand its sum is

$$\frac{1}{1-\dfrac{z-a}{t-a}}=\frac{t-a}{t-z}.$$

$$\therefore \qquad 1+\frac{z-a}{t-a}+\left(\frac{z-a}{t-a}\right)^2+\ldots=\frac{t-a}{t-z}.$$

Multiplying both sides by $\dfrac{1}{2\pi i}\dfrac{f(t)}{t-a}$,

$$\frac{1}{2\pi i}\frac{f(t)}{t-a}+\frac{1}{2\pi i}\frac{f(t)}{(t-a)^2}(z-a)+\frac{1}{2\pi i}\frac{f(t)}{(t-a)^3}\times(z-a)^2+\ldots=\frac{1}{2\pi i}\frac{f(t)}{t-z}.$$

By term-by-term integration along Γ_2',

$$\sum_{n=0}^{\infty}(z-a)^n\frac{1}{2\pi i}\int_{\Gamma_2'}\frac{f(t)}{(t-a)^{n+1}}dt=\frac{1}{2\pi i}\int_{\Gamma_2'}\frac{f(t)}{t-z}dt \qquad (9.3.2)$$

The series on the left is absolutely convergent.

Evaluation of the integral along Γ_1'. Now for the same z in D_2 and for any t on Γ_1'

$$|t-a|<|z-a| \quad \text{or} \quad \left|\frac{t-a}{z-a}\right|<1.$$

Hence the geometric series

$$1+\frac{t-a}{z-a}+\left(\frac{t-a}{z-a}\right)^2+\ldots$$

is absolutely convergent and its sum function is

$$\frac{1}{1-\dfrac{t-a}{z-a}}=\frac{z-a}{z-t}=-\frac{z-a}{t-z}$$

$$\therefore \qquad 1+\frac{t-a}{z-a}+\left(\frac{t-a}{z-a}\right)^2+\ldots=-\frac{z-a}{t-z}.$$

Multiplying both sides by $\dfrac{1}{2\pi i}\dfrac{f(t)}{z-a}$,

$$\sum_{n=-1}^{-\infty} \frac{1}{2\pi i} \frac{f(t)}{(t-a)^{n+1}} (z-a)^n = -\frac{1}{2\pi i} \frac{f(t)}{t-z}$$

By term-by-term integration along Γ_1'

$$\sum_{n=-1}^{-\infty} (z-a)^n \frac{1}{2\pi i} \int_{\Gamma_1'} \frac{f(t)}{(t-a)^{n+1}} dt = -\frac{1}{2\pi i} \int_{\Gamma_1'} \frac{f(t)}{t-z} dt. \qquad (9.3.3)$$

The series on the left is absolutely convergent.

Therefore, by (9.3.2) and (9.3.3), (9.3.1) becomes

$$f(z) = \sum_{n=0}^{\infty} (z-a)^n \frac{1}{2\pi i} \int_{\Gamma_2'} \frac{f(t)}{(t-a)^{n+1}} dt + \sum_{n=-1}^{-\infty} (z-a)^n \frac{1}{2\pi i} \int_{\Gamma_1'} \frac{f(t)}{(t-a)^{n+1}} dt. \qquad (9.3.4)$$

Now, C is a scro curve in-between Γ_1' and Γ_2'. Hence, by extension to Cauchy's fundamental theorem

$$\int_{\Gamma_2'} = \int_C \text{ and } \int_C = \int_{\Gamma_1'}$$

Hence (9.3.4) becomes

$$\begin{aligned} f(z) &= \sum_{n=-\infty}^{\infty} (z-a)^n \frac{1}{2\pi i} \int_C \frac{f(t)}{(t-a)^{n+1}} dt \\ &= \sum_{n=-\infty}^{\infty} a_n (z-a)^n, \text{ where } a_n = \frac{1}{2\pi i} \int_C \frac{f(t)}{(t-a)^{n+1}} dt. \end{aligned}$$

Since r_1 and r_2 are arbitrary, this result is true for all z in D_1 and for all C in D_1.

Remark. If a Taylor's expansion is about $z = a$, then the coefficients of the powers of $z - a$ are $f(a), f'(a)/1!, f''(a)/2!, \ldots$ which can be calculated by differentiation and substitution.

But the coefficient of $(z - a)^n$ in the Laurent's expansion, namely,

$$a_n = \frac{1}{2\pi i} \int_C \frac{f(t)}{(t-a)^{n+1}} dt$$

has to be evaluated by complex integration. However, in practice, the Laurent's expansion is obtained as the binomial expansion, namely,

$$(1+z)^n = 1 + \frac{n}{1!} z + \frac{n(n-1)}{2!} z^2 + \ldots, \quad |z| < 1$$

or as any other known expansion such as exponential, logarithmic, etc.

STEPS TO FIND LAURENT'S SERIES FOR *F* (*Z*)

1. Find the point $z = a$ about which $f(z)$ is to be expanded. Here the expansion will be in negative integral powers of $z - a$ (with or without positive integral powers of $z - a$).
2. Find the singular points of $f(z)$.
3. Consider all the concentric circles having their centres at $z = a$ and each passing through a singular point.
4. Expand $f(z)$ for all z in each annular region in-between every two consecutive circles.

9.4. CAUCHY PRODUCT AND DIVISION

When two series

$$a_0 + a_1 z + a_2 z^2 + a_3 z^3 + ...$$
$$b_0 + b_1 z + b_2 z^2 + b_3 z^3 + ...$$

are given, the power series $c_0 + c_1 z + c_2 z^2 + c_3 z^3 + ...$, where

$$c_0 = a_0 b_0, \quad c_1 = a_0 b_1 + a_1 b_0, \quad c_2 = a_0 b_2 + a_1 b_1 + a_2 b_0, \text{ etc.}$$

is called the *Cauchy product* of the given series. Suppose $|z| = r$ and $|z| = R$ are the circles of convergence of the series and the series converge to $f(z)$ and $g(z)$ in $|z| < r$ and $|z| < R$ and, if $r < R$, then the Cauchy product of the series converges to the product $f(z)\, g(z)$ in $|z| < r$.

Division of a series by another. If $g(z) \neq 0$ in $|z| < r$, then $f(z)/g(z)$ is analytic in $|z| < r$ and $f(z)/g(z)$ can be expanded as a power series in z directly or a few of the terms of the power series can be obtained by dividing the series for $f(z)$ by the series for $g(z)$.

9.5. SINGULAR POINT OR SINGULARITY

A singular point of a function is a point at which the function ceases to be analytic.

For example, 0 and 1 are the singular points of the function $1/z(z-1)$because the function is not analytic there since the function attains the value ∞ there. A function which is not defined at a point is not analytic there. So such a point is a singular point of the function.

Singular points of cosec z are $\pm n\pi$, $n = 0, 1, 2, 3, ...$ and ∞.

The point $z = a$ is a singular point of $f(z)$

(*i*) if $f(z) = \infty$ or

(*ii*) if $f(z)$ is not defined at $z = a$ or

(*iii*) if $f(z)$ is not differentiable at $z = a$.

The following are examples for the first two cases:

(*i*) The singular points of the fuction

$$f(z) = \frac{z}{(z-1)(z+2)}$$

are $z = 1$ and $z = -2$ because $f(1) = \infty$ and $f(-2) = \infty$.

(*ii*) The entire function

$$f(z) = e^z$$

is not defined at $z = \infty$ as shown earlier and $z = \infty$ is the only singular point of e^z.

(*iii*) The function

$$f(z) = \frac{e^z}{z}$$

has two singular points $z = 0$ and $z = \infty$ because

(*i*) $f(0) = \infty$

(*ii*) e^z is not defined at $z = \infty$ and so e^z/z is not defined at $z = \infty$.

Remark. The point $z = \infty$ is a singular point of all the trigonometric functions and hyperbolic functions because they are functions of e^z.

Singular points in certain illustrative cases including $z = \infty$ are tabulated below.

$f(z)$	Singular points $z = a$	$f(a)$
$\dfrac{1}{z(1-z)}$	0, 1	∞
$\dfrac{z}{(z+1)(z-2)}$	$-1, 2$	∞
$\dfrac{1+z}{1+z^2}$	$i, -i$	∞
$1 + e^z, e^{-z}$	∞	undefined
$e^{1/z}, e^{-1/z}$	0	undefined
$\sin z$, $\cos z$	∞	undefined
$1/\sin z$	$\infty; \pm n\pi, n = 0, 1, 2, 3, ...$	undefned; ∞
$1/\cos z$	$\infty; \pm (2n+1)\frac{1}{2}\pi, n = 0, 1, 2, ...$	undefined; ∞
$1/\sinh z$	$\infty; \pm n\pi i, n = 0, 1, 2, 3, ...$	undefined; ∞
$1/\cosh z$	$\infty; \pm (2n+1)\frac{1}{2}\pi i, n = 0, 1, 2, ...$	undefined; ∞

Illustration. The function $e^{1/(z-a)}$ is an illustrative example for an underfined function at the singularity $z = a$. The function is analytic in $|z - a| > 0$ and it is not defined at $z = a$ because, if $a = \alpha + i\beta$ and $z = (\alpha + i\beta) + h$, then as $h \to 0$, $z \to a$ and

$$f(z) = e^{1/h} \to \infty \quad \text{if } h \text{ is positive}$$
$$\to 0 \quad \text{if } h \text{ is negative.}$$

i.e., the limit of the function as $z \to a$ does not exist.

Isolated singualrity. If $z = a$ is a singular point of a function $f(z)$ and if there exists a neighbourhood of $z = a$ containing no other singular point of $f(z)$, then $z = a$ is said to be an *isolated singular point* of $f(z)$.

Non-isolated singularity. If $f(z)$ has infinity of singular points in every neighbourhood of $z = a$, then $z = a$ is a non-isolated singular point but it is a limit point of the set of singular points of $f(z)$.

Example 1. 0 and 1 are the singularities of $1/z(z-1)$. They are isolated.

Example 2. All the singular points of $1/\sin z$, namely, ... $-2\pi, -\pi, 0, \pi, 2\pi, 3\pi$, ... except ∞ are isolated singular points. ∞ is a limit point of the set of the singular points because every neighbourhood $|z| > R$ of ∞ contains infinity of singular points. As such, ∞ is not an isolated singular point.

Example 3. The function $1/\{\sin(1/z)\}$ is not analytic at the points

$$z = \pm 1/\pi, \pm 1/(2\pi), \pm 1/(3\pi), ... \tag{9.5.1}$$

because sin $(1/z)$ vanishs there. Further the function is not defined at $z = 0$ since sin z is not defined at ∞. Thus $z = 0$ also is a singular point. Further it is the limit point of the sequence of points (9.5.1). Hence $z = 0$ is a non-isolated singular point. However the singular point ∞ is an isolated point.

9.6. ISOLATED SINGULARITIES

Isolated singularities are classified into three different groups. This classification is done with reference to the negative powers of the Laurent's expansion of the function about the singular point. If $z = a$ is an isolated singular point of a function $f(z)$, then the singularity is called (*i*) *a removable singularity* or

(*ii*) a *pole* or

(*iii*) an *essential singularity*

according as the Laurent's series about $z = a$ of $f(z)$, valid in a deleted neighbourhood of $z = a$, has

(*i*) no negative powers or

(*ii*) a finite number of negative powers or

(*iii*) an infinite number of negative powers.

9.7. REMOVABLE SINGULARITY

Suppose $z = a$ is a removable singularity of $f(z)$. Then, by definition, the Laurent's series of $f(z)$ about $z = a$ is

$$a_0 + a_1(z-a) + a_2(z-a)^2 + \ldots = f(z).$$

The limit of $f(z)$ as $z \to a$ is $a_0 = f(a)$. But $f(z)$ is not analytic at $z = a$. It may be so because it might not have been defined there. Thus, if the function is defined to have a value a_0 at $z = a$, then the function is analytic there. As such, the singularity is called a removable singularity.

Alternative definition for removable singularity. If $z = a$ is a singular point of $f(z)$ and if

$$\lim_{z \to a} f(z)$$

exists and is finite, then $z = a$ is a *removable singularity* of $f(z)$.

***L′* Hospital's rule.** In finding limits we can use L′ Hospital's rule which is true for complex functions also.

Example 1. Consider the function

$$\frac{\sin z}{z}, \quad z \neq \infty \qquad \textbf{(1978, 79)}$$

In this case the function is not defined at $z = 0$ since it takes the form 0/0. So $z = 0$ is a singular point of the function. But

$$\frac{\sin z}{z} = \frac{1}{z}\left(z - \frac{z^3}{3!} + \frac{z^5}{5!} - \ldots\right) = 1 - \frac{z^2}{3!} + \frac{z^4}{5!} - \ldots$$

in $0 < |z| < \infty$. Therefore the singularity can be removed if the value of $\sin z/z$ at $z = 0$ is defined as 1 because

$$\lim_{z \to 0} \frac{\sin z}{z} = 1.$$

Since the limit exists and is finite, the singularity at $z = 0$ is a removable singularity.

Example 2. Consider the function

$$\frac{z - \sin z}{z^3} = \frac{1}{z^3}\left(\frac{z^3}{3!} - \frac{z^5}{5!} + \frac{z^7}{7!} - \ldots\right), \ 0 < |z| < \infty$$

Its singularity at $z = 0$ is a removable singularity because the function becomes analytic at $z = 0$ if its value at $z = 0$ is defined to be 1/3!

Alternative method. The function $f(z)$ is not defined at $z = 0$. But, by *L′* Hospital's rule.

$$\lim_{z \to 0} \frac{\sin z - z}{z^3} = \lim_{z \to 0} \frac{\cos z - 1}{3z^2} = \lim_{z \to 0} \frac{-\sin z}{6z} = -\frac{1}{6}.$$

Since the limit exists and is finite, the singularity at $z = 0$ is a removable singularity.

A removal singularity of a function can also be defined as follows:

Removable singularity. If $z = a$ is a singular point of a function and if the limit of the function as z tends to a exists and is finite, then $z = a$ is removable singularity of the function.

9.8. POLE

Given a function $f(z)$ and a point $z = a$, if the Laurent's expansion of $f(z)$ in a deleted neighbourhood of $z = a$ has a finite number of negative powers of $z - a$, then the point $z = a$ is said to be a pole of $f(z)$.

If $z = a$ is a pole of $f(z)$, then the Laurent's expansion for $f(z)$ about $z = a$ is of the form

$$\frac{a_{-m}}{(z-a)^m} + \frac{a_{-(m-1)}}{(z-a)^{m-1}} + \ldots + \frac{a_{-1}}{z-a} + a_0 + a_1(z-a) + \ldots,$$

where $a_{-m} \neq 0$. Here $z = a$ is said to be a pole of order m or a pole of multiplicity m.

Principal part. The rational function formed by the negative powers, namely

$$\frac{a_{-m}}{(z-a)^m} + \frac{a_{-(m-1)}}{(z-a)^{m-1}} + \ldots + \frac{a_{-1}}{z-a}$$

is called the principal part of $f(z)$ at the pole $z = a$. This part tends to ∞ as $z \to a$.

Alternative definition. Given an analytic function $f(z)$ with a singularity at $z = a$, if

$$\lim_{z \to a} f(z) = \infty,$$

then $z = a$ is a pole of $f(z)$.

Simple pole. When the order of a pole is 1, the pole is said to be a *simple pole.*

Double pole. When the order of a pole is 2, the pole is said to be a *double pole.*

Example 1. The singularity $z = 1$ of the function

$$\frac{3}{z-1} + 4 + 5(z-1) + 6(z-1)^2$$

is a simple pole. Its principal part is $3/(z-1)$.

Example 2. $z = 0$ is a simple pole of the function e^z/z because its expansion about $z = 0$ is

$$\frac{e^z}{z} = \frac{1}{z} + \frac{1}{1!} + \frac{z}{2!} + \frac{z^2}{3!} + \ldots$$

Its principal part is $\frac{1}{z}$.

Example 3. The singularity $z = 1$ of the function

$$\frac{1}{(z-1)^2} + \frac{1}{z-1} + 1 + (z-1) + (z-1)^2$$

is a double pole. Its principal part is $1/(z-1)^2 + 1/(z-1)$.

Example 4. $z = 0$ is a double pole of the function e^z/z^2 because its expansion about $z = 0$ is

$$\frac{e^z}{z^2} = \frac{1}{z^2} + \frac{1}{z} + \frac{1}{2!} + \frac{z}{3!} + \frac{z^2}{4!} + \ldots$$

Example 5. $z = 1$ is a triple pole of $\frac{e^{2z}}{(z-1)^3}$ because

$$\frac{e^{2z}}{(z-1)^3} = \frac{e^2 e^{2(z-1)}}{(z-1)^3} = \frac{e^2}{(z-1)^3}\left[1+\frac{2(z-1)}{1!}+\frac{2^2(z-1)^2}{2!}+\ldots\right].$$

To find the order of a pole. One method of finding the order of a pole $z = a$ is to expand the function about $z = a$ and then to find the order. However, it is evident from the definition that, if $z = a$ is a pole of $f(z)$, then there exists a positive integer m such that

$$\lim_{z \to a} [(z-a)^m f(z)]$$

exists and is equal to a nonzero finite complex number and this m is the order of $z = a$.

From the concept the following tabulated results are evident:

Function		Pole	Order
$\frac{z+1}{z^2-2z}$	$\frac{z+1}{z(z-2)}$	$z=0$ $z=2$	1 1
$\frac{1}{z^2+1}$	$\frac{1}{(z+i)(z-i)}$	$z=i$ $z=-i$	1 1
$\frac{1}{z^4+2z^2+1}$	$\frac{1}{(z+i)^2(z-i)^2}$	$z=i$ $z=-i$	2 2
$\frac{1}{z^2(z-2)^3}$	$\frac{1}{z^2(z-2)^3}$	$z=0$ $z=0$	2 3

9.9. ESSENTIAL SINGULARITY

If $z = a$ is an essential singularity of $f(z)$ then the Laurent's expansion for $f(z)$ about $z = a$ is of the form

$$f(z) = \sum_{n=1}^{\infty} \frac{a_{-n}}{(z-a)^n} + a_0 + a_1(z-a) + \ldots$$

In this, the principal part is an infinite series of negative powers of $z - a$ while the remaining part is a finite or infinite series of positive powers of $z - a$.

Example 1. $z = 0$ is an essential singularity of the function $e^{1/z} + 1 + z + z^2$ because its expansion aboaut $z = 0$ is $\sum_{n=1}^{\infty} \frac{1}{n!\,z^n} + 2 + z + z^2$.

Example 2. $z = -1$ is an essential singularity of the function

$$e^{1/(z+1)} + \frac{1}{2-z} \quad \text{or} \quad e^{1/(z+1)} + \frac{1}{3-(z+1)}$$

because its expansion about $z = -1$ is

$$\left(\sum_{n=1}^{\infty} \frac{1}{n!\,(z+1)^n} + 1\right) + \frac{1}{3}\sum_{n=0}^{\infty}\left(\frac{z+1}{3}\right)^n.$$

Isolation of a singularity. Now it is evident that the removable singularities and the poles are isolated singularities. But the essential singularity is either an isolated or non-isolated singularity which is evident from the following examples.

9.10. BEHAVIOUR OF A FUNCTION AT AN ISOLATED SINGULARITY

The results which pertain to the behaviour of functions at their singularities are given below.

1. If $z = a$ is a removable singularity of $f(z)$, then $f(z)$ is bounded in every deleted neighbourhood of $z = a$.
2. If $z = a$ is a pole of $f(z)$, then $f(z)$ is not bounded in every deleted neighbourhood of $z = a$.
3. If $z = a$ is an essential singularity of $f(z)$, then the behaviour of $f(z)$, close to $z = a$, is complicated and at $z = a$ the function is not defined. This behaviour is embodied in the theorem of Weierstrass.

Suppose $z = a$ is an isolated singularity of a function $f(z)$. If it is a removable singularity, the limit of $f(z)$ as $z \to a$ is finite and $f(z)$ is bounded in a small deleted neighbourhood of $z = a$. However, if $z = a$ is a pole, the limit of $f(z)$ as $z \to a$ is ∞. If $z = a$ is an essential singularity, the behaviour of $f(z)$ close to $z = a$ is complicated and at $z = a$, $f(z)$ is not defined. These results are embodied in theorems 9.6, 9.7, 9.8, 9.9, 9.12.

BEHAVIOUR AT A REMOVABLE SINGULARITY

Theorem 9.6. If $z = a$ is a removable singularity of a function $f(z)$, then there exists a deleted neighbourhood of $z = a$ in which $f(z)$ is bounded.

Proof: Since a removable singularity is an isolated singularity, there is a deleted neighbourhood

$$0 < |z - a| < R$$

in which $f(z)$ is analytic. By the definition of removable singularity, the Laurent's series for $f(z)$ in this deleted neighbourhood contains no negative powers of $z - a$. So the expansion is of the form

$$f(z) = a_0 + a_1 (z - a) + a_2 (z - a)^2 + \ldots$$

Now $f(z)$ is analytic in the deleted neighbourhood. But, if we define the value of $f(z)$ at $z = a$ as a_0 (or zero if $a_0 = 0$), then the new $f(z)$ is analytic at $z = a$ and hence in the whole of the neighbourhood. Therefore $f(z)$ is bounded in some closed neighbourhood of $z = a$. But, actually, $f(z)$ is not defined at $z = a$ or is discontinuous at $z = a$. Thus, leaving out the point $z = a$, we get that $f(z)$ is bounded in some deleted neighbourhood of $z = a$.

Theorem 9.7. Converse of theorem 9.6. A function has an isolated singularity at $z = a$ but is analytic in the deleted neighbourhood $0 < |z - a| < R$. If the function is bounded in the deleted neighbourhood, then the singularity is a removable singularity.

Proof: Since $f(z)$ is bounded, suppose $|f(z)| < M$ in the deleted neighbourhood. Let the Laurent's series of $f(z)$ about $z = a$ be

$$\sum_{-\infty}^{\infty} a_n (z - a)^n, \quad a_n = \frac{1}{2\pi i} \int_C \frac{f(t)}{(t - a)^{n+1}} dt,$$

where C is a circle $|t - a| = \rho$, $\rho < R$. Then

$$|a_n| = \left| \frac{1}{2\pi i} \int_C \frac{f(t)}{(t - a)^{n+1}} dt \right|$$

$$\leq \frac{1}{2\pi} \underset{\text{on } C}{\text{Max}} \frac{|f(t)|}{|t - a|^{n+1}} \, l(C)$$

$$< \frac{1}{2\pi} \frac{M}{\rho^{n+1}} (2\pi\rho) = \frac{M}{\rho^n}.$$

Letting ρ approach zero, we find that

$$|a_n| = 0 \quad \text{or} \quad a_n = 0 \text{ if } n = -1, -2, -3, \ldots$$

Therefore, in the Laurent's expansion of $f(z)$, all the negative powers vanish. Hence

$$f(z) = a_0 + a_1(z-a) + a_2(z-a)^2 + \ldots$$

Thus, by definition, $z = a$ is a removable singularity.

BEHAVIOUR AT A POLE

Theorem 9.8. If $z = a$ is a pole of a function $f(z)$, then

$$\lim_{z \to a} f(z) = \infty.$$

Proof: Let the order of the pole be m and the Laurent's expansion about $z = a$ in the deleted neighbourhood $0 < |z-a| < \rho$ be

$$f(z) = \frac{a_{-m}}{(z-a)^m} + \ldots + \frac{a_{-1}}{z-a} + a_0 + a_1(z-a) + \ldots$$

$$= \frac{a_{-m} + \ldots + a_{-1}(z-a)^{m-1}}{(z-a)^m} + a_0 + a_1(z-a) + \ldots$$

$$= \frac{\phi(z)}{(z-a)^m} + a_0 + a_1(z-a) + a_2(z-a)^2 + \ldots, \text{ say.}$$

Now, as $z \to a$, taking any path,

(*i*) $\phi(z)$, a polynomial in $z - a$, tends to a_{-m},

(*ii*) $1/(z-a)^m$ tends to ∞,

(*iii*) $a_0 + a_1(z-a) + a_2(z-a)^2 + \ldots$ tends to a_0. Hence $f(z) \to \infty$ as $z \to a$.

Remark. If z is a pole, $|f(z)| \to \infty$ as $z \to a$.

Theorem 9.9. (Converse of 9.8.) If a function $f(z)$ is analytic in a deleted neighbourhood of $z = a$ and if

$$\lim_{z \to a} f(z) = \infty,$$

then $z = a$ is a pole of $f(z)$.

Proof: Since $f(z) \to \infty$ as $z \to a$, given a positive number A, there exists a δ such that

$$|f(z)| > A \qquad \text{for all } z \text{ in } |z-a| < \delta.$$

Consider the function $g(z)$ defined by $g(z) = 1/f(z)$. Then

$$|g(z)| = \frac{1}{|f(z)|} < \frac{1}{A} \qquad \text{for all } z \text{ in } |z-a| < \delta.$$

i.e., in $|z-a| < \delta$, $g(z)$ is bounded and analytic. This implies that $g(z)$ has a removable singularity at $z = a$. So, expressing $g(z)$ in Laurent's series about $z = a$,

$$g(z) = b_0 + b_1(z-a) + b_2(z-a)^2 + \ldots$$

But, as $z \to a$, $f(z) \to \infty$ and hence $g(z) \to 0$. This implies that either $b_0 = 0$ or the first few coefficeints are zero. Suppose

$$b_0 = b_1 = \ldots = b_{m-1} = 0; \; b_m \neq 0.$$

Then

$$g(z) = (z-a)^m [b_m + b_{m+1}(z-a) + \ldots.]$$

$$= (z-a)^m \phi(z), \text{ say.}$$

Here $\phi(z)$ is analytic in a neighbourhood of $z = a$ and $\phi(a) = b_m \neq 0$. So $1/\phi(z)$ is analytic in a neighbourhood of $z = a$. Expanding $1/\phi(z)$ as a Taylor's series about $z = a$,

$$1/\phi(z) = c_0 + c_1(z-a) + c_2(z-a)^2 + \ldots$$

$$\therefore \quad f(z) = \frac{1}{g(z)} = \frac{1}{(z-a)^m \phi(z)} = \frac{1}{(z-a)^m}[c_0 + c_1(z-a) + \ldots]$$

$$= \frac{c_0}{(z-a)^m} + \frac{c_1}{(z-a)^{m-1}} + \frac{c_2}{(z-a)^{m-2}} + \ldots$$

Therefore $z = a$ is a pole of $f(z)$.

Theorem. 9.10. If $z = a$ is a zero of order m of an analytic function $f(z)$, then $z = a$ is a pole of order m of the function $1/f(z)$.

Proof: Since $z = a$ is a zero of order m of $f(z)$, suppose

$$f(z) = (z-a)^m \phi(z).$$

Then $\phi(a) \neq 0$ and $\phi(z)$ is analytic in a neighbourhood of a. Consequently the function $1/\phi(z)$ is analytic in that neighbourhood. Hence, expanding $1/\phi(z)$ as a Taylor's series about $z = a$,

$$\frac{1}{\phi(z)} = b_0 + b_1(z-a) + b_2(z-a)^2 + \ldots$$

$$\therefore \quad \frac{1}{f(z)} = \left[\frac{1}{(z-a)^m}\right]\left[\frac{1}{\phi(z)}\right] = \frac{1}{(z-a)^m}[b_0 + b_1(z-a) + \ldots.]$$

$$= \frac{b_0}{(z-a)^m} + \frac{b_1}{(z-a)^{m-1}} + \frac{b_2}{(z-a)^{m-2}} + \ldots$$

Therefore $z = a$ is a pole of order m of the function $1/f(z)$.

Corollary. Since zeros are isolated points (Theorem 8.13), poles are also isolated points.

Theorem 9.11. If $z = a$ is a pole of order m of a function $f(z)$, then $z = a$ is a zero of order m of the function $1/f(z)$.

Proof: Since $z = a$ is a pole of order m of the function $f(z)$, there exists a deleted neighbourhood of $z = a$ in which the Laurent's expansion for $f(z)$ about $z = a$, is

$$f(z) = \frac{a_{-m}}{(z-a)^m} + \ldots + \frac{a_{-1}}{z-a} + a_0 + a_1(z-a) + \ldots$$

Let $\psi(z) = (z-a)^m f(z)$

$$= a_{-m} + a_{-(m-1)}(z-a) + a_{-(m-2)}(z-a)^2 + \ldots$$

Then $\psi(a) = a_{-m}$ and $\dfrac{1}{\psi(a)} \neq 0.$

Now $\dfrac{1}{f(z)} = (z-a)^m \left[\dfrac{1}{\psi(z)}\right], \dfrac{1}{\psi(a)} \neq 0.$

Therefore $1/f(z)$ has a zero of order m.

BEHAVIOUR AT AN ESSENTIAL SINGULARITY

Behaviour in a neighbourhood of an essential singularity. The complicated behaviour of an analytic function in a neighbourhood of an essential singularity is embodied in the following theorem of Weierstrass. The complicated behaviour is that an analytic function comes arbitrarily close to any complex value in every neighbourhood of an essential singularity, however small it may be.

Theorem 9.12. (Weierstrass theorem). If $f(z)$ has an isolated essential singularity at $z = a$ and if c is any complex constant, then, for any positive $\in$, however small, there exists an z in every deleted neighbourhood of $z = a$ such that

$$|f(z) - c| < \in.$$

Proof: Let $$D: 0 < |z - a| < \rho$$

be an arbitrarily chosen small deleted neighbourhood of $z = a$. Given an $\in > 0$, suppose it is not possible to find an z in D such that $|f(z) - c| < \in$. Then, for all z in D,

$$|f(z) - c| \geq \in \quad \text{or} \quad \left|\frac{1}{f(z) - c}\right| \leq \frac{1}{\in}.$$

Here $1/\{f(z) - c\}$ is bounded and analytic in D. And it is not analytic at $z = a$ because $f(z)$ is not analytic there. So, by theorem 9.7, $z = a$ is a removable singularity of $1/\{f(z) - c\}$. Therefore its Laurent's expansion about $z = a$ is of the form, either

$$a_0 + a_1 (z - a) + a_2 (z - a)^2 + ..., \quad a_0 \neq 0$$

or $$a_m (z - a)^m + a_{m+1} (z - a)^{m+1} + ..., \quad a_m \neq 0.$$

In other words, $1/\{f(z) - c\}$ has a removable singularity at $z = a$ and

either $z = a$ is not a zero of $1/\{f(z) - c\}$

or $z = a$ is a zero of order m of $1/\{f(z) - c\}$.

Consequently,

either $f(z) - c$ has a removable singularity at $z = a$

or $f(z) - c$ has a pole $z = a$ of order m.

∴ Either $f(z)$ has a removable singularity at $z = a$

or $f(z)$ has a pole $z = a$ of order m.

This contradicts the hypothesis that $z = a$ is an essential singularity of $f(z)$. Hence the theorem, as stated.

9.11. DETERMINATION OF THE NATURE OF SINGULARITIES

We defined a removable singularity, a pole and an essential singularity of a function $f(z)$ at $z = a$ with reference to the number of negative powers in the Laurent's expansion of $f(z)$ about $z = a$. But, from the foregoing theorems, we see that we can define the singularity with reference to the limit of the function as $z \to a$. If $z = a$ is a singularity of $f(z)$, then it is

(*i*) a removable singularity if $\lim_{z \to a} f(z)$ exists and is finite,

(*ii*) a pole if $\lim_{z \to a} f(z)$ exists and equals ∞, and

(*iii*) an essential singularity if $\lim_{z \to a} f(z)$ does not exist.

In finding limits we can use L' Hospital's rule which is true for complex functions also.

Example 1. $\lim_{z \to 0} \frac{\sin z}{z} = 1$ and so the singularity of $(\sin z)/z$ at $z = 0$ is a removable one. So also $(\sin \sqrt{z})/\sqrt{z}$ has a removable singularity at $z = 0$ since limit of this function at $z \to 0$ is 1.

Example 2. $\lim_{z \to 0} \left(\operatorname{cosec} z - \frac{1}{z}\right) = 0.$. So $z = 0$ is a removable singularity of $\operatorname{cosec} z - (1/z)$.

Example 3. $\lim_{z \to 1} \frac{e^z}{(z-1)^2} = \infty.$. So $z = 1$ is a pole of $\frac{e^z}{(z-1)^2}$.

Example 4. $\lim_{z \to 0} e^{1/z}$ does not exist. So $z = 0$ is an essential singularity of $e^{1/z}$.

Limit point of zeros and limit point of poles. From theorem 9.3 we see that, if a function $f(z)$ vanishes at points

$$z_n, \quad n = 1, 2, 3, \ldots$$

where the sequence $\{z_n\}$ converges to a limit a and if $f(z)$ does not vanish identically, then $f(z)$ is not analytic at $z = a$. That is $z = a$ is a singular point and it is isolated. Obviously it is not a removable singularity or a pole. Therefore it should be an essential singularity; more specifically, it should be an isolated essential singularity. On the other hand the limit point of poles is a non-isolated singularity because the limit point cannot be a pole since poles are isolated singularities.

Example 1. The zeros of $f(z) = \sin \{1/(z - a)\}$ are given by $\sin \dfrac{1}{z-a} = 0$ as $z = a + \dfrac{1}{n\pi}$, $n = \pm 1, \pm 2, \ldots$

Now $z = a$ is the limit point of the zeros. It can be seen that it is an isolated essential singularity of $f(z)$.

Example 2. The poles of $f(z) = 1/\left\{\sin \dfrac{1}{z-a}\right\}$ are given by $\sin \dfrac{1}{z-a} = 0$ as $z = a + \dfrac{1}{n\pi}$, $n = \pm 1, \pm 2, \ldots$

Now $z = a$ is the limit point of the poles. It can be seen that it is a non-isolated essential singularity of $f(z)$.

Isolation of a singularity. Now it is evident that removable singularities and poles are isolated singularities while the essential singularity is either an isolated or a non-isolated singularity.

9.12. NATURE OF SINGULARITY AT ∞

Suppose $f(z)$ has an isolated singularity at ∞. Then there is a deleted neighbourhood of $z = \infty$, $R < |z| < \infty$, in which $f(z)$ is analytic. Now, through the transformation $w = 1/z$, we get that $f(1/z)$ has an isolated singularity at $z = 0$ and $f(1/z)$ is analytic in the deleted neighbourhood of $z = 0$, $0 < |z| < 1/R$. So to examine the nature of singularity of $f(z)$ at $z = \infty$ we should examine the nature of singularity of $f(1/z)$ at $z = 0$. Thus, if $f(z)$ is expanded in a power series in z valid in $R < |z| < \infty$, then $z = \infty$ is a removable singularity or a pole or an essential singularity according as the series has no positive powers or a finite number of positive powers or an infinite number of positive powers.

9.13. EXAMPLES

9.1. Taylor's Series

1. $f(z) = \sum_{n=0}^{\infty} \dfrac{f^n(a)}{n!}(z-a)^n$ is a ... series.

2. Expand e^z about $z = 2$.

[*Hint* : $e^z = \sum_{n=0}^{\infty} \dfrac{f^n(2)}{n!}(z-2)^n = \sum_{n=0}^{\infty} \dfrac{e^2}{n!}(z-2)^n$]

3. Give the Taylor's series for $\sin z$ about $z = \pi/4$.

[**Ans.** $\sum_{n=0}^{\infty} \dfrac{\sin(\pi/4 + n\pi/2)}{n!}\left(z - \dfrac{\pi}{4}\right)^n$]

4. Find the Taylor's series for ze^z about $z = 1$.
[*Hint* : If $f(z) = ze^z, f^n(z) = (z+n)e^z, f^n(1) = (1+n)e$. So

$$ze^z = \sum_{n=0}^{\infty} \frac{f^n(1)}{n!}(z-1)^n \sum_{n=0}^{\infty} \frac{(1+n)\,e}{n!}(z-1)^n$$]

5. **Solved sum.** Expand $\cos z$ as Taylor's series about the points given below.
(*i*) $z = 0$ (*ii*) $z = \pi/4$ (*iii*) $z = \pi/3$.

✍ $\cos z$ is an entire function, that is, a function analytic everywhere except at $z = \infty$. We shall denote $\cos z$ by $f(z)$. Then, by successive differentiation, we have

$$f^n(z) = \cos(z + n\pi/2).$$

$$\therefore \quad f^n(0) = \cos(n\pi/2),$$

$$f^n(\pi/4) = \cos(\pi/4 + n\pi/2),$$

$$f^n(\pi/3) = \cos(\pi/3 + n\pi/2).$$

(*i*) Here the expansion is about $z = 0$. So the expansion is

$$f(z) = \sum_{n=0}^{\infty} \frac{f^n(0)}{n!}(z-0)^n$$

$$= \sum_{n=0}^{\infty} \frac{\cos(0 + n\pi/2)}{n!} z^n = \sum_{n=0}^{\infty} \frac{\cos n\pi/2}{n!} z^n$$

$$= 1 - \frac{z^2}{2!} + \frac{z^4}{4!} - \frac{z^6}{6!} + \ldots$$

(*ii*) Here the expansion is about $z = \pi/4$. So it is

$$f(z) = \sum_{n=0}^{\infty} \frac{f^n(\pi/4)}{n!}\left(z - \frac{\pi}{4}\right)^n$$

$$\therefore \quad \cos z = \sum_{n=0}^{\infty} \frac{\cos(\pi/4 + n\pi/2)}{n!}\left(z - \frac{\pi}{4}\right)^n$$

(*iii*) Here the expansion is about $z = \pi/3$. So it is

$$\cos z = \sum_{n=0}^{\infty} \frac{\cos(\pi/3 + n\pi/2)}{n!}\left(z - \frac{\pi}{3}\right)^n.$$

6. **Solved sum.** Expand $\log(1+z)$ as a Taylor's series about $z = 0$.

✍ Now $\log(1+z)$ is analytic everywhere in the finite plane except at $z = -1$ (given by $z + 1 = 0$). Now we want the expansion about $z = 0$. So consider the circle C with centre $z = 0$ and passing through the singular point $z = -1$. Now the distance beetween $z = 0$ and $z = -1$ is 1. So C is the circle

$$C: \quad |z| = 1.$$

Thus we shall expand $\log(1+z)$ in C_i. Denoting $\log(1+z)$ by $f(z)$, we have

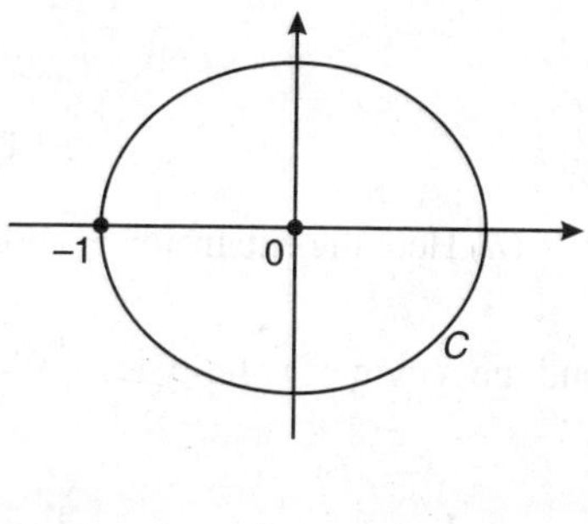

$$f(z) = \log(1+z) \qquad f(0) = 0$$

$$f'(z) = \frac{1}{1+z} \qquad f'(0) = 1$$

$$f''(z) = -\frac{1}{(1+z)^2} \qquad f''(0) = -1$$

$$f'''(z) = \frac{2!}{(1+z)^3} \qquad f'''(0) = 2!$$

$$f^4(z) = -\frac{3!}{(1+z)^4} \qquad f^4(0) = -3!$$

....

So the Taylor's series for log (1 + z) about $z = 0$ is

$$\log(1+z) = z - \frac{z^2}{2} + \frac{z^3}{3} - \frac{z^4}{4} + \ldots \quad |z| < 1.$$

Note. The expansion is valid for $z = 1$ also.

7. **Solved sum.** Find the Taylor's series for $1/z$ about

(*i*) $z = 1$ (*ii*) $z = 2$ (*iii*) $z = i$.

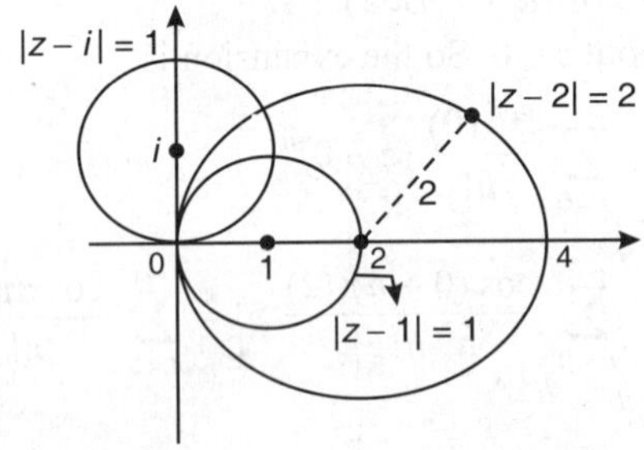

(*i*) $f(z) = 1/z$ is not analytic only at $z = 0$. Now we want its expansion about $z = 1$. So we have to choose the largest circle with centre $z = 1$ such that $1/z$ is analytic in its interior. So its radius should be the distance between $z = 0$ and $z = 1$. The distance is 1. Thus the circle is

$$C : |z - 1| = 1.$$

Therefore the region of validity of the Taylor's expansion in powers of $z - 1$ is

$$C_i : |z - 1| < 1.$$

Now the required Taylor's series about $z = 1$ is

$$f(z) = f(1) + \frac{f'(1)}{1!}(z-1) + \frac{f''(1)}{2!}(z-1)^2 + \ldots$$

Since $f(z) = 1/z$, $f(1) = 1$ and

$$f'(z) = -\frac{1}{z^2} \qquad f'(1) = -1$$

$$f''(z) = \frac{2!}{z^3} \qquad f''(1) = 2!$$

$$f'''(z) = -\frac{3!}{z^4} \qquad f'''(1) = -3!$$

....

$$\therefore \quad \frac{1}{z} = 1 - (z-1) + (z-1)^2 - (z-1)^3 + \ldots$$

(*ii*) Here the expansion is about $z = 2$ and the distance between $z = 2$ and $z = 0$ is 2. Hence

$$C : |z - 2| = 2$$

and the region of validity is

$$C_i : |z - 2| < 2.$$

So the required Talor's series is

$$f(z) = f(2) + \frac{f'(2)}{1!}(z-2) + \frac{f''(2)}{2!}(z-2)^2 + \dots$$

But $f(2) = 1/2$ and

$$f'(2) = -\frac{1}{2^2}, \quad f''(2) = \frac{2!}{2^3}, \quad f'''(2) = -\frac{3!}{2^4}, \dots$$

$$\therefore \qquad \frac{1}{z} = \frac{1}{2} - \frac{1}{2^2}(z-2) + \frac{1}{2^3}(z-2)^2 - \frac{1}{2^4}(z-2)^3 + \dots$$

(*iii*) Here the expansion is about $z = i$ and the distance between $z = i$ and $z = 0$ is 1. Hence

$$C : |z - i| = 1$$

and the region of validity is

$$C_i : |z - i| < 1$$

So the required Taylor's series is

$$f(z) = f(i) + \frac{f'(i)}{1!}(z-i) + \frac{f''(i)}{2!}(z-i)^2 + \dots$$

$$= \frac{1}{i} - \frac{1}{i^2}(z-i) + \frac{1}{i^3}(z-i)^2 - \frac{1}{i^4}(z-i)^3 + \dots$$

Second method (*i*). To get the series we can use binomial theorem also. For any z in the circular disc $z - 1| < 1$,

we have

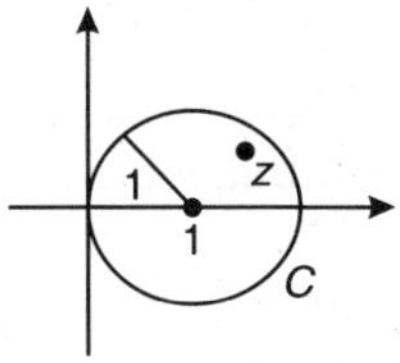

$$\frac{1}{z} = \frac{1}{1+(z-1)} = [1 + (z-1)]^{-1}$$

$$= 1 - (z-1) + (z-1)^2 - (z-1)^3 + (z-1)^4 - \dots$$

(***ii***) For any z in the disc $|z - 2| < 2$, (1)

$$\frac{1}{z} = \frac{1}{(z-2)+2} = \frac{1}{2}\frac{1}{1+\dfrac{z-2}{2}}.$$

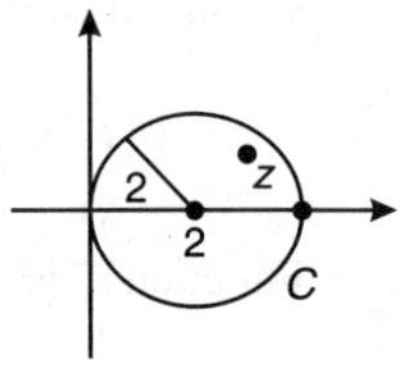

From (1) we have, $\left|\dfrac{z-2}{2}\right| < 1.$

So, by binomial theorem,

$$\frac{1}{z} = \frac{1}{2}\left[1 + \frac{z-2}{2}\right]^{-1}$$

$$= \frac{1}{2}\left[1 - \left(\frac{z-2}{2}\right) + \left(\frac{z-2}{2}\right)^2 - \left(\frac{z-2}{2}\right)^3 + \dots\right]$$

(***iii***) For any z in the disc, $\quad |z - i| < 1$ (1)

$$\frac{1}{z} = \frac{1}{(z-i)+i} = \frac{1}{i}\frac{1}{1+\dfrac{z-i}{i}} = \frac{1}{i}\left[1 + \frac{z-i}{i}\right]^{-1}$$

Now by (1), $\left|\dfrac{z-i}{i}\right| = \dfrac{|z-i|}{|i|} < 1$. Therefore

$$\frac{1}{z} = \frac{1}{i}\left[1-\left(\frac{z-i}{i}\right)+\left(\frac{z-i}{i}\right)^2-\left(\frac{z-i}{i}\right)^3+\ldots\right]$$

8. **Solved sum.** Find the Taylor's expansion about $z = 0$ of

$$f(z) = \frac{z}{(z+1)(z-3)}.$$

✍ $f(z)$ is not analytic at

$$z = -1, \quad z = 3.$$

Now we have to find the expansion about $z = 0$. So the largest circle C with centre $z = 0$ such that $f(z)$ is analytic in C_i is the circle which passes through $z = -1$. So it is $|z| = 1$. Thus, for all z in C_i, we get

$$|z| < 1.$$

Splitting $f(z)$ into partial fractions,

$$f(z) = \frac{1}{4}\cdot\frac{1}{z+1}+\frac{3}{4}\cdot\frac{1}{z-3}$$

$$= \frac{1}{4}\cdot\frac{1}{1+z}+\frac{3}{4}\cdot\frac{1}{-3}\frac{1}{1-z/3}$$

$$= \frac{1}{4}(1+z)^{-1}-\frac{1}{4}(1-z/3)^{-1}.$$

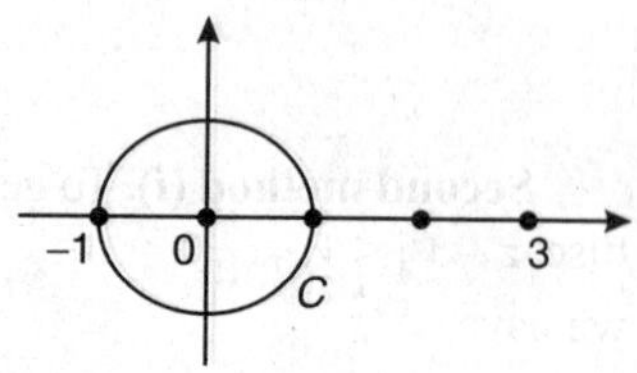

But, for $|z| < 1$, we get that $|z/3| < 1$. Thus

$$f(z) = \frac{1}{4}(1-z+z^2-z^3+z^4-\ldots)-\frac{1}{4}\left(1+\frac{z}{3}+\frac{z^2}{3^2}+\frac{z^3}{3^3}+\frac{z^4}{3^4}+\ldots\right)$$

$$= \frac{1}{4}\sum_{n=1}^{\infty}\left[(-1)^n-\frac{1}{3^n}\right]z^n.$$

9. **Solved sum.** Show that, for every finite z,

$$e^z = e\left[1+\frac{z-1}{1!}+\frac{(z-1)^2}{2!}+\frac{(z-1)^3}{3!}+\ldots\right] \qquad \textbf{(1974)}$$

Function e^z is analytic in the finite plane and the given series is an expansion about $z = 1$. Now

$$f(z) = f'(z) = f''(z) = f'''(z) = \ldots = e^z.$$

$$\therefore \quad f(1) = f'(1) = f''(1) = f'''(1) = \ldots = e.$$

By Taylor's theorem, for every z in a circular disc with centre at $z = 1$, say $|z-1| < \rho$,

$$e^z = e+\frac{e}{1!}(z-1)+\frac{e}{2!}(z-1)^2+\ldots$$

But ρ is arbitrary. So this expansion is valid for all finite z.

10. Find the Taylor's series for e^z about $z = 1$. State the region of validity of expansion. **(1979)**

[*Hint* : Same as the previous sum]

11. Find the Taylor's series for ze^z about $z = 1$.

[*Hint* : $f(z) = ze^z$; $f(1) = e$, $f'(1) = 2e$, $f''(1) = 3e$, $f'''(1) = 4e$, ... For all finite z,

$$ze^z = e\left\{1 + \frac{2(z-1)}{1!} + \frac{3(z-1)^2}{2!} + \frac{4(z-1)^3}{3!} + \ldots\right\}].$$

12. Find the Taylor's series for the functions

(*i*) $\log(1-z)$ **(1981)** (*ii*) $\log(1+z)$

about $z = 0$. State the regions in which the expansions are valid.

[*Hint* : (*i*) $z = 1$ is a singular point of $\log(1-z)$ and its distance from 0 is 1. So $\log(1-z) = -(z + z^2/2 + z^3/3 + \ldots)$ for all z in $|z - 0| < 1$ or $|z| < 1$ (*ii*) Refer Solved sum No. 6, 9.13. Examples. $z = -1$ is a singular point. So $\log(1+z) = z - z^2/2 + z^3/3 - z^4/4 + \ldots$ for all z in $|z| < 1$]

13. Show that $1/z^2 = 1 + 2(z+1) + 3(z+1)^2 + \ldots$ for all z in $|z+1| < 1$. **(1975)**

[*Hint* : The expansion is about $z = -1$. Distance of the point $z = -1$ from the singular point $z = 0$ is 1. So expansion is valid in $|z - (-1)| < 1$].

14. Show that $1/z^2 = 1 - 2(z-1) + 3(z-1)^2 - 4(z-1)^3 + \ldots$ for all z in $|z-1| < 1$.

15. Show that $\dfrac{1}{z^2} = \dfrac{1}{4}\left[1 - 2\left(\dfrac{z-2}{2}\right) + 3\left(\dfrac{z-2}{2}\right)^2 - \ldots\right]$ for all z in $|z-2| < 2$.

16. Find the Taylor's series for $(z-1)/z^2$ about $z = 1$. State the region of validity. **(1980)**

[*Hint* : Vide sum 6; $(z-1)(1/z^2) = (z-1)\{1 - 2(z-1) + 3(z-1)^2 - \ldots\}$. Expansion is valid in $|z-1| < 1$].

17. Obtain the Taylor's series for $z/(z+2)$ about $z = 1$. State the region of validity.

[*Hint* : $f(z) = 1 - \dfrac{2}{z+2} = \dfrac{1}{3} + 2\sum\limits_{n=1}^{\infty}(-1)^{n+1}\dfrac{(z-1)^n}{3^{n+1}}$. Region of validity is $|z-1| < 3$ because distance between 1 and -2 is 3].

18. Obtain the Taylor's series to represent $1/\{(z+1)(z+3)\}$ in $|z| < 1$. **(1981)**

[*Hint* : Expansion needed is about $z = 0$. Function is analytic in $|z| < 1$ since its singular points are -1 and -3.

$$\frac{1}{(z+1)(z+3)} = \frac{1}{2}\frac{1}{1+z} - \frac{1}{2}\frac{1}{3+z}$$

$$= \frac{1}{2}\frac{1}{1+z} - \frac{1}{6}\frac{1}{1+(z/3)}$$

$$= \frac{1}{2}(1 - z + z^2 - z^3 + \ldots) - \frac{1}{6}\left(1 - \frac{z}{3} + \frac{z^2}{3^2} - \frac{z^3}{3^3} + \ldots\right)$$

$= \dfrac{1}{2}\sum\limits_{n=0}^{\infty}(-1)^n\left[1 - \dfrac{1}{3^{n+1}}\right]z^n$. This is so either by Taylor's expansion or by binomial expansion]

19. Find the Taylor's series to represent $\dfrac{z^2-1}{(z+2)(z+3)}$ in $|z| < 2$. **(1976, 80, 81)**

[*Hint* : Expansion needed is about $z = 0$. Function is analytic in $|z| < 2$ since its singular points are $z = -2$ and, $z = -3$.

$$\frac{z^2-1}{(z+2)(z+3)} = 1 + \frac{3}{z+2} - \frac{8}{z+3}$$

$$= 1 + \frac{3}{2}\,\frac{1}{1+z/2} - \frac{8}{3}\,\frac{1}{1+z/3}$$

$$= -\frac{1}{6} + \sum_{n=1}^{\infty} (-1)^n \left\{\frac{3}{2^{n+1}} - \frac{8}{3^{n+1}}\right\} z^n \;]$$

20. Show that $-1 + \sum_{n=1}^{\infty} (-1)^{n+1} \left[1 + \frac{1}{4^n}\right] z^n$ represents $\frac{(z-2)(z+2)}{(z+1)(z+4)}$ in $|z| < 1$. **(1982)**

[*Hint*: $\frac{(z-2)(z+2)}{(z+1)(z+4)} = 1 - \frac{1}{z+1} - \frac{4}{z+4}$]

21. Show that

(*i*) $\frac{1}{1+z} = 1 - z + z^2 - z^3 + \ldots$ for $|z| < 1$

(*ii*) $\frac{1}{1+z^2} = 1 - z^2 + z^4 - z^6 + \ldots$ for $|z| < 1$

(*iii*) $\sin z = z - \frac{z^3}{3!} + \frac{z^5}{5!} - \frac{z^7}{7!} + \ldots$ for $|z| < \infty$

(*iv*) $\sin z^2 = z^2 - \frac{z^6}{3!} + \frac{z^{10}}{5!} - \frac{z^{14}}{7!} + \ldots$ for $|z| < \infty$

(*v*) $\cos z = 1 - \frac{z^2}{2!} + \frac{z^4}{4!} - \frac{z^6}{6!} + \ldots$ for $|z| < \infty$

(*vi*) $z \cosh z^2 = z + \frac{z^5}{2!} + \frac{z^9}{4!} + \frac{z^{13}}{6!} + \ldots$ for $|z| < \infty$

9.3. Laurent's Series

(*i*) *In powers of z*

22. **Solved sum.** Obtain the series for $1/(z-3)$ valid in

(*i*) $|z| < 3$ (*ii*) $|z| > 3$. **(1978)**

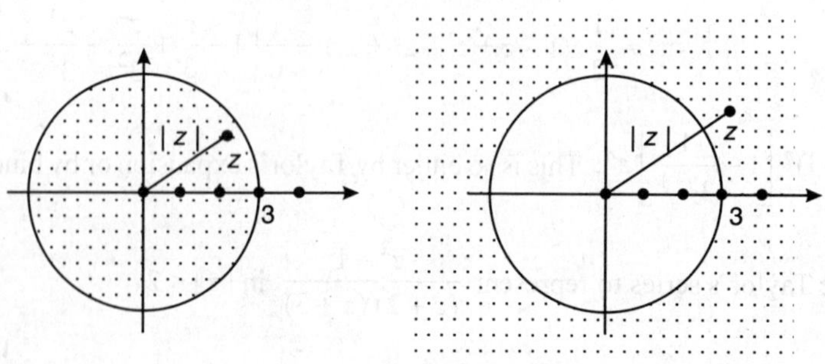

Fig. 9.13.1

(*i*) $1/(z-3)$ is not analytic at $z = 3$ but analytic in the disc $|z| < 3$ (Fig. 9.13.1). Hence the Taylor's series about $z = 0$ which is valid in $|z| < 3$ is

$$\frac{1}{z-3} = -\frac{1}{3}\frac{1}{1-(z/3)} = (-1/3)\,[1 + (z/3) + (z/3)^2 + ...]$$

(*ii*) $1/(z - 3)$ is analytic in $|z| > 3$. But, for all finite z in $|z| > 3$, $|3/z| < 1$. So, using binomial series, the Laurent's expansion is

$$\frac{1}{z-3} = \frac{1}{z\,(1-3/z)} = \frac{1}{z}\left[1+\frac{3}{z}+\frac{3^2}{z^2}+...\right] = \frac{1}{z}+\frac{3}{z^2}+\frac{3^2}{z^3}+\frac{3^3}{z^4}+...$$

23. **Solved sum.** Obtain two different Laurent's series for $1/z^2\,(1 - z)$ about $z = 0$ and state the regions of validity. **(1980)**

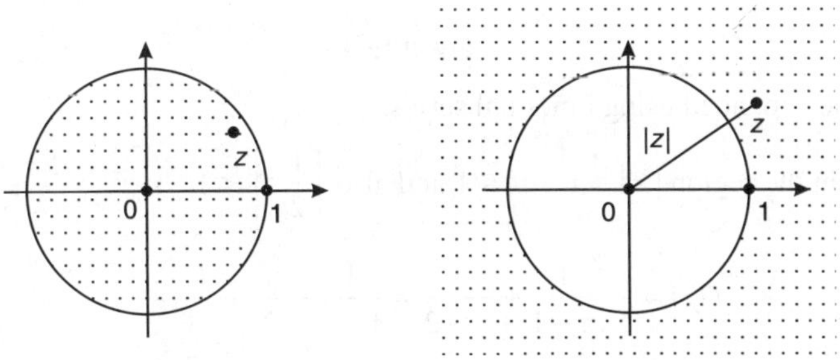

Fig. 9.13.2

The given function is not analytic at $z = 0$ and $z = 1$ but is analytic in the deleted disc (*i*) $0 < |z| < 1$ and (*ii*) in $|z| > 1$ (Fig. 9.13.2). Therefore there are two Laurent's series, one is valid in $0 < |z| < 1$ and the other is valid in $|z| > 1$.

(*i*)
$$\frac{1}{z^2\,(1-z)} = \frac{1}{z^2}\,(1-z)^{-1}$$

$$= \frac{1}{z^2}\,(1 + z + z^2 + z^3 + ...),\ 0 < |z| < 1$$

$$= 1/z^2 + 1/z + 1 + z + z^2 + ...,\ 0 < |z| < 1$$

(*ii*)
$$\frac{1}{z^2\,(1-z)} = \frac{1}{z^3}\frac{-1}{1-1/z}$$

$$= -\frac{1}{z^3}\left(1+\frac{1}{z}+\frac{1}{z^2}+...\right),\ |1/z| < 1$$

$$= -(1/z^3 + 1/z^4 + 1/z^5 + ...),\ |z| > 1$$

24. **Solved sum.** Expand $1/\{(z - 1)(z - 2)\}$ as a power series in z valid in

(*i*) $|z| < 1$ (*ii*) $1 < |z| < 2$ (*iii*) $|z| > 2$ **(1980, 81, 84)**

Let the given function be $f(z)$. The points $z = 1$ and $z = 2$ are its singular points (Fig. 9.13.3). So $f(z)$ is analytic

(*i*) in the disc $|z| < 1$

(*ii*) in the annulus $1 < |z| < 2$ and

(*iii*) in the multiply connected region $|z| > 2$.

Therefore $f(z)$ can be expanded, about $z = 0$, as a Taylor's series valid in $|z| < 1$, as a Laurent's series valid in $1 < |z| < 2$ and as a Laurent's series valid in $|z| > 2$. Now

$$f(z) = \frac{1}{(z-1)(z-2)} = \frac{-1}{z-1} + \frac{1}{z-2}.$$

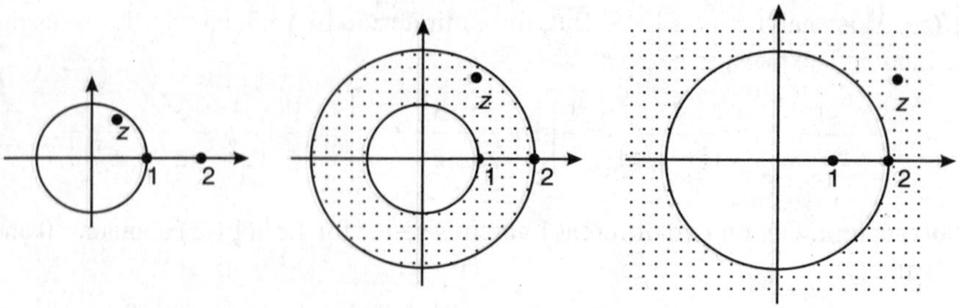

Fig. 9.13.3

So $f(z)$ can be expanded using binomial series.

(*i*) For any z in the region $|z| < 1$, $|z| < 1$ and also $\left|\frac{1}{2}z\right| < 1$.

$$\therefore \qquad f(z) = -\frac{1}{z-1} + \frac{1}{z-2} = \frac{1}{1-z} - \frac{1}{2}\frac{1}{1-\frac{1}{2}z}$$

$$= (1 + z + z^2 + ...) - \frac{1}{2}\left[1 + \left(\frac{1}{2}z\right) + \left(\frac{1}{2}z\right)^2 + ...\right]$$

$$= \sum_{n=0}^{\infty}\left(1 - \frac{1}{2^{n+1}}\right)z^n$$

(*ii*) In the annulus region $1 < |z| < 2$, $|1/z| < 1$ and $\left|\frac{1}{2}z\right| < 1$. Therefore

$$f(z) = \frac{-1}{z-1} + \frac{1}{z-2} = -\frac{1}{z}\frac{1}{1-(1/z)} - \frac{1}{2}\frac{1}{1-\frac{1}{2}z}$$

$$= -\frac{1}{z}\left[1 + \frac{1}{z} + \frac{1}{z^2} + ...\right] - \frac{1}{2}\left[1 + \left(\frac{1}{2}z\right) + \left(\frac{1}{2}z\right)^2 + ...\right]$$

$$= -\sum_{n=0}^{\infty}\frac{1}{z^{n+1}} - \sum_{n=0}^{\infty}\frac{z^n}{2^{n+1}} = -\sum_{n=0}^{\infty}\left(\frac{1}{z^{n+1}} + \frac{z^n}{2^{n+1}}\right).$$

(*iii*) In $|z| > 2$, $|1/z| < 1$ and $|2/z| < 1$. So

$$f(z) = -\frac{1}{z-1} + \frac{1}{z-2} = -\frac{1}{z}\frac{1}{1-(1/z)} + \frac{1}{z}\frac{1}{1-(2/z)}$$

$$= -\frac{1}{z}\left[1 + \frac{1}{z} + \frac{1}{z^2} + ...\right] + \frac{1}{z}\left[1 + \frac{2}{z} + \frac{2^2}{z^2} + ...\right]$$

$$= \sum_{n=1}^{\infty}\frac{2^n - 1}{z^{n+1}}.$$

25. Find the Laurent's series for

$$f(z) = \frac{1}{z^2 - z - 2}$$

in the domain $1 < |z| < 2$.

[*Hint* : In $1 < |z| < 2$, for any z, $|z/2| < 1$, $|1/z| < 1$. So

$$f(z) = \frac{1}{(z+1)(z-2)} = \frac{1/3}{z-2} - \frac{1/3}{z+1} = -\frac{1}{6} \cdot \frac{1}{1-z/2} - \frac{1}{3z} \cdot \frac{1}{1+1/z}$$

$$= -\frac{1}{6}\left(1 + \frac{z}{2} + \frac{z^2}{2^2} + \ldots\right) - \frac{1}{3z}\left(1 - \frac{1}{z} + \frac{1}{z^2} - \frac{1}{z^3} + \ldots\right)]$$

26. **Solved sum.** Find the Laurent's expansion for

$$f(z) = \frac{z^2 - 1}{(z+2)(z+3)}$$

(*i*) in $2 < |z| < 3$ (*ii*) $|z| > 3$

✍ Now the centre of the given domains are $z = 0$. So the required expansion is about $z = 0$, that is, in powers of z. The function $f(z)$ is not analytic at $z = -2$, $z = -3$.

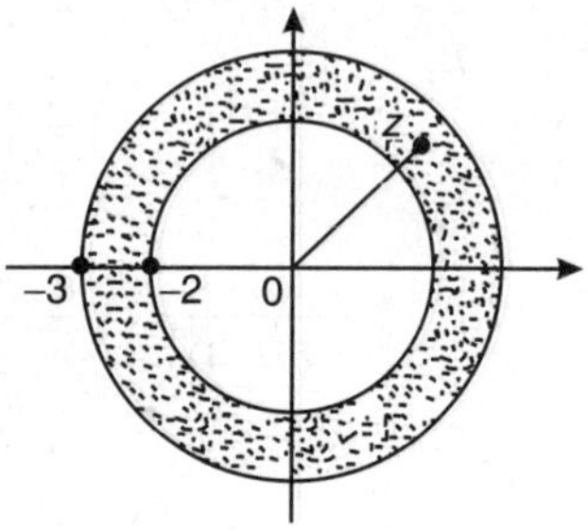

(*i*) Annular domain $2 < |z| < 3$. For all z in the annular domain

$$2 < |z| < 3,$$

z is such that

$$2 < |z|, \quad |z| < 3.$$

$\therefore$ $$|2/z| < 1, \quad |z/3| < 1.$$

So, rewriting $f(z)$ as one which is expandable as binomial series,

$$f(z) = \frac{z^2 - 1}{(z+2)(z+3)} = 1 + \frac{3}{z+2} - \frac{8}{z+3}$$

$$= 1 + \frac{1}{z} \cdot \frac{3}{1+2/z} - \frac{1}{3} \cdot \frac{8}{1+z/3} = 1 + \frac{3}{z}\left(1 + \frac{2}{z}\right)^{-1} - \frac{8}{3}\left(1 + \frac{z}{3}\right)^{-1}$$

$$= 1 + \frac{3}{z}\left(1 - \frac{2}{z} + \frac{2^2}{z^2} - \ldots\right) - \frac{8}{3}\left(1 - \frac{z}{3} + \frac{z^2}{3^2} - \ldots\right)$$

$$= \left(1 - \frac{8}{3}\right) + 8\left(\frac{z}{3^2} - \frac{z^2}{3^3} + \frac{z^3}{3^4} - \ldots\right) + 3\left(\frac{1}{z} - \frac{2}{z^2} + \frac{2^2}{z^3} - \frac{2^3}{z^4} + \ldots\right)$$

(*ii*) Domain $|z| > 3$. For all z in the domain

$$|z| > 3,$$

z is such that

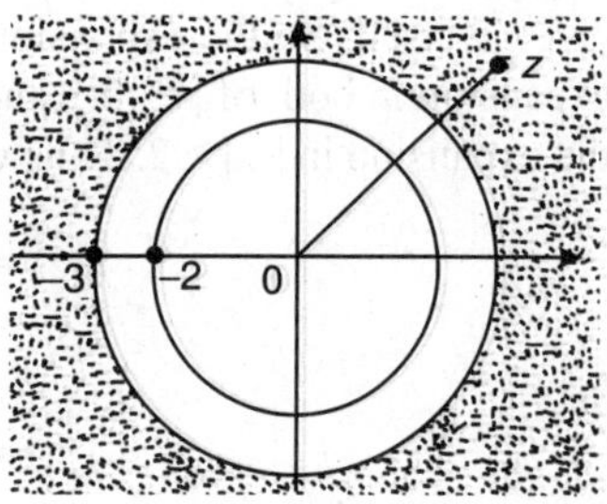

$$2 < |z|, \quad 3 < |z|$$

$\therefore$ $$|2/z| < 1, \quad |3/z| < 1.$$

So, rewriting $f(z)$ as one which is expandable as binomial series,

$$f(z) = 1 + \frac{3}{z+2} - \frac{8}{z+3} = 1 + \frac{1}{z} \cdot \frac{3}{1+2/z} - \frac{1}{z} \cdot \frac{8}{1+3/z}$$

$$= 1+\frac{3}{z}\left(1+\frac{2}{z}\right)^{-1} - \frac{8}{z}\left(1+\frac{3}{z}\right)^{-1}$$

$$= 1+\frac{3}{z}\left(1-\frac{2}{z}+\frac{2^2}{z^2}-\frac{2^3}{z^3}+...\right)-\frac{8}{z}\left(1-\frac{3}{z}+\frac{3^2}{z^2}-\frac{3^3}{z^3}+...\right)$$

$$= 1+(3-8)\frac{1}{z} - (3\times 2 - 8\times 3)\frac{1}{z^2} + (3\times 2^2 - 8\times 3^2)\frac{1}{z^3} - ...$$

27. Find the Laurent's series for the following functions valid in $|z| > 1$:

(*i*) $\dfrac{1}{z^2(z-1)}$ (*ii*) $\dfrac{1}{z(z-1)^2}$

[*Hint* : Since $|z| > 1$, we have $|1/z| < 1$. Therefore

(*i*) $$\frac{1}{z^2(z-1)} = \frac{1}{z^2}\cdot\frac{1}{z(1-1/z)} = \frac{1}{z^3}\left(1-\frac{1}{z}\right)^{-1}$$

$$= \frac{1}{z^3}\left(1+\frac{1}{z}+\frac{1}{z^2}+\frac{1}{z^3}+...\right)$$

(*ii*) $$\frac{1}{z(z-1)^2} = \frac{1}{z}\cdot\frac{1}{z^2(1-1/z)^2} = \frac{1}{z^3}\left(1-\frac{1}{z}\right)^{-2}$$

$$= \frac{1}{z^3}\left(1+\frac{2}{z}+\frac{3}{z^2}+\frac{4}{z^3}+...\right)]$$

28. **Solved sum.** Find the Laurent's expansion of

$$\frac{7z-2}{z(z-2)(z+1)}$$

in the domains

(*i*) $|z| < 2$ (*ii*) $|z| > 3$ (*iii*) $2 < |z| < 3$.

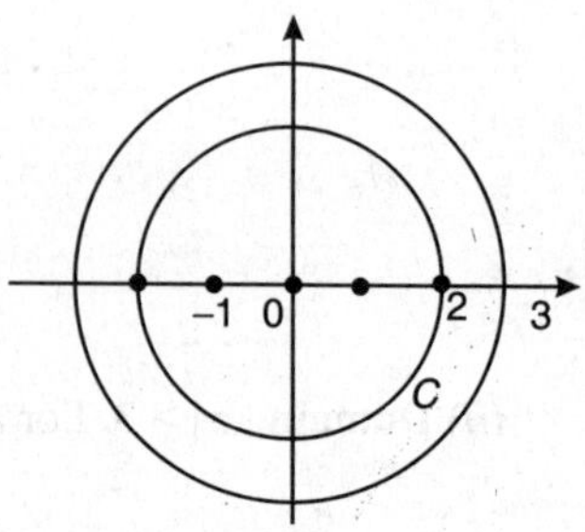

✍ We shall name the circle $|z| = 2$ as C. Now $0, -1, 2$ are the singular points of $f(z)$.

1. Domain $|z| < 2$. This region is C_i. This contains the singular points $z = -1$, $z = 0$. So the expansion in this region as a Taylor's series or as a Laurent's series is not possible.

2. Domains $|z| > 3$ and $2 < |z| < 3$. None of $-1, 0, 2$ lies in $|z| > 2$. But the domains $|z| > 3$ and $2 < |z| < 3$ are contained in $|z| > 2$ itself. So the expansion about $z = 0$ in $|z| > 2$ is also the expansion in both of $|z| > 3$ and $2 < |z| < 3$. So we shall now find the expansion in $|z| > 2$. Thus we have $|2/z| < 1$.

$$\therefore \quad f(z) = \frac{1}{z}+\frac{2}{z-2}-\frac{3}{z+1} = \frac{1}{z}+\frac{2}{z(1-2/z)}-\frac{3}{z(1+1/z)}$$

$$= \frac{1}{z}+\frac{2}{z}\left(1-\frac{2}{z}\right)^{-1} - \frac{3}{z}\left(1+\frac{1}{z}\right)^{-1}.$$

Since $|2/z| < 1$, we get that $|1/z| < 1$.

$$f(z) = \frac{1}{z} + \frac{2}{z}\left(1 + \frac{2}{z} + \frac{2^2}{z^2} + \frac{2^3}{z^3} + \ldots\right) - \frac{3}{z}\left(1 + \frac{1}{z} + \frac{1}{z^2} - \frac{1}{z^3} + \ldots\right)$$

$$= \sum_{n=2}^{\infty} [2^n + 3(-1)^n]\frac{1}{z^n}.$$

This is valid in both of $|z| > 3$ and $2 < |z| < 3$.

29. Expand $1/\{z(z^2 - 3z + 2)]$ as a power series valid in $1 < |z| < 2$. **(1981)**

[*Hint* : $$\frac{1}{z(z^2 - 3z + 2)} = \frac{1}{2}\frac{1}{z} - \frac{1}{z-1} + \frac{1}{2}\frac{1}{z-2}$$

$$= \frac{1}{2}\frac{1}{z} - \frac{1}{z}\frac{1}{1-(1/z)} - \frac{1}{2^2}\frac{1}{1-(z/2)}$$

$$= \frac{1}{2z} - \frac{1}{z}\left[1 + \frac{1}{z} + \frac{1}{z^2} + \ldots\right] - \frac{1}{4}\left[1 + \frac{z}{2} + \frac{z^2}{2^2} + \ldots\right]$$

$$= -\sum_{n=1}^{\infty}\left(\frac{1}{z^{n+2}} + \frac{z^{n-1}}{2^{n+1}}\right)]$$

30. Expand $1/\{(z+1)(z+3)\}$ in the regions

(*i*) $|z| < 1$ (*ii*) $1 < |z| < 3$ (*iii*) $|z| > 3$ **(1982)**

[*Hint* : Function: $\frac{1}{2}\frac{1}{z+1} - \frac{1}{2}\frac{1}{z+3}$; singularities: $-1, -3$;

(*i*) $\frac{1}{2}(1 - z + z^2 - \ldots) - \frac{1}{6}\{1 - (z/3) + (z/3)^2 - \ldots\}$

(*ii*) $\frac{1}{2z}\left\{1 - \frac{1}{z} + \frac{1}{z^2} - \ldots\right\} - \frac{1}{6}\left\{1 - \frac{z}{3} + \frac{z^2}{3^2} - \ldots\right\}$

(*iii*) $\frac{1}{2z}\left\{1 - \frac{1}{z} + \frac{1}{z^2} - \ldots\right\} - \frac{1}{2z}\left\{1 - \frac{3}{z} + \frac{3^2}{z^2} - \ldots\right\}$]

31. Show that the Taylor's or the Laurent's series in z for $z/\{(z-1)(2-z)\}$, respectively in $|z| < 1, 1 < |z| < 2, |z| > 2$, are $-[1 + z + z^2 + \ldots] + [1 + (z/2) + (z/2)^2 + \ldots]$, $|z| < 1$

$$\frac{1}{z}\left[1 + \frac{1}{z} + \frac{1}{z^2} + \ldots\right] + [1 + (z/2) + (z/2)^2 + \ldots], \quad 1 < |z| < 2$$

$$\frac{1}{z}\left[1 + \frac{1}{z} + \frac{1}{z^2} + \ldots\right] - \frac{2}{z}\left[1 + \frac{2}{z} + \frac{2^2}{z^2} + \ldots\right], \quad |z| > 2.$$

(ii) Laurent's series in powers of z – a

32. **Solved sum.** Expand $1/\{z(z-1)\}$ as Laurent's series in powers of z and in Laurent's series in powers of $z - 1$ and state the regions of validity. **(1978)**

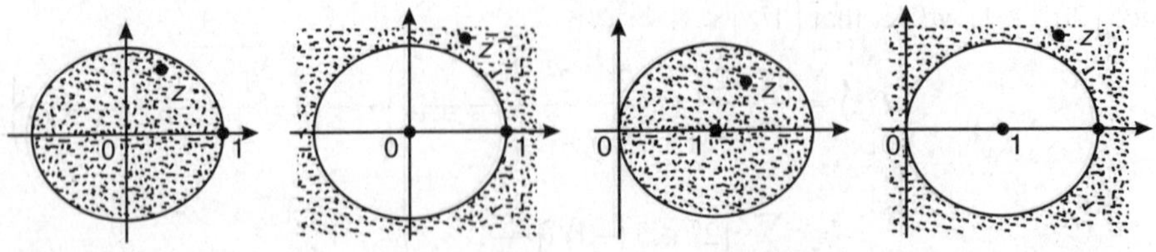

Fig. 9.13.4

The function $1/\{z(z-1)\}$ is not analytic at $z = 0$ and at $z = 1$, and it is analytic in the following regions :

(*i*) Deleted disc $0 < |z| < 1$ (Fig. 9.13.4a)

(*ii*) Region $|z| > 1$ (Fig. 9.13.4b)

(*iii*) Deleted disc $0 < |z-1| < 1$ (Fig. 9.13.4c)

(*iv*) Region $|z-1| > 1$ (Fig. 9.13.4d)

Case 1 : For all z in the deleted disc (*i*), $|z| < 1$. So in this disc

$$\frac{1}{z(z-1)} = \frac{1}{z}\left(\frac{-1}{1-z}\right) = \frac{-1}{z}(1 + z + z^2 + \dots)$$

$$= -(1/z) - 1 - z - z^2 - \dots$$

Case 2 : For all z in the region (*ii*), $|z| > 1$. So in this region

$$\frac{1}{z(z-1)} = \frac{1}{z^2}\frac{1}{1-(1/z)} = \frac{1}{z^2}\left(1 + \frac{1}{z} + \frac{1}{z^2} + \dots\right)$$

$$= 1/z^2 + 1/z^3 + 1/z^4 + \dots$$

Case 3 : For all z in the deleted disc (*iii*), $|z-1| < 1$. Therefore in this disc

$$\frac{1}{z(z-1)} = \frac{1}{(z-1)\{1+(z-1)\}} = \frac{1}{z-1}\{1 - (z-1) + (z-1)^2 - \dots\}.$$

Case 4 : For all z in the region (*iv*), $|z-1| > 1$. Therefore in this region

$$\frac{1}{z(z-1)} = \frac{1}{(z-1)\{1+(z-1)\}} = \frac{1}{(z-1)^2}\frac{1}{1+\{1/(z-1)\}}$$

$$= \frac{1}{(z-1)^2}\left\{1 - \frac{1}{z-1} + \frac{1}{(z-1)^2} - \dots\right\}.$$

33. **Solved sum.** For $f(z) = \dfrac{z+4}{(z+3)(z-1)^2}$, find the Laurent's expansions in the domains

(*i*) $0 < |z-1| < 4$ (*ii*) $|z-1| > 4$

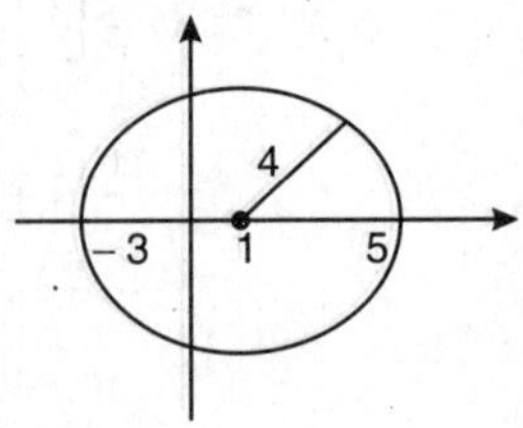

✍ From the given domains it is seen that the required expansions is about $z = 1$ so that it is in powers of $z - 1$. The function is not analytic at

$$z = 1, \quad z = -3$$

and, from the figure, it is seen that it is analytic in the given domains namely,

(*i*) $0 < |z-1| < 4$ (*ii*) $|z-1| > 4$.

For brevity, setting

$$z - 1 = u \quad \text{or} \quad z = u + 1,$$

$$f(z) = \frac{z+4}{(z+3)(z-1)^2} = \frac{(u+1)+4}{[(u+1)+3]u^2} = \frac{u+5}{(u+4)u^2}$$

$$= \frac{1}{u^2}\frac{(u+4)+1}{u+4} = \frac{1}{u^2}\left[1+\frac{1}{u+4}\right].$$

Now we want $f(z)$ in powers of u.

Case (*i*) $0 < |z-1| < 4$. In this case

$$0 < |u| < 4 \quad \text{or} \quad |u/4| < 1.$$

$$\therefore \quad f(z) = \frac{1}{u^2}\left[1+\frac{1}{u+4}\right] = \frac{1}{u^2}\left[1+\frac{1}{4(1+u/4)}\right]$$

$$= \frac{1}{u^2} + \frac{1}{4u^2(1+u/4)} = \frac{1}{u^2} + \frac{1}{4u^2}(1+u/4)^{-1}$$

$$= \frac{1}{u^2} + \frac{1}{4u^2}\left(1 - \frac{u}{4} + \frac{u^2}{4^2} - \frac{u^3}{4^3} + \ldots\right)$$

$$= \frac{5}{4u^2} - \frac{1}{4^2 u} + \frac{1}{4^3} - \frac{u}{4^4} + \frac{u^2}{4^5} - \frac{u^3}{4^6} + \ldots, \text{ where } u = z-1.$$

Case (*ii*) $|z-1| > 4$. In this case

$$|u| > 4 \quad \text{or} \quad |4/u| < 1.$$

$$\therefore \quad f(z) = \frac{1}{u^2}\left[1+\frac{1}{u+4}\right] = \frac{1}{u^2}\left[1+\frac{1}{u(1+4/u)}\right]$$

$$= \frac{1}{u^2} + \frac{1}{u^3(1+4/u)} = \frac{1}{u^2} + \frac{1}{u^3}\left(1+\frac{4}{u}\right)^{-1}$$

$$= \frac{1}{u^2} + \frac{1}{u^3}\left(1 - \frac{4}{u} + \frac{4^2}{u^2} - \frac{4^3}{u^3} + \ldots\right)$$

$$= \frac{1}{u^2} + \frac{1}{u^3} - \frac{4}{u^4} + \frac{4^2}{u^5} - \frac{4^3}{u^6} + \ldots, \text{ where } u = z-1.$$

34. **Solved sum.** Expand $f(z) = \dfrac{z}{(z+1)(z+2)}$ as a Laurent's series about $z = -2$.

The function $f(z)$ is not anlaytic at

$$z = -1, \quad z = -2.$$

The expansion is about $z = -2$. So we should consider the domains bounded by circles with centres at $z = -2$. From the figure we see that the domains in which $f(z)$ is analytic are

(*i*) $0 < |z+2| < 1$ (*ii*) $|z+2| > 1$.

For brevity, setting

$$z+2 = u \quad \text{or} \quad z = u-2,$$

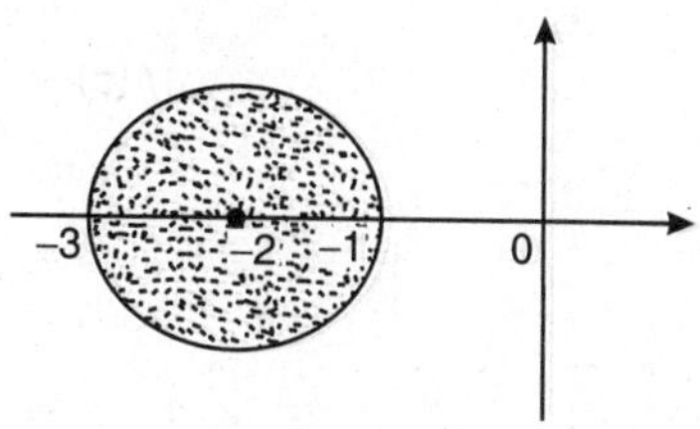

$$f(z) = \frac{u-2}{(u-2+1)u} = \frac{1}{u} \cdot \frac{u-2}{u-1}$$

$$= \frac{1}{u} \cdot \frac{(u-1)-1}{u-1} = \frac{1}{u}\left[1 - \frac{1}{u-1}\right].$$

Now we want $f(z)$ in powers of u.

Case (*i*) $0 < |z+2| < 1$. In this case

$$0 < |u| < 1 \text{ i.e., } \quad |u| < 1.$$

$$\therefore \quad f(z) = \frac{1}{u}\left[1 - \frac{1}{u-1}\right] = \frac{1}{u}\left[1 + \frac{1}{1-u}\right]$$

$$= \frac{1}{u} + \frac{1}{u} \cdot \frac{1}{1-u} = \frac{1}{u} + \frac{1}{u}(1-u)^{-1}$$

$$= \frac{1}{u} + \frac{1}{u}(1 + u + u^2 + u^3 + u^4 + \ldots)$$

$$= \frac{2}{u} + 1 + u + u^2 + u^3 + u^4 + \ldots, \text{ where } u = z + 2.$$

Case (*ii*) $|z+2| > 1$. In this case

$$|u| > 1 \quad \text{or} \quad |1/u| < 1.$$

$$\therefore \quad f(z) = \frac{1}{u}\left[1 - \frac{1}{u-1}\right] = \frac{1}{u} - \frac{1}{u} \cdot \frac{1}{u-1}$$

$$= \frac{1}{u} - \frac{1}{u} \cdot \frac{1}{u(1-1/u)} = \frac{1}{u} - \frac{1}{u^2}\left(1 - \frac{1}{u}\right)^{-1}$$

$$= \frac{1}{u} - \frac{1}{u^2}\left(1 + \frac{1}{u} + \frac{1}{u^2} + \frac{1}{u^3} + \ldots\right)$$

$$= \frac{1}{u} - \left(\frac{1}{u^2} + \frac{1}{u^3} + \frac{1}{u^4} + \frac{1}{u^5} + \ldots\right), \text{ where } u = z + 2.$$

35. **Solved sum.** Find the Laurent's expansion for $f(z) = \dfrac{1}{z(1-z)^2}$ in the domains

(*i*) $0 < |z| < 1$ (*ii*) $0 < |z-1| < 1$

✍ The function $f(z)$ is not analytic at

$$z = 0, \quad z = 1.$$

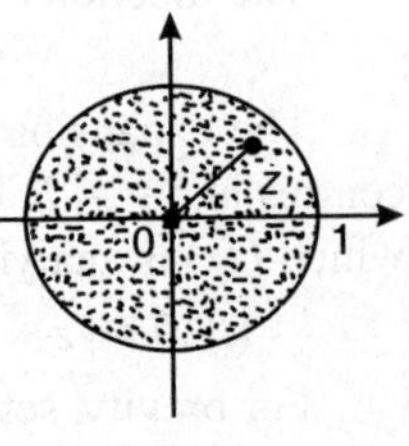

(*i*) In $0 < |z| < 1$, $f(z)$ is analytic and $|z| < 1$. So

$$f(z) = \frac{1}{z(1-z)^2} = \frac{1}{z}(1-z)^{-2}$$

$$= \frac{1}{z}(1 + 2z + 3z^2 + 4z^3 + 5z^4 + \ldots)$$

$$= \frac{1}{z} + 2 + 3z + 4z^2 + 5z^3 + 6z^4 + \ldots$$

(*ii*) In $0 < |z-1| < 1, f(z)$ is analytic and z is such that

$$|z-1| < 1.$$

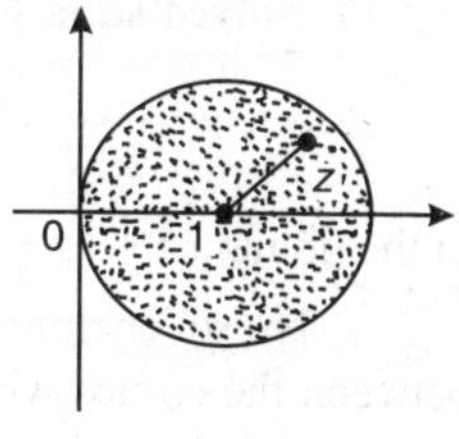

Set $z - 1 = u$ or $z = 1 + u$. Then we have

$$|u| < 1.$$

$$\therefore \quad f(z) = \frac{1}{(1+u)u^2} = \frac{1}{u^2}(1+u)^{-1}$$

$$= \frac{1}{u^2}\,[1 - u + u^2 - u^3 + u^4 - u^5 + ...]$$

$$= \frac{1}{u^2} - \frac{1}{u} + 1 - u + u^2 - u^3 + u^4 - ..., \text{ where } u = z - 1.$$

36. **Solved sum.** Find the Laurent's expansion of

$$f(z) = \frac{3z^2 - 7}{(z-1)(z-2)(z+3)} \text{ in the domain } |z+3| > 5.$$

✍ The given domain $D: |z+3| > 5$ is the region exterior to the circle

$$|z+3| = 5$$

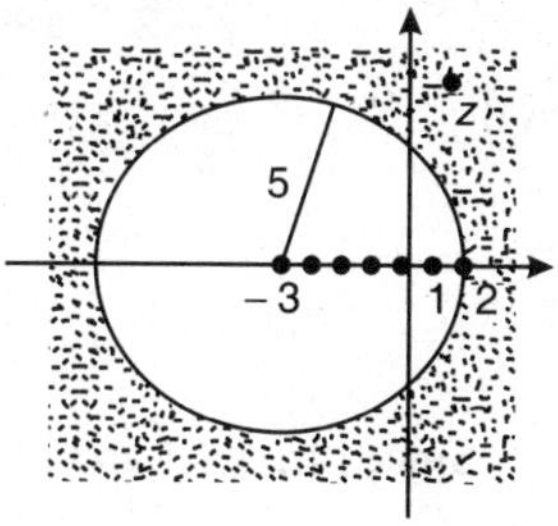

whose centre is $z = -3$ and radius is 5. From the figure it is evident that the singular points $1, 2, -3$ of $f(z)$ are not in R. Set

$$z + 3 = u, \quad z = u - 3$$

for brevity. Then, for all z in $|z+3| > 5$, we have

$$|u| > 5 \quad \text{or} \quad |5/u| < 1.$$

Splitting $f(z)$ into partial fractions,

$$f(z) = \frac{1}{z-1} + \frac{1}{z-2} + \frac{1}{z+3} = \frac{1}{(u-3)-1} + \frac{1}{(u-3)-2} + \frac{1}{u}$$

$$= \frac{1}{u-4} + \frac{1}{u-5} + \frac{1}{u} = \frac{1}{u}\cdot\frac{1}{1-4/u} + \frac{1}{u}\cdot\frac{1}{1-5/u} + \frac{1}{u}$$

$$= \frac{1}{u}\left(1 - \frac{4}{u}\right)^{-1} + \frac{1}{u}\left(1 - \frac{5}{u}\right)^{-1} + \frac{1}{u}.$$

Since $|5/u| < 1$, we get $|4/u| < 1$. Therefore

$$f(z) = \frac{1}{u}\left[1 + \frac{4}{u} + \frac{4^2}{u^2} + ...\right] + \frac{1}{u}\left[1 + \frac{5}{u} + \frac{5^2}{u^2} + ...\right] + \frac{1}{u}$$

$$= \left[\frac{1}{u} + \frac{4}{u^2} + \frac{4^2}{u^3} + \frac{4^3}{u^4} + ...\right] + \left[\frac{1}{u} + \frac{5}{u^2} + \frac{5^2}{u^3} + \frac{5^3}{u^4} + ...\right] + \frac{1}{u}$$

$$= \frac{3}{u} + \left[(4+5)\frac{1}{u^2} + (4^2+5^2)\frac{1}{u^3} + ...\right]$$

$$= \frac{3}{u} + \sum_{n=1}^{\infty} \frac{4^n + 5^n}{u^{n+1}}, \text{ where } u = z + 3.$$

37. **Solved sum.** Find the Laurent's expansion of

$$f(z) = \frac{7z-2}{z(z-2)(z+1)}$$

in the domain $1 < |z+1| < 3$.

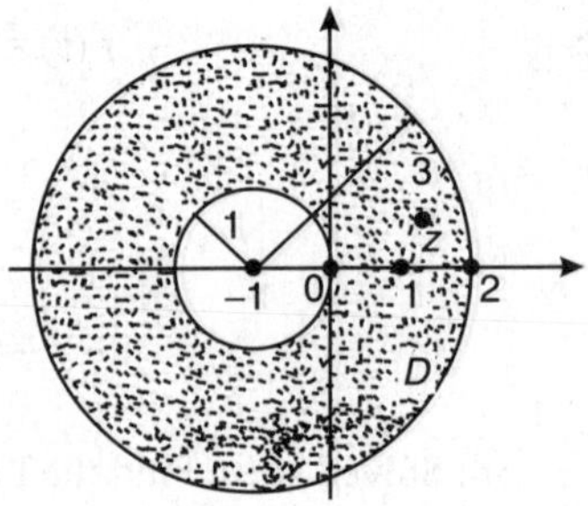

✍ The given domain $D : 1 < |z+1| < 3$ is the annular domain between the circles whose centres are $z = -1$ and radii are 1 and 3. From the figure it is evident that the singular points 0, 2, – 1 of $f(z)$ are not in D. Set

$$z + 1 = u, \quad z = u - 1$$

for brevity. Then, for all z in $1 < |z+1| < 3$, we have

$$1 < |u| < 3.$$

i.e., $$1 < |u|, \quad |u| < 3.$$

$\therefore$ $$|1/u| < 1, \quad |u/3| < 1.$$

Splitting $f(z)$ into partial fractions,

$$f(z) = \frac{1}{z} + \frac{2}{z-2} - \frac{3}{z+1} = \frac{1}{u-1} + \frac{2}{u-3} - \frac{3}{u}$$

$$= \frac{1}{u}\left(\frac{1}{1-\frac{1}{u}}\right) - \frac{2}{3}\left(\frac{1}{1-\frac{u}{3}}\right) - \frac{3}{u}$$

$$= \frac{1}{u}(1-\tfrac{1}{u})^{-1} - \frac{2}{3}(1-\tfrac{u}{3})^{-1} - \frac{3}{u}.$$

Since $|1/u| < 1$ and $|u/3| < 1$,

$$f(z) = \frac{1}{u}\left[1 + \frac{1}{u} + \frac{1}{u^2} + \frac{1}{u^3} + \ldots\right] - \frac{2}{3}\left[1 + \frac{u}{3} + \frac{u^2}{3^2} + \ldots\right] - \frac{3}{u}$$

$$= \left[\frac{1}{u} + \frac{1}{u^2} + \frac{1}{u^3} + \frac{1}{u^4} \ldots\right] - \frac{2}{3}\left[1 + \frac{u}{3} + \frac{u^2}{3^2} + \ldots\right] - \frac{3}{u}$$

$$= -\frac{2}{u} + \left[\frac{1}{u^2} + \frac{1}{u^3} + \frac{1}{u^4} \ldots\right] - \frac{2}{3}\left[1 + \frac{u}{3} + \frac{u^2}{3^2} + \frac{u^3}{3^3} + \ldots\right]$$

$$= -\frac{2}{u} + \sum_{n=2}^{\infty} \frac{1}{u^n} - \frac{2}{3}\sum_{n=0}^{\infty} \frac{u^n}{3^n}, \text{ where } u = z + 1.$$

38. **Solved sum.** Find the Laurent's expansion of

$$f(z) = \frac{z+5}{z(z+2)(z+3)}$$

in (*i*) $|z+1| < 1/2$ (*ii*) $1 < |z+1| < 2$.

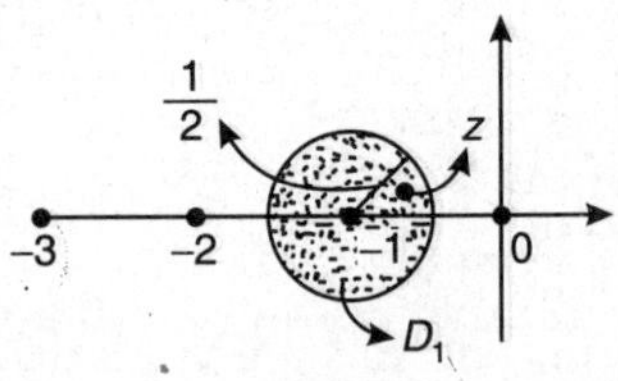

✍ Now the singular points of $f(z)$ are 0, – 2, – 3. From the domains, it is seen that the expansions should be in powers of $z + 1$.

Case (*i*) $D_1 : |z+1| < 1/2$.

This domain contains no singular point. So the expansion will be a Taylor's series. Set

$$z + 1 = u \quad \text{or} \quad z = u - 1$$

for brevity. Then, in $|z+1|<1/2$, $|u|<1/2$. Splitting $f(z)$ into partial fractions,

$$f(z) = \frac{5}{6}\cdot\frac{1}{z} - \frac{3}{2}\cdot\frac{1}{z+2} + \frac{2}{3}\cdot\frac{1}{z+3}$$

$$= \frac{5}{6}\cdot\frac{1}{u-1} - \frac{3}{2}\cdot\frac{1}{u+1} + \frac{2}{3}\cdot\frac{1}{u+2}$$

$$= -\frac{5}{6}\cdot\frac{1}{1-u} - \frac{3}{2}\cdot\frac{1}{1+u} + \frac{1}{3}\cdot\frac{1}{1+u/2}$$

$$= -\frac{5}{6}(1-u)^{-1} - \frac{3}{2}(1+u)^{-1} + \frac{1}{3}(1+u/2)^{-1}.$$

Since $|u|<1$ and $|u/2|<1$,

$$f(z) = -\frac{5}{6}(1+u+u^2+u^3+...) - \frac{3}{2}(1-u+u^2-u^3+...) + \frac{1}{3}\left(1-\frac{u}{2}+\frac{u^2}{2^2}-\frac{u^3}{2^3}+...\right)$$

$$= -2 + \sum_{n=1}^{\infty}\left[-\frac{5}{6}-\frac{3}{2}(-1)^n + \frac{1}{3}\frac{(-1)^n}{2^n}\right]u^n, \text{ where } u = z+1.$$

Case (*ii*) The given

$$D_2 : 1 < |z+1| < 2$$

is the annular domain between the circles whose centres are $z=-1$ and radii are 1 and 2. From the figure it is evident that the singular points are not in D_2. Now, for all z in

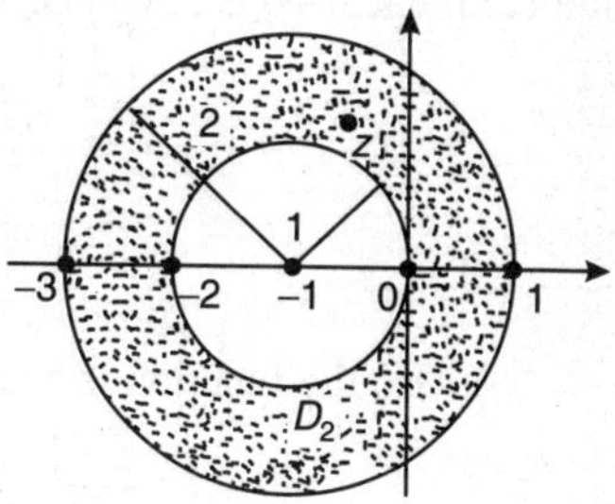

$$1 < |z+1| < 2$$

$$1 < |u|, \quad |u| < 2.$$

$$\therefore \quad |1/u| < 1, \quad |u/2| < 1.$$

$$\therefore f(z) = \frac{5}{6}\cdot\frac{1}{z} - \frac{3}{2}\cdot\frac{1}{z+2} + \frac{2}{3}\cdot\frac{1}{z+3}$$

$$= \frac{5}{6}\cdot\frac{1}{u-1} - \frac{3}{2}\cdot\frac{1}{u+1} + \frac{2}{3}\cdot\frac{1}{u+2}$$

$$= \frac{5}{6}\cdot\frac{1}{u(1-1/u)} - \frac{3}{2}\cdot\frac{1}{u(1+1/u)} + \frac{2}{3}\cdot\frac{1}{2(1+u/2)}.$$

Now $|1/u|<1$, $|u/2|<1$. Therefore

$$f(z) = \frac{5}{6}\cdot\frac{(1-1/u)^{-1}}{u} - \frac{3}{2}\cdot\frac{(1+1/u)^{-1}}{u} + \frac{1}{3}\cdot(1+u/2)^{-1}$$

$$= \frac{5}{6u}\left(1+\frac{1}{u}+\frac{1}{u^2}+\frac{1}{u^3}+\frac{1}{u^4}+...\right)$$

$$- \frac{3}{2u}\left(1-\frac{1}{u}+\frac{1}{u^2}-\frac{1}{u^3}+\frac{1}{u^4}-...\right)$$

$$+ \frac{1}{3}\left(1-\frac{u}{2}+\frac{u^2}{2^2}-\frac{u^3}{2^3}+\frac{u^4}{2^4}-...\right)$$

$$= \sum_{n=1}^{\infty}\left[\frac{5}{6}+\frac{3}{2}(-1)^n\right]\frac{1}{u^n}+\frac{1}{3}\sum_{n=0}^{\infty}(-1)^n\left(\frac{u}{2}\right)^n.$$

39. Find the Laurent's series for

$$f(z) = \frac{6z+5}{z(z-2)(z+1)} \text{ in the domain } 1 < |z+1| < 3.$$

[*Hint* : $f(z) = -\frac{5}{2}\cdot\frac{1}{z} - \frac{1}{3}\cdot\frac{1}{z+1} + \frac{17}{6}\cdot\frac{1}{z-2}$.

If $u = z + 1$, then $|1/u|, |u/3| < 1$;

$$f(z) = -\frac{5}{2}\cdot\frac{1}{u}\left(1-\frac{1}{u}\right)^{-1} - \frac{1}{3}\cdot\frac{1}{u} - \frac{17}{18}\left(1-\frac{u}{3}\right)^{-1}]$$

40. **Solved sum.** Expand $(z - 1)/z^2$ about $z = 1$ (*i*) in Taylor's series and (*ii*) in Laurent's series. **(1982)**

(*i*) The given function is not analytic at $z = 0$. But we want an expansion about $z = 1$. But the function is analytic in the disc $|z - 1| < 1$ because the distance between the centre $z = 1$ of the disc and the point $z = 0$ is 1. Hence, for all z in this disc,

$$\frac{z-1}{z^2} = \frac{z-1}{\{1+(z-1)\}^2}$$
$$= (z-1)[1-2(z-1)+3(z-1)^2 - \ldots]$$
$$= (z-1) - 2(z-1)^2 + 3(z-1)^3 + \ldots$$

(*ii*) Also the function is analytic in the region $|z - 1| > 1$. Hence, for all z in this region,

$$\frac{z-1}{z^2} = \frac{z-1}{\{1+(z-1)\}^2} = \frac{z-1}{(z-1)^2\{1+1/(z-1)\}^2}$$

$$= \frac{1}{z-1}\left[1-\frac{2}{z-1}+\frac{3}{(z-1)^2}-\ldots\right]$$

41. Show that the expansion of $f(z) = z/\{(z - 1)(z - 3)\}$ valid in $0 < |z - 1| < 2$, is

$$-\frac{1}{2}\frac{1}{z-1} - \frac{3}{4}\left[1+\frac{z-1}{2}+\frac{(z-1)^2}{2^2}+\ldots\right]$$ **(1981)**

[*Hint* : $f(z) = -\frac{1}{2}\frac{1}{z-1}+\frac{3}{2}\frac{1}{z-3} = -\frac{1}{2}\frac{1}{z-1}-\frac{3}{4}\frac{1}{1-\frac{1}{2}(z-1)}$

or $f(z) = \frac{1}{z-1}\frac{(z-1)+1}{(-2)\{1-\frac{1}{2}(z-1)\}}$]

42. Show that the Laurent's series for $z/\{(z - 1)(2 - z)\}$, valid in $|z - 1| > 1$, is

$$-(z-1)^{-1} - 2(z-1)^{-2} - 2(z-1)^{-3} - 2(z-1)^{-4} - \ldots$$

43. Find the Laurent's series for the following functions about the indicated singularities:

(*i*) $\frac{e^{2z}}{(z-1)^3}$ about $z = 1$ (*ii*) $(z-3)\sin\frac{1}{z+2}$ about $z = -2$

[*Hint* : (*i*) $\dfrac{e^2\, e^{2(z-1)}}{(z-1)^3} = \dfrac{e^2}{(z-1)^3}\left\{1+\dfrac{2(z-1)}{1!}+...\right\}$

(*ii*) $\{(z+2)-5\}\left\{\dfrac{1}{z+2}-\dfrac{1}{3!\,(z+2)^3}+\dfrac{1}{5!\,(z+2)^5}-...\right\}$]

44. **Solved sum.** Prove that, for any finite $z \neq 0$,

$$\cosh\left(z+\frac{1}{z}\right) = a_0 + \sum_{n=1}^{\infty} a_n\left(z^n+\frac{1}{z^n}\right),$$

where $$a_n = \frac{1}{2\pi}\int_0^{2\pi} \cos n\theta \cosh(2\cos\theta)\, d\theta.$$ **(1974, 80)**

$\cosh(z+1/z)$ is analytic everywhere except at $z=0$ and $z=\infty$. So the Laurent's series representing $\cosh(z+1/z)$ in the annulus $0<|z|<\infty$ is

$$\sum_{-\infty}^{\infty} a_n z^n,\; a_n = \frac{1}{2\pi i}\int_C \frac{\cosh(z+1/z)}{(z-0)^{n+1}}\, dz,$$

where C is any circle with centre at $z=0$. In particular, if C is the positively oriented unit circle, $z=e^{i\theta}$, $0\le\theta\le 2\pi$, then

$$a_n = \frac{1}{2\pi i}\int_0^{2\pi} \frac{\cosh(2\cos\theta)}{e^{(n+1)i\theta}}\, e^{i\theta}\, i d\theta$$

$$= \frac{1}{2\pi}\int_0^{2\pi} \cosh(2\cos\theta)\{\cos n\theta - i\sin n\theta\}\, d\theta$$

$$= \frac{1}{2\pi}\int_0^{2\pi} \cosh(2\cos\theta)\cos n\theta\, d\theta - \frac{i}{2\pi}\int_0^{2\pi} \cosh(2\cos\theta)\sin n\theta\, d\theta$$

$$= \frac{1}{2\pi} I_1 - \frac{i}{2\pi} I_2 \text{ say.}$$

Then $$I_2 = \int_0^{2\pi} \cosh(2\cos\theta)\sin n\theta\, d\theta$$

$$= \int_0^{2\pi} \cosh\{2\cos(2\pi-\theta)\}\sin n(2\pi-\theta)\, d\theta$$

$$\left(\because \int_0^a f(x)\,dx = \int_0^a f(a-x)\,dx\right)$$

$$= -\int_0^{2\pi} \cosh(2\cos\theta)\sin n\theta\, d\theta = -I_2.$$

So $I_2 = 0$ and hence

$$a_n = \frac{1}{2\pi}\int_0^{2\pi} \cosh(2\cos\theta)\cos n\theta\, d\theta$$

Further
$$a_{-n} = \frac{1}{2\pi}\int_0^{2\pi} \cosh(2\cos\theta)\cos(-n\theta)\, d\theta$$

$$= \frac{1}{2\pi}\int_0^{2\pi} \cosh(2\cos\theta)\cos n\theta\, d\theta = a_n.$$

So the Laurent's series representing $\cosh(z + 1/z)$ is

$$\sum_{-\infty}^{\infty} a_n z^n = a_0 + \sum_{n=1}^{\infty} a_n (z^n + z^{-n}).$$

45. **Solved sum.** Show that $f(z) = e^{\frac{1}{2}c(z-1/z)} = \sum_{-\infty}^{\infty} a_n z^n$,

where
$$a_n = \frac{1}{2\pi}\int_0^{2\pi} \cos(n\theta - c\sin\theta)\, d\theta.. \qquad \textbf{(1977, 81)}$$

Function $f(z)$ is analytic everywhere except at $z = 0$ and $z = \infty$. So, expanding it in Laurent's series about $z = 0$,

$$f(z) = \sum_{-\infty}^{\infty} a_n (z-0)^n,\ a_n = \frac{1}{2\pi i}\int_C \frac{f(t)}{(t-0)^{n+1}}\, dt.$$

In this, if we take C to be the circle $|t| = 1$, then

$$t = e^{i\theta},\ 0 \le \theta \le 2\pi.$$

$$\therefore \quad a_n = \frac{1}{2\pi i}\int_C e^{\frac{1}{2}c(t-1/t)} \frac{dt}{t^{n+1}} = \frac{1}{2\pi i}\int_0^{2\pi} \frac{e^{ci\sin\theta}\,(ie^{i\theta}\, d\theta)}{e^{i(n+1)\theta}}$$

$$= \frac{1}{2\pi}\int_0^{2\pi} e^{-i}\,(n\theta - c\sin\theta)\, d\theta$$

$$= \frac{1}{2\pi}\int_0^{2\pi} [\cos(n\theta - c\sin\theta) - i\sin(n\theta - c\sin\theta)]\, d\theta$$

of which the imaginary part vanishes because it is

$$I = \int_0^{2\pi} \sin(n\theta - c\sin\theta)\, d\theta$$

$$= \int_0^{2\pi} \sin\{n(2\pi - \theta) - c\sin(2\pi - \theta)\}\, d\theta = -I.$$

46. Show that $e^{\frac{1}{2}i(z-1/z)} = \sum_{-\infty}^{\infty} a_n z^n$, where

$$a_n = \frac{1}{2\pi}\int_0^{2\pi} (\cos n\theta - i\sin\theta)\, d\theta. \quad \textbf{(1979)}$$

[*Hint* : Set $c = i$ in the previous sum]

9.4. Cauchy Product and Division

47. **Solved sum.** Obtain the first four terms of the Laurent's expansion

$$\frac{e^z}{z(z^2+1)} = \frac{1}{z} + 1 - \frac{1}{2}z - \frac{5}{6}z^2 + \ldots \quad 0 < |z| < 1 \quad \textbf{(1975)}$$

The singularities of the function are 0, i, $-i$, ∞. So the function is analytic in $0 < |z| < 1$. But

$$\frac{e^z}{z} = \frac{1}{z} + 1 + \frac{z}{2} + \frac{z^2}{6} + \ldots \qquad 0 < |z| < 1$$

$$\frac{1}{z^2+1} = 1 - z^2 + z^4 - z^6 + \ldots \qquad 0 < |z^2| < 1$$

$$\therefore \quad \frac{e^z}{z(z^2+1)} = \left(\frac{1}{z} + 1 + \frac{z}{2} + \frac{z^2}{6} + \ldots\right)(1 - z^2 + z^4 - \ldots)$$

$$= \frac{1}{z} + 1 + \left(-1 + \frac{1}{2}\right)z + \left(\frac{1}{6} - 1\right)z^2 + \ldots$$

$$= \frac{1}{z} + 1 - \frac{z}{2} - \frac{5z^2}{6} + \ldots \qquad 0 < |z| < 1$$

48. **Solved sum.** Obtain the first few terms of the Laurent's expansion of $1/(e^z - 1)$, valid in $0 < |z| < 2\pi$.

The singular points of $(1/e^z - 1)$ are 0, $2\pi i$, $4\pi i$, ... So

$$\frac{1}{e^z - 1} = \frac{1}{z + \frac{z^2}{2!} + \frac{z^3}{3!} + \ldots} \qquad 0 < |z| < 2\pi$$

$$= \frac{1}{z}\left[1 + \left(\frac{z}{2!} + \frac{z^2}{3!} + \frac{z^3}{4!} + \ldots\right)\right]^{-1}$$

$$= \frac{1}{z}\left[1 - \left(\frac{z}{2!} + \frac{z^2}{3!} + \ldots\right) + \left(\frac{z}{2!} + \frac{z^2}{3!} + \ldots\right)^2 - \ldots\right]$$

$$= \frac{1}{z}\left[1 - \frac{z}{2!} + \left(-\frac{1}{3!} + \frac{1}{(2!)^2}\right)z^2 + \ldots\right]$$

$$= \frac{1}{z} - \frac{1}{2} - \frac{1}{12}z + (0)z^2 - \frac{1}{720}z^3 + \ldots$$

49. Find the first few terms of the Laurent's expansion for $1/\{z\,(e^z - 1)\}$ valid in $0 < |z| < 2\pi$ as

$$\frac{1}{z^2} - \frac{1}{2z} + a_0 + a_2 z^2 + \ldots \qquad \textbf{(1976)}$$

[*Hint* : Use the previous sum; $a_0 = 1/12$, $a_2 = -1/720$]

50. Find the first few terms of the Laurent's expansion of tan z, sec z, cosec z as

(*i*) $\tan z = z + \dfrac{z^3}{3} + \dfrac{2z^5}{15} + \ldots \qquad |z| < \dfrac{1}{2}\pi$

(*ii*) $\sec z = 1 + \dfrac{z^2}{2!} + \left(-\dfrac{1}{4!} + \dfrac{1}{(2!)^2}\right) z^4 + \ldots \qquad |z| < \dfrac{1}{2}\pi$

(*iii*) $\operatorname{cosec} z = \dfrac{1}{z} + \dfrac{1}{3!} z - \left(\dfrac{1}{5!} - \dfrac{1}{(3!)^2}\right) z^3 + \ldots \qquad 0 < |z| < \pi$

9.7 To 9.12: Singularities

51. **Solved sum.** Find the zeros and singularities of

$$f(z) = \frac{z^2 + 1}{1 - z^2}.$$

✍ Now $f(z) = \dfrac{(z+i)(z-i)}{(1+z)(1-z)}$. The zeros of $f(z)$ are given by $f(z) = 0$ as

$$z = -i, \quad z = i.$$

The poles are given by $f(z) = \infty$ as

$$z = -1, \quad z = 1.$$

52. Find the zeros of $\dfrac{z^3 - 1}{z^3 + 1}$.

[**Ans.** Cube roots of 1, namely, 1, ω, ω^2]

53. **Solved sum.** Find the zeros of $f(z) = \sin \dfrac{1}{z-a}$.

✍ The zeros are given by

$$\frac{1}{z-a} = n\pi, \quad n = \pm 1, \pm 2, \pm 3, \ldots$$

that is, by $(z - a)\, n\pi = 1$. Thus the zeros are

$$z = a + \frac{1}{n\pi}, \quad n = \pm 1, \pm 2, \pm 3, \ldots$$

54. **Solved sum.** Find the nature of the singularities of

$$f(z) = \frac{\tan z}{z}.$$

The finite singularities of $f(z)$ are

$$z = 0, \quad z = \pm\frac{\pi}{2}, \pm\frac{3\pi}{2}, \pm\frac{5\pi}{2}, \ldots$$

Singularity $z = 0$. At $z = 0, f(z)$ is not defined since it is of the form "0 divided by 0". But

$$\lim_{z \to 0} \frac{\tan z}{z} = 1.$$

So $z = 0$ is a removable singularity of $f(z)$.

Singularity $z = \pi/2$. Now $f(\pi/2) = \infty$. So $z = \pi/2$ is a pole of $f(z)$. Consider

$$\lim_{z \to \pi/2} (z - \pi/2) \frac{\tan z}{z} = \lim_{z \to \pi/2} \frac{(z - \pi/2) \sin z}{z \cos z}$$

$$= \lim_{z \to \pi/2} \frac{\sin z + (z - \pi/2) \cos z}{\cos z - z \sin z} \text{ by } L' \text{ Hospital rule.}$$

$$= \frac{1 + 0}{0 - \pi/2} = -\frac{1}{\pi/2}$$

which is finite not equal to 0. So $z = \pi/2$ is a simple pole of $f(z)$. Similarly all the other singularities are simple poles.

55. Find the nature of the singularities of

$$\frac{z^2}{(z-1)^2 (z-2)}.$$

[**Ans.** $z = 1$ is a double pole and $z = 2$ is a simple pole]

56. Find the nature of the singularity of

$$f(z) = (z - 3) \sin \frac{1}{z + 2}.$$

[*Hint* : $f(z) = [(z + 2) - 5] \left[\frac{1}{z + 2} - \frac{1}{3!\,(z + 2)^3} + \ldots \right]$ has an infinite number of negative powers of $z + 2$. So $z = -2$ is an essential singularity]

57. Find the nature of the singularities of $\dfrac{z}{\sin z}$.

[**Ans.** $z = 0$ is a removanble singularity. $z = n\pi$, $n = 1, 2, 3, \ldots$, are poles]

58. State the nature of singularity of $f(z)$ at $z = 0$ in the following cases : **(1979, 82)**

(*i*) $(e^z - 1)/z$

(*ii*) $(\sin z)/z$

(*iii*) $(1 - \cos z)/z^2$

(*iv*) $\dfrac{\tan z}{z}$

(*v*) $\operatorname{cosec} z - \dfrac{1}{z}$

[*Hint* : Removable singularities (0/0 form); The limits of the functions as $z \to 0$ are (*i*) 1 (*ii*) 1 (*iii*) 1/2 (*iv*) $\lim\limits_{z \to 0} \dfrac{\tan z}{z} = 1$ (*v*) $\lim\limits_{z \to 0} \dfrac{z - \sin z}{z \sin z} = 0$, a finite limit].

59. Is $z = 0$ a singular point of $z^2 e^{-z}$? What is the expansion of this function about $z = 0$?

[*Hint* : No; $z^2 - z^3/1! + z^4/2! - \ldots$]

60. Find the poles of $\dfrac{z^3 - 1}{z^3 + 1}$.

[*Hint* : $z^3 + 1 = 0$ gives the poles as $-1, \dfrac{1}{2} \pm i \dfrac{\sqrt{3}}{2}$]

61. Find the poles of $\dfrac{z}{\cos z}$.

[**Ans.** $\pm (2n + 1)(\pi/2)$, $n = 0, 1, 2, 3, ...$]

62. **Solved sum.** Find the poles of

(*i*) cosec z (*ii*) cot z (*iii*) tan z

✍ (*i*) Now cosec $z = \dfrac{1}{\sin z}$. The zeros of sin z are

$$\pm n\pi, \quad n = 0, 1, 2, 3, ...$$

which are the poles of cosec z.

(*ii*) Now cot $z = \dfrac{\cos z}{\sin z}$. The zeros of sin z are

$$\pm n\pi, \quad n = 0, 1, 2, 3, ...$$

which are the poles of cot z.

(*iii*) Now tan $z = \dfrac{\sin z}{\cos z}$. The zeros of cos z are

$$\pm (2n + 1)(\pi/2), \quad n = 0, 1, 2, ...$$

which are the poles of tan z.

63. **Solved sum.** Find the poles of $f(z) = \dfrac{1}{\sin \dfrac{1}{z - a}}$.

✍ The poles are given by

$$\frac{1}{z - a} = n\pi, \quad n = \pm 1, \pm 2, \pm 3, ..$$

that is, by $(z - a)\, n\pi = 1$. Thus the poles are $z = a + \dfrac{1}{n\pi}$, $n = \pm 1, \pm 2, \pm 3 ...$

64. **Solved sum.** Find the orders of the poles of

$$f(z) = \frac{1}{z\,(e^z - 1)}.$$

✍ It is evident that the denominator vanishes when

$$z = 0, \quad z = \pm 2\pi i, \ \pm 4\pi i, \ \pm 6\pi i, \ ...$$

So they are the poles of $f(z)$. Here we know that

$$\lim_{z \to 0} \frac{z}{e^z - 1} = 1.$$

So, multiplying $f(z)$ by z^2 and taking limit as $z \to 0$,

$$\lim_{z \to 0} z^2 f(z) \lim_{z \to 0} \frac{z}{e^z - 1} = 1$$

which is finite, not equal to 0. So $z = 0$ is a pole of order 2. To find the order of $z = 2\pi i$, consider

$$\lim_{z \to 2\pi i} (z - 2\pi i)\, f(z) = \lim_{z \to 2\pi i} \frac{z - 2\pi i}{z\,(e^z - 1)}$$

$$= \lim_{z \to 2\pi i} \frac{1}{(e^z - 1) + ze^z} \text{ by } L' \text{ Hospital's rule}$$

$$= \frac{1}{(1-1)+2\pi i\,(1)} = \frac{1}{2\pi i}$$

which is finite and not equal to 0. So $z = 2\pi i$ and similarly the other poles are simple poles.

65. Find the orders of the pole $z = 0$ of the following functions:

(*i*) $\frac{e^z}{z}$ (*ii*) $\frac{e^z}{z^2}$ (*iii*) $\frac{1-\sin z}{z^3}$

[*Hint* : (*i*) $\frac{e^z}{z} = \frac{1}{z} + \frac{1}{1!} + \frac{z}{2!} + \ldots$ So $z = 0$ is a simple pole. (*ii*) Order: 2 (*iii*) Order: 3]

66. Find the order of the pole $z = 0$ in relation to

(*i*) e^z/z (*ii*) e^z/z^2 (*iii*) $(1 - \sin z)/z^3$,

[**Ans.** 1, 2, 3]

67. Show that $z = 0$ is an essential singularity of the following functions:

$e^{1/z}$, $ze^{1/z}$, $z^2 e^{-1/z}$, e^{-1/z^2}, $\sin (1/z)$ **(1978)**, $\cosh (1/z)$.

[*Hint* : The Laurent's expansions about $z = 0$ has an infinite number of negative powers of z]

68. **Solved sum.** Show that $z = 1$ is a double pole of $f(z) = 1/(z-1)^2$. Investigate the behaviour of this function at $z = \infty$.

Now $f(z) = 1/(z-1)^2$. It tends to ∞ as $z \to 1$. So $z = 1$ is a pole. But $1/(z-1)^2$ is the Laurent's expansion of the functio about $z = 1$. So $z = 1$ is a pole of order 2.

Further $f(1/z) - z^2/(1-z)^2$ which is analytic at $z = 0$. Also its expansion about $z = 0$ is $z^2 + 2z^3 + 3z^4 + \ldots$ So $z = 0$ is not a singular point of $f(1/z)$ and consequently $z = \infty$ is not a singular point of $f(z)$.

69. Show that the singularities of the following functions are poles and find their orders. Investigate the behaviour of the functions at $z = \infty$.

(*i*) $\frac{z+1}{z^2-2z}$ **(1984)** (*ii*) $\frac{1}{z^2+1}$ **(1981)** (*iii*) $\frac{1}{z-z^3}$ **(1984)**

(*iv*) $\frac{2z+3}{z(z^2+1)}$ **(1980)** (*v*) $\frac{1}{z^4+2z^2+1}$ (*vi*) $\frac{1}{z^2(z-3)^2}$.

[*Hint*: (*i*) $f(z) = \frac{z+1}{z(z-2)} = \frac{A}{z} + \frac{B}{z-2}$. Expanding about $z = 0$, $\frac{A}{z} + B\,(a_0 + a_1 z + \ldots)$. Expanding about $z = 2$, $\frac{B}{z-2} + A\,\{b_0 + b_1 (z-2) + \ldots\}$. So $z = 0$ and $z = 2$ are S.P's.

Second method. Expanding $f(z)$ in $0 < |z| < 2$ about $z = 0$,

$$f(z) = \frac{1+z}{z(-2)}\,\frac{1}{1-\frac{1}{2}z} = -\frac{1}{2z}\,(1+z)\left(1 + \frac{z}{2} + \frac{z^2}{2^2} + \ldots\right)$$

(*ii*) $i, -i$ are S.P's (*iii*) $0, 1, -1$ are S.P's (*iv*) $0, i, -i$ are S.P's (*v*) $i, -i$ are double poles (*vi*) 0, 3 are double poles. (*i*) to (*vi*) $f(1/z)$ is analytic at $z = 0$; so also is $f(z)$ at $z = \infty$].

70. **Solved sum.** Show that, if x and y are real, then

$$\sin z = \sin (x + iy) = \sin x \cosh y + i \cos x \sinh y$$

and deduce that $\sin z$ vanishes only for certain real values of z and find the singular points of $\sin z$.

$$\sin z = \sin (x + iy) = \sin x \cos iy + \sin iy \cos x$$
$$= \sin x \cosh y + (i \sinh y) \cos x.$$

So $\sin z$ vanishes only when both

$$\sin x \cosh y \text{ and } \cos x \sinh y$$

vanish. But the first expression $\sin x \cosh y$ vanishes only when $\sin x$ vanishes, that is, when

$$x = n\pi, \quad n = 0, \pm 1, \pm 2, \pm 3, \ldots$$

because $\cosh y$ is always greater than or equal to 1. So, when $x = n\pi$, we shall search for the situation under which the second expression $\cos x \sinh y$ vanishes. But, when $x = n\pi$, $\cos x \neq 0$. So we shall find out when $\sinh y$ vanishes. But $\sinh y$ vanishes only when $y = 0$. Thus, we get that $\sin z = 0$ when

$$z = n\pi + 0i \text{ or } n\pi, \; n = 0, \pm 1, \pm 2, \pm 3, \ldots$$

$\sin z$ is analytic in the finite plane, and since it is a function of exponential, it is not defined at $z = \infty$. So $z = \infty$ is the only singularity of $\sin z$. But $z = \infty$ is the limit point of the zeros of $\sin z$. Hence it is an isolated essential singularity.

71. **Solved sum.** Find the singular point of

$$f(z) = \sin \frac{1}{z-a}.$$

State the nature of singularity.

✍ $z = a$ is the only singular point in the finite plane. Since

$$\sin \frac{1}{z-a} = \frac{1}{z-a} - \frac{1}{3!\,(z-a)^3} + \frac{1}{5!\,(z-a)^5} - \ldots,$$

$z = a$ is an essential singularity. It is an isolated essential singularity.

72. **Solved sum.** Find the singular points of

$$f(z) = \frac{1}{\sin \dfrac{1}{z-a}}.$$

State their nature.

✍ $f(z)$ has an infinite number of poles which are given by

$$\frac{1}{z-a} = n\pi, \quad n = \pm 1, \pm 2, \pm 3, \ldots$$

that is, by $(z-a)\,n\pi = 1$. Thus the poles are

$$z = a + \frac{1}{n\pi}, \quad n = \pm 1, \pm 2, \pm 3, \ldots$$

But $z = a$ also is a singular point. It is an essential singularity. Further it is a limit point of the poles. So it is a non-isolated essential singularity.

73. Show that $\cos z$ vanishes only when

$$z = \pm (2n+1)\,\frac{1}{2}\pi, \quad n = 0, 1, 2, 3, \ldots$$

and find its singular points.

[*Hint* : Consider $\cos z = \cos(x + iy) = \cos x \cosh y - i \sin x \sinh y$. $z = \infty$ is the only singularity of $\cos z$ and it is an isolated essential singularity.

74. Show that $\sinh z$ vanishes only when

$$z = (n\pi)\,i, \quad n = 0, \pm 1, \pm 2, \pm 3, \ldots$$

and argue that these points are the singular points of $\coth z$.

[*Hint* : $\sinh z = (1/i) \sin iz$.]

75. Show that cosh z vanishes only when

$$z = \pm (2n + 1)\, \frac{1}{2} \pi i, \quad n = 0, 1, 2, 3, \ldots$$

and argue that these points are singular poins of tanh z.

[*Hint* : cosh z = cos iz.]

76. Obtain the singularities of cot z and tan z and their nature. **(1978)**

[*Hint* : (*i*) cot z = cos z/sin z ; zeros of sin z are $0, \pm \pi, \pm 2\pi, \ldots$ which are the poles of cot z. So $z = \infty$, the limit point of these poles, is a non-isolated E.S. (*ii*) Similarly $z = \infty$ is a non-isolated E.S. of tan z.]

77. Find the zeros of sin z – cos z and thus find its singularity and its nature. **(1980)**

[*Hint* : Zeros are $\pm n\pi + \frac{1}{4}\pi$; The limit of these zeros, $z = \infty$, is an isolated E.S.]

78. Find the poles of 1/(sin z – cos z) and thus find its singularity and its nature.

[*Hint* : Poles are $\pm n\pi + \frac{1}{4}\pi$; The limit of these poles $z = \infty$, is a non-isolated E.S.]

79. Show that $z = \infty$ is a non-isolated singularity of $1/(1 - e^z)$. **(1982)**

[*Hint* : Zeros of $1 - e^z$, or poles of $1/(1 - e^z)$, are $0, 2\pi i, 4\pi i, \ldots$]

80. Obtain the nature of singularities of the following functions :

(*i*) $\sin \dfrac{1}{z}$ at $z = 0$ (*ii*) $\sin \dfrac{1}{1 - z}$ at $z = 1$.

[*Hint* : (*i*) Zeros of sin $(1/z)$ are given by $1/z = 0, n\pi$, or by $z = \infty, 1/n\pi$; The limit point of these zeros, namely $z = 0$, is an isolated E.S. (*ii*) The limit point of the zeros $1 - (1/n\pi)$ is 1 and it is an isolated E.S.]

81. Classify the points $z = 0, z = 1, z = \infty$ in relation to the function

$$f(z) = \frac{z - 2}{z^2} \sin \frac{1}{1 - z}. \qquad (1981)$$

[*Hint* : $z = 0$ is a pole of order 2; Zeros of sin $1/(1 - z)$ are given by $1/(1 - z) = n\pi$ as $1 - (1/n\pi)$. The limit of these zeros, namely $z = 1$, is an isolated singularity. $z = \infty$ is not a singular point because $f(1/z)$ is analytic at $z = 0$.]

82. Show that $z = 0$ is a non-isolated essential singularity of cot $(1/z)$ and cosec $(1/z)$.

[*Hint* : Zeros of sin $(1/z)$ are poles of cot $(1/z)$ and cosec $(1/z)$; $z = 0$ is the limit point of these poles.]

83. Determine the singularities of the following functions and state their nature.

(*i*) $\dfrac{z}{e^z - 1}$ **(1980)** (*ii*) $\dfrac{z}{e^{1/z} - 1}$ (*iii*) $\dfrac{1}{(2 \sin z - 1)^2}$

[*Hint* : (*i*) $z = 0$ is an R.S. since the limit of $f(z)$ as $z \to 0$ is 1. Poles: $\pm 2\pi i, \pm 4\pi i, \ldots$; Limit of these poles, $z = \infty$, is an E.S. (*ii*) Poles: $\pm 1/2\, \pi i, \pm 1/4\, \pi i, \ldots$; Limit of these poles, $z = 0$, is a non-isolated E.S.; $z = \infty$ is a pole of order 2 since $f(1/z) = 1/z^2 \quad 1/2z + 1/12 + \ldots$ has $z = 0$ as a pole of order 2 (*iii*) Poles are given by $2 \sin z - 1 = 0$ as $z = \pm n\pi + (-1)^n (\pi/6)$; Limit of these poles, $z = \infty$, is a non-isolated E.S.]

84. Obtain the singularities of the following functions and state their nature:

(*i*) $e^{z/(z-2)}$ **(1981)** (*ii*) $\cos (z^2 + z^{-2})$.

[*Hint* : (*i*) $z = 2$ is an E.S. since the only singularity of e^z is ∞ and it is essential (*ii*) $0, \infty$ are E.S's.]

CHAPTER 10

RESIDUES

10.1. RESIDUE

First we define residue. If $z = a$ is an isolated singularity of $f(z)$ and if the Laurent's expansion for $f(z)$ about $z = a$ is

$$\sum_{-\infty}^{\infty} a_n (z-a)^n,$$

then the coefficient of $(z-a)^{-1}$, namely a_{-1}, is called the residue of $f(z)$ at $z = a$.
We know that, in this expansion,

$$a_n = \frac{1}{2\pi i}\int_C \frac{f(z)}{(z-a)^{n+1}}\,dz,$$

where C is a scro curve encircling z = a and lying in the annulus region in which the expansion has been made. Thus, setting $n = -1$, residue of $f(z)$ at $z = a$ is

$$a_{-1} = \frac{1}{2\pi i}\int_C f(z)\,dz, \tag{10.1.1}$$

RESIDUE AT A REMOVABLE SINGULARITY

If $z = a$ is a removable singularity of $f(z)$, then the Laurent's series about $z = a$ has no negative powers of $z - a$. As such, the coefficient of $(z-a)^{-1}$ is 0. So the residue of $f(z)$ at $z = a$ is 0.

RESIDUE AT A POLE

The following are methods of calculation of residues under different situations.

Method 1.

Expansion of $f(z)$ about $z = a$ in Laurent's series to find the coefficient of $\frac{1}{z-a}$.

Method 2.

The residue at $z = a$ of a function $f(z)$ can be calculated directly by evaluating

$$\frac{1}{2\pi i}\int_C f(z)\,dz$$

choosing suitably a small scro curve C around $z = a$. (10.2.1)

10.2. CALCULATION OF RESIDUES

The following are the various methods of calculation of residues under different situations :

Example 1. If $f(z) = e^{z-a}/(z-a)^2$, the Laurent's expansion about the singularity $z = a$, is

$$f(z) = \frac{1}{(z-a)^2} + \frac{1}{1!\,(z-a)} + \frac{1}{2!} + \frac{z-a}{3!} + \frac{(z-a)^2}{4!} + \ldots$$

The coefficient of $(z-a)^{-1}$ is $1/1!$. So the residue of $f(z)$ at the *double pole* $z = a$ is $1/1!$. From this example it is evident that it is not necessary to evaluate the integral (10.1.1) to find the residue.

Example 2. Suppose $f(z) = \dfrac{2z+1}{z^2 - z - 2} = \dfrac{2z+1}{(z+1)(z-2)}$. Then $z = -1$ and $z = 2$ are poles. But

$$\frac{2z+1}{(z+1)(z-2)} = \frac{1}{3}\frac{1}{z+1} + \frac{5}{3}\frac{1}{(z-2)}.$$

So, if C_1 is the circle $|z+1| = 1$ and C_2 is the circle $|z-2| = 1$,

$$\left.\begin{matrix}\text{Res. of } f(z)\\ \text{at } z=-1\end{matrix}\right\} = \frac{1}{2\pi i}\int_{C_1}\left(\frac{1}{3}\frac{1}{z+1} + \frac{5}{3}\frac{1}{z-2}\right)dz$$

$$= \frac{1}{2\pi i}\left\{\frac{1}{3}(2\pi i) + \frac{5}{3}(0)\right\} = \frac{1}{3} \text{ by (8.3.3).}$$

$$\left.\begin{matrix}\text{Res. of } f(z)\\ \text{at } z=2\end{matrix}\right\} = \frac{1}{2\pi i}\int_{C_2} f(z)\,dz = \frac{1}{2\pi i}\left\{0 + \frac{5}{3}(2\pi i)\right\} = \frac{5}{3}.$$

Residue at a simple pole.

1. If $z = a$ is a *simple pole* of $f(z)$, then

$$\left.\begin{matrix}\text{Res. of } f(z)\\ \text{at } z=a\end{matrix}\right\} = \lim_{z\to a}(z-a)f(z) \qquad (10.2.2)$$

since $\displaystyle\lim_{z\to 0}(z-a)f(z) = \lim_{z\to a}(z-a)\left[\frac{a_{-1}}{z-a} + a_0 + a_1(z-a) + \ldots\right] = a_{-1}.$

Example 1. If $f(z) = \dfrac{2z+1}{z^2 - z - 2} = \dfrac{2z+1}{(z+1)(z-2)}$,

$$\left.\begin{matrix}\text{Res. at}\\ z=-1\end{matrix}\right\} = \lim_{z\to -1}(z+1)\frac{2z+1}{(z+1)(z-2)} = \frac{-1}{-3} = \frac{1}{3}$$

$$\left.\begin{matrix}\text{Res. at}\\ z=2\end{matrix}\right\} = \lim_{z\to 2}(z-2)\frac{2z+1}{(z+1)(z-2)} = \frac{5}{3}.$$

Example 2. The residue of $\cot z$ at the simple pole $z = 0$ is

$$\lim_{z\to 0} z\frac{\cos z}{\sin z} = \left(\lim_{z\to 0}\frac{z}{\sin z}\right)\left(\lim_{z\to 0}\cos z\right) = (1)(1) = 1.$$

2. If $z = a$ is a *simple pole* of $f(z)$ and if $f(z)$ can be expressed as a quotient of two analytic functions $h(z)$ and $g(z)$ as

$$f(z) = \frac{h(z)}{g(z)}, \text{ where } h(a) \neq 0,\ g(a) = 0,\ g'(a) \neq 0,$$

$$\left.\begin{array}{l}\text{Res. at}\\ z=a\end{array}\right\} = \lim_{z\to a}(z-a)\frac{h(z)}{g(z)}$$

$$= \lim_{z\to a}\frac{z-a}{g(z)-g(a)}h(z) \quad \text{since } g(a)=0$$

$$= \frac{h(a)}{g'(a)}. \tag{10.2.3}$$

Remark: In the above case the result can also be written as

$$\left.\begin{array}{l}\text{Res. of } f(z)\\ \text{at } z=a\end{array}\right\} = \left[\frac{h(z)}{g'(z)}\right]_{z=a}$$

Example. If $f(z) = \cot z = \cos z/\sin z$, then $z = 0$ is a simple pole of $f(z)$ and

$$h(z) = \cos z,\ g(z) = \sin z,\ h(0) \neq 0,\ g(0) = 0,\ g'(0) \neq 0.$$

$$\therefore \quad \left.\begin{array}{l}\text{Res. at}\\ z=0\end{array}\right\} = \frac{h(0)}{g'(0)} = \frac{\cos 0}{\cos 0} = 1.$$

3. If $z = a$ is a *pole of order m* of (z), suppose $(z-a)^m f(z) = \phi(z)$.

Then $\phi(z)$ is analytic in a neighbourhood of $z = a$. So $\phi(z)$ can be expanded about $z = a$ as a Taylor's series as

$$\phi(z) = \phi(a) + \frac{\phi'(a)}{1!}(z-a) + \frac{\phi''(a)}{2!}(z-a)^2 + \ldots$$

$$\therefore \quad f(z) = \frac{\phi(z)}{(z-a)^m} = \phi(a)\frac{1}{(z-a)^m} + \frac{\phi'(a)}{1!}\frac{1}{(z-a)^{m-1}} + \ldots + \frac{\phi^{(m-1)}(a)}{(m-1)!}\frac{1}{z-a} + \ldots$$

Hence the residue of $f(z)$ at $z = a$ is

$$\frac{\phi^{(m-1)}(a)}{(m-1)!} \quad \text{or} \quad \frac{1}{(m-1)!}\left[\frac{d^{m-1}}{dz^{m-1}}\phi(z)\right]_{z=a}$$

i.e.,

$$\frac{1}{(m-1)!}\left[\frac{d^{m-1}}{dz^{m-1}}(z-a)^m f(z)\right]_{z=a} \tag{10.2.3}$$

The following theorem is an application of residues.

Theorem 10.1. (Residue theorem), Suppose

(*i*) D is a simiply-connected region,

(*ii*) $f(z)$ is analytic in D except at its singularities which are finite in number,

(*iii*) C is a scro curve in D, not passing through any singularity.

Then $\int_C f(z)\,dz = 2\pi i$ (sum of the residue of $f(z)$ in C_i).

Proof: Let the singular points of $f(z)$ in C_i be $a_1, a_2, a_3, \ldots a_n$. Let $\Gamma_1, \Gamma_2 \ldots, \Gamma_n$ be positively oriented small circles,

$$|z-a_1| = r_1, |z-a_2| = r_2, \ldots, |z-a_n| = r_n$$

in C_i such that no one intersects the other (Fig. 10.2.1). Then, by extension to Cauchy's fundamental theorem,

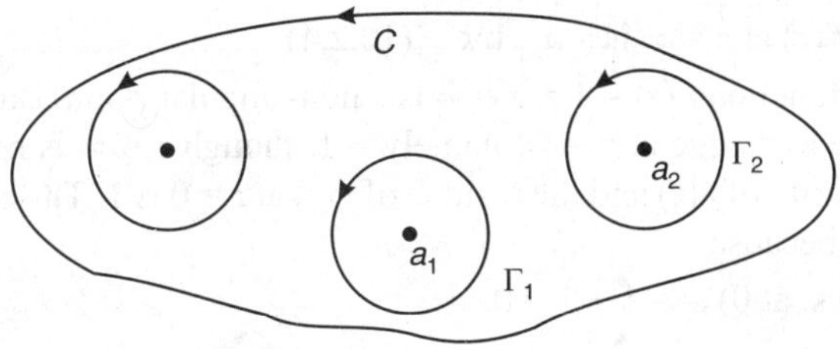

Fig. 10.2.1

$$\begin{aligned}\int_C f(z)\,dz &= \int_{\Gamma_1} f(z)\,dz + \int_{\Gamma_2} f(z)\,dz + \ldots + \int_{\Gamma_n} f(z)\,dz \\ &= 2\pi i\,(\text{Residue of } f(z) \text{ at } z = a_1) \\ &\quad + 2\pi i\,(\text{Residue of } f(z) \text{ at } z = a_2) \\ &\quad \ldots\ldots\ldots\ldots\ldots\ldots\ldots\ldots \\ &\quad + 2\pi i\,(\text{Residue of } f(z) \text{ at } z = a_n) \\ &= 2\pi i\,(\text{Sum of the residues of } f(z) \text{ in } C_i).\end{aligned}$$

Residue at $z = \infty$. To define the residue at $z = \infty$ which is an isolated singularity, we use (10.1.1). If C is a positively oriented circle $|z| = R$, then it is negatively oriented with respect to $z = \infty$ and $-C$ is positively oriented with respect to $z = \infty$. Hence

$$\left.\begin{array}{l}\text{Res. of } f(z) \\ \text{at } z = \infty\end{array}\right\} = \frac{1}{2\pi i}\int_{-C} f(z)\,dz = \frac{1}{2\pi i}\int_C f(z)\,dz, \tag{10.2.4}$$

where the exterior of the circle C contains no singularity of $f(z)$ other than $z = \infty$.

With this definition for residue of $f(z)$ at $z = \infty$ we have the following theorem :

Theorem 10.2. If a function $f(z)$ is analytic in the extended plane except at a finite number of singularities including $z = \infty$, then the sum of the residues of $f(z)$ is zero.

Proof: Suppose the singularities are $a_1, a_2, a_3, \ldots, a_{n-1}$ and $a_n = \infty$. If C is a scro curve encircling $a_1, a_2, \ldots, a_{n-1}$, then by the residue theorem,

$$\left.\begin{array}{l}\text{Sum of the residue of} \\ f(z) \text{ at } a_1, a_2, \ldots, a_{n-1}\end{array}\right\} = \frac{1}{2\pi i}\int_C f(z)\,dz$$

$$\text{Residue of } f(z) \text{ at } z = \infty = -\frac{1}{2\pi i}\int_C f(z)\,dz, \text{ by} \tag{10.2.4}$$

Therefore the sum of all the residues is zero.

Theorem 10.3. If $f(z)$ is analytic in a deleted neighbourhood of $z = \infty$ and if its Laurent's expansion in $R < |z| < \infty$ is

$$f(z) = \ldots + \frac{a_{-2}}{z^2} + \frac{a_{-1}}{z} + a_0 + a_1 z + a_2 z^2 + \ldots,$$

then the residue of $f(z)$ at $z = \infty$ is $-a_{-1}$.

Proof: Integrating $f(z)$, and the series term by term, along, a circle $C : |z| = r$ which lies in $R < |z| < \infty$,

$$\int_C f(z)\,dz = \int_C \frac{a_{-1}}{z}\,dz$$

since all the other integrals vanish. Thus, by (8.3.3),

$$\int_C f(z)\,dz = a_{-1}(2\pi i) \quad \text{or} \quad -\frac{1}{2\pi i}\int_C f(z)\,dz = -a_{-1}.$$

$\therefore$ Residue of $f(z)$ at $z = \infty$ is $-a_{-1}$ by (10.2.4)

Remark. For the function $f(z) = 1/z$, $z = \infty$ is a non-singular point. But in this case $-a_{-1} = -1$. So $f(z)$ is said to have a residue at $z = \infty$, namely -1, though $z = \infty$ is not a singular point. Now $z = 0$ is the only singularity of $f(z)$ and the residue of $f(z)$ at $z = 0$ is 1. This situation is in conformity with the theorem 10.2 because

(Res. at ∞) + (Res. at 0) $= -1 + 1 = 0$.

10.3. REAL DEFINITE INTEGRALS

The residue theorem is useful to evaluate certain types of real definite integrals. To illustrate the usage we shall consider here the following four types only;

Type 1: Integrals of the form

$$\int_0^{2\pi} f(\cos\theta, \sin\theta)\, d\theta,$$

where f is a rational function in $\cos\theta$ and $\sin\theta$.

Type 2: Integrals of the form

$$\int_{-\infty}^{\infty} \frac{P(x)}{Q(x)}\, dx,$$

where $P(x)$ and $Q(x)$ are polynomials in x such that the degree of Q exceeds that of P at least by two and $Q(x)$ does not vanish for any x.

Type 3: Integrals of the form

$$\int_{-\infty}^{\infty} (\sin ax)\, f(x)\, dx \quad \text{or} \quad \int_{-\infty}^{\infty} (\cos ax)\, f(x)\, dx,\ a > 0,$$

where $f(z)$ is such that $f(z) \to 0$ as $z \to \infty$ and it does not have a pole on the real axis.

Type 4: Integrals of the form

$$\int_{-\infty}^{\infty} f(x)\, dx,$$

where $f(z)$ has a finite number of poles on the real axis.

Contour integration. The complex integration along the scro curve used in evaluating the definite integral is called contour integration.

Type 1: The method to evaluate the integrals of this type is simply to consider the equivalent integral

$$\int_0^{2\pi} f(\cos\theta, \sin\theta) d\theta = \int_{C;|z|=1} f\left\{\frac{1}{2}\left(z + \frac{1}{z}\right), \frac{1}{2i}\left(z - \frac{1}{z}\right)\right\} \frac{dz}{iz}, \tag{10.3.1}$$

where C is the positively oriented unit circle, $z = e^{i\theta}$, $0 \le \theta \le 2\pi$.

By (8.2.3) this integral reduces to the integral

$$\int_0^{2\pi} f(\cos\theta, \sin\theta)\, d\theta, \tag{10.3.2}$$

because the substitution $z = e^{i\theta}$ renders $z = \cos\theta + i\sin\theta$,

$$\frac{1}{z} = \cos\theta - i\sin\theta, \quad dz = (ie^{i\theta})d\theta = i\, z\, d\theta,$$

$$\frac{1}{2}\left(z+\frac{1}{z}\right)=\cos\theta,\ \frac{1}{2i}\left(z-\frac{1}{z}\right)=\sin\theta,\ \frac{dz}{iz}=d\theta.$$

Illustration. We evaluate the integral

$$I=\int_0^{2\pi}\frac{1}{a+b\cos\theta}\,d\theta,\quad a>|b|>0.$$

Now $\displaystyle\int_0^{2\pi}\frac{1}{a+b\cos\theta}\,d\theta = \int_{|z|=1}\frac{1}{a+b\left\{\frac{1}{2}\left(z+\frac{1}{z}\right)\right\}}\frac{dz}{iz}$

$$=\frac{2}{i}\int_{|z|=1}\frac{1}{bz^2+2az+b}\,dz$$

$$=\frac{2}{i}\int_{|z|=1}\frac{1}{b(z-z_1)(z-z_2)}\,dz,$$

where z_1 and z_2 are the zeros of $bz^2+2az+b$ so that

$$z_1=-\frac{a}{b}+\frac{\sqrt{(a^2-b^2)}}{b},\ z_2=-\frac{a}{b}-\frac{\sqrt{(a^2-b^2)}}{b}$$

But z_1z_2 = product of the zeros (or $\dfrac{\text{Constant term}}{\text{Coeff. of } z^2}$ of the quadratic equation) = 1, that is $|z_1|\cdot|z_2|=1$. So, of z_1 and z_2, one lies inside the unit circle and the other lies outside the unit circle. But $(z_1)^2<(z_2)^2$, $|z_1|<|z_2|$. Hence z_1 lies inside the circle. That is, z_1 is the only singular point of the integrand lying in the interior of the unit circle C and z_1 is a simple pole of the integrand. So, by the Residue theorem,

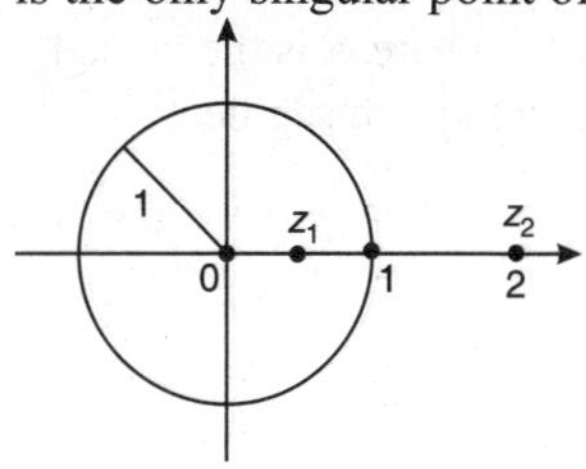

$$I=\frac{2}{i}(2\pi i)\,(\text{Residue of the integrand at } z_1)$$

$$=4\pi\left\{\lim_{z\to z_1}(z-z_1)\frac{1}{b(z-z_1)(z-z_2)}\right\}$$

$$=\frac{4\pi}{b(z_1-z_2)}=\frac{4\pi}{b}=\frac{b}{2\sqrt{(a^2-b^2)}}=\frac{2\pi}{\sqrt{(a^2-b^2)}}.$$

Type 2. Suppose we want to evaluate the improper integral

$$\int_{-\infty}^{\infty}\frac{P(x)}{Q(x)}\,dx$$

wherein the degree of the denominator polynomial $Q(x)$ is greater than the degree of the numerator polynomial $P(x)$ at least by 2. The contour to be used is the semicircle

$$\Gamma:|z|=R,\ \ \operatorname{Im} z\geq 0$$

which lies in the first and second quadrants and the line L joining the points

$$(-R,0),\ \ (R,0).$$

The complex integral along Γ,

$$\int_{\Gamma}\frac{P(z)}{Q(z)}\,dz,$$

becomes zero as $R \to \infty$ because

$$\left| \int_\Gamma f(z)\, dz \right| \le \begin{bmatrix} \max |f(z)| \\ \text{on } \Gamma \end{bmatrix} l(\Gamma)$$

$$\le \frac{1}{R^2} (\pi R) \qquad (1)$$

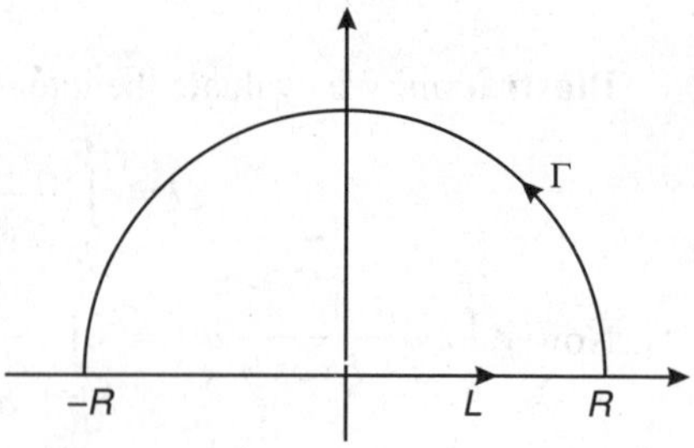

which vanishes when $R \to \infty$ and

$$\int_L \frac{P(z)}{Q(z)}\, dz = \int_{-\infty}^{\infty} \frac{P(x)}{Q(x)}\, dx$$

since, on L, $z = x$.

***Note*:** To obtain the inequality (1), we should use in general, for the numerator $|P(z)|$,

$$|z_1 + z_2| \le |z_1| + |z_2|$$

and for the denominator $|Q(z)|$,

$$|z_1 - z_2| \ge |z_1| - |z_2|.$$

As an example we evaluate the integral

$$\text{I} = \int_0^\infty \frac{x^2}{1 + x^4}\, dx \qquad (1980)$$

Cosider the integral

$$\text{I}' = \int_c \frac{z^2}{1 + z^4}\, dz,$$

where C is the closed curve consisting the positively oriented semicircle $\Gamma : |z| = R$, **Im** $z \ge 0$, and the line segment

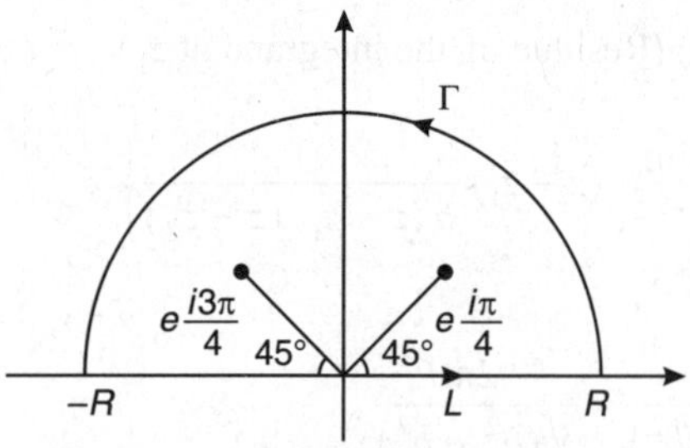

Fig. 10.3.1

$L : [-R, \; R]$ (Fig. 10.3.1). The denominator $1 + z^4$ of the integrand vanishes at $z^4 = -1 = \cos(2k\pi + \pi) + i \sin(2k\pi + \pi)$

$$z = e^{i(2k\pi + \pi)/4}, \quad k = 0, 1, 2, 3.$$

These points are simple poles of the integrand $z^2/(1 + z^4)$.

Further, of these, only the points

$$e^{i\pi/4}, \quad e^{i3\pi/4}$$

lie in C_i for $R > 1$. So, from "$a_{-1} = h(a)/g'(a)$" the residues of the integrand at these points are

$$\left(\frac{z^2}{4z^3}\right)_{z = e^{i\pi/4}}, \left(\frac{z^2}{4z^3}\right)_{z = e^{i3\pi/4}}$$

The sum of these residues is

$$\frac{e^{-i\pi/4}}{4}+\frac{e^{-i3\pi/4}}{4} = \frac{1}{4}\left(\cos\frac{\pi}{4}-i\sin\frac{\pi}{4}\right)+\frac{1}{4}\left(\cos\frac{3\pi}{4}-i\sin\frac{3\pi}{4}\right)$$

$$= \frac{1}{4}\left(\cos\frac{\pi}{4}-i\sin\frac{\pi}{4}\right)+\frac{1}{4}\left(-\cos\frac{\pi}{4}-i\sin\frac{\pi}{4}\right)$$

$$= -\frac{i}{2\sqrt{2}}.$$

So, by the Residue theorem, the value of I' is

$$\int_C \frac{z^2}{1+z^4}\,dz = 2\pi i\left[\frac{-i}{2\sqrt{2}}\right]$$

i.e., $$\int_\Gamma \frac{z^2}{1+z^4}\,dz+\int_L \frac{z^2}{1+z^4}\,dz = \frac{\pi}{\sqrt{2}}$$

i.e., $$\int_\Gamma \frac{z^2}{1+z^4}\,dz+\int_{-R}^{R} \frac{x^2}{1+x^4}\,dx = \frac{\pi}{\sqrt{2}}$$

since, on L, $z = x$. Now, taking limit as $R \to \infty$,

$$\int_{-\infty}^{\infty} \frac{x^2}{1+x^4}\,dx = \frac{\pi}{\sqrt{2}}$$

if the integral along Γ tends to 0 as $R \to \infty$. But this integral is such that its absolute value

$$\left|\int_\Gamma \frac{z^2}{1+z^4}\,dz\right| \le \left[\underset{z\in\Gamma}{\text{Max}}\left|\frac{z^2}{1+z^4}\right|\right] l(\Gamma)$$

$$\le \left[\underset{z\in\Gamma}{\text{Max}}\frac{|z|^2}{|z^4|-1}\right] l(\Gamma) \quad \text{since } |z^4-(-1)| \ge |z^4|-|-1|$$

$$= \frac{R^2}{R^4-1}(\pi R) = \pi\frac{R^3}{R^4-1}.$$

But $\pi R^3/(R^4-1) \to 0$ as $R \to \infty$. So the integral vanishes as $R \to \infty$.

Now the integrand is even. Therefore

$$2\int_0^{\infty} \frac{x^2}{x^4+1}\,dx = \frac{\pi}{\sqrt{2}} \quad \text{or} \quad \int_0^{\infty} \frac{x^2}{x^4+1}\,dx = \frac{\pi}{2\sqrt{2}}.$$

Type 3. This type is similar to the type 2 in all aspects except for the fact that, here, to show that the integral along Γ vanishes as $R \to \infty$, a result called Jordan's lemma, is used. We just state this result without giving its proof.

Jordan's lemma. If $m > 0$ and $f(z)$ is a continuous function such that max $|f(z)|$ on Γ tends to 0 as $R \to \infty$, then

$$\lim_{R\to\infty}\int_\Gamma f(z)\,e^{imz}\,dz = 0$$

where Γ is the semicircular arc, $|z| = R$, **Im** $z \ge 0$.

Illustration. As an example of type 3, we evaluate

(*i*) $\displaystyle\int_0^\infty \frac{\cos mx}{x^2+a^2}\,dx,\ m>0,\ a>0$ **(1982)**

(*ii*) $\displaystyle\int_0^\infty \frac{\cos mx}{x^2+1}\,dx,\ m>0$ **(1980, 81)**

(*i*) Let $C = \Gamma + L$, as specified in type 2, and consider

$$\int_C \frac{e^{imz}}{z^2+a^2}\,dz\,.$$

Here the singularities of the integrand are the simple poles ai and $-ai$ of which ai lies in the upper half-plane. The residue of the integrand at ai as

$$\lim_{z\to ai}\left[(z-ai)\frac{e^{imz}}{(z+ai)(z-ai)}\right] = \frac{e^{-ma}}{2ai}\,.$$

So, for an $R > a$, ai lies in C_i and, by Residue theorem,

$$\int_C \frac{e^{imz}}{z^2+a^2}\,dz = 2\pi i\left(\frac{e^{-ma}}{2ai}\right) = \frac{\pi e^{-ma}}{a}$$

i.e.,
$$\int_\Gamma \frac{e^{imz}}{z^2+a^2}\,dz + \int_L \frac{e^{imz}}{z^2+a^2}\,dz = \frac{\pi e^{-ma}}{a}$$

or
$$\int_\Gamma \frac{e^{imz}}{z^2+a^2}\,dz + \int_{-R}^{R} \frac{e^{imx}}{x^2+a^2}\,dx = \frac{\pi e^{-ma}}{a}\,.$$

As $R\to\infty$, by Jordan's lemma, the first integral vanishes because $m > 0$ and, taking $R > a$,

$$\lim_{R\to\infty}\left[\max_{\text{on }\Gamma}\left|\frac{1}{z^2+a^2}\right|\right] = \lim_{R\to\infty}\left[\max_{\text{on }\Gamma}\frac{1}{|z^2|-a^2}\right]$$

$$= \lim_{R\to\infty}\frac{1}{R^2-a^2} = 0$$

$$\therefore \quad \int_{-\infty}^{\infty} \frac{e^{imx}}{x^2+a^2}\,dx = \frac{\pi e^{-ma}}{a}$$

Thus, splitting the integral into real and imaginary parts,

$$\int_{-\infty}^{\infty} \frac{\cos mx}{x^2+a^2}\,dx = \frac{\pi e^{-ma}}{a}, \quad \int_{-\infty}^{\infty} \frac{\sin mx}{x^2+a^2}\,dx = 0\,.$$

Since $\cos mx/(x^2+a^2)$ is even, the first result becomes

$$\int_0^\infty \frac{\cos mx}{x^2+a^2}\,dx = \frac{\pi e^{-ma}}{2a}$$

(*ii*) This integral can be evaluated as a corollary to the result in (*i*) by setting $a = 1$ or, otherwise, independently.

Type 4. This type is a more general case of type 3. In type 3 the integrand $f(z)$ does not have a pole on the real axis while in this type the integrand has a finite number of poles on the real axis. The integrals of this type can be evaluated using the result of the following lemma.

Lemma : Let $f(z)$, an analytic function in a region D, have a simple pole at $z = a$ with residue k there. Let $\widehat{AB}$ a circular arc in D (Fig. 10.3.2) specified by

$$|z-a| = \rho, \quad \alpha \le \text{arc}\,(z-a) \le \beta.$$

Then $$\lim_{\rho \to 0}\left[\int_{\widehat{AB}} f(z)\,dz\right] = ik(\beta - \alpha). \qquad (10.3.3)$$

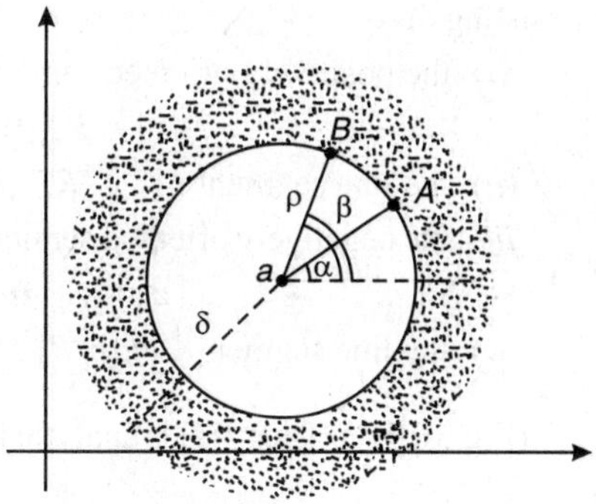

Fig. 10.3.2

Proof: Now $z = a$ is a simple pole of $f(z)$. Therefore

$$(z-a) f(z) \to k \quad \text{as} \quad z \to a.$$

Hence, for a given $\in > 0$, there exists a $\delta = \delta(\in)$ such that

$|(z-a) f(z) - k| < \in$ for all z in $|z-a| < \delta$.

Let $\phi(z)$ denote the function $(z-a) f(z) - k$. Then

$$|\phi(z)| < \in \quad \text{for all } z \text{ in } |z-a| < \delta.$$

Now

$$\int_{\widehat{AB}} f(z)\,dz = \int_{\widehat{AB}} \frac{k + \phi(z)}{z-a}\,dz = k \int_{\widehat{AB}} \frac{1}{z-a}\,dz + \int_{\widehat{AB}} \frac{\phi(z)}{z-a}\,dz$$

$$= k \int_{\alpha}^{\beta} \frac{1}{(a + \rho e^{i\theta} - a)} (i\rho e^{i\theta})\,d\theta + \int_{\widehat{AB}} \frac{\phi(z)}{z-a}\,dz$$

$$= i(\beta - \alpha) k + \int_{\widehat{AB}} \frac{\phi(z)}{z-a}\,dz \qquad (10.3.4)$$

But $$\left|\int_{\widehat{AB}} \frac{\phi(z)}{z-a}\,dz\right| \le \left[\max_{z \in \widehat{AB}} \left|\frac{\phi(z)}{z-a}\right|\right] l(\widehat{AB})$$

$$< \left(\frac{\in}{\rho}\right)[(\beta - \alpha)\rho] \qquad \text{if } \rho < \delta$$

Thus, for any $\in > 0$, there exists a δ such that

$$\int_{\widehat{AB}} \frac{\phi(z)}{z-a}\,dz < \in (\beta - \alpha), \quad \text{if} \quad \rho < \delta. \qquad (10.3.5)$$

Since $\in$ is arbitrary, this implies that

$$\lim_{\rho \to 0} \int_{\widehat{AB}} \frac{\phi(z)}{z-a}\,dz = 0;$$

otherwise (10.3.5) will be contradicted. Thus, from (10.3.4), we get, by taking limit as $\rho \to 0$, that

$$\lim_{\rho \to 0} \int_{\widehat{AB}} f(z)\,dz = i(\beta - \alpha) k.$$

Illustration 1. First we show that

$$\int_0^{\infty} \frac{\sin x}{x}\,dx = \frac{\pi}{2}. \qquad \textbf{(1981, 83)}$$

Consider the function $f(z) = e^{iz}/z$ which has the only simple pole $z = 0$ in the finite plane. Let C be the scro curve consisting of

(*i*) the positively oriented semicircle C_R specified by
$$|z| = R, \quad \mathbf{Im}\, z \geq 0,$$
(*ii*) the line segment $L_1 : [-R, -r]$,

(*iii*) the negatively oriented semicircle $-C_r$ specified by
$$|z| = r, \quad \mathbf{Im}\, z \geq 0,$$
(*iv*) the line segment $L_2 : [r, R]$

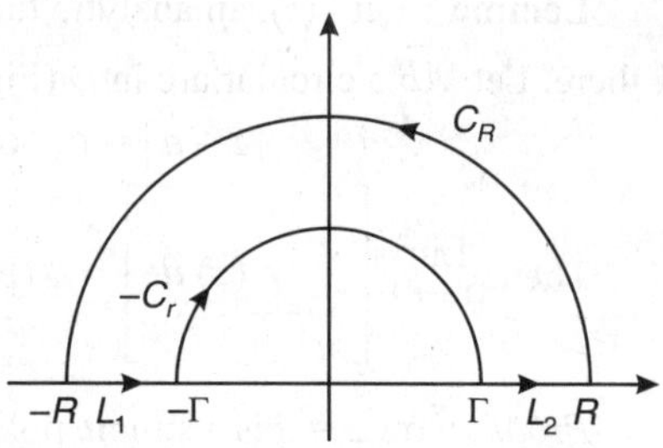

Fig. 103.3

(Fig. 10.3.3). Then $f(z)$ is analytic inside and on C. So we get, by Cauchy's fundamental theorem,

$$\int_C \frac{e^{iz}}{z}\, dz = 0$$

i.e.,
$$\int_{C_R} + \int_{L_1} + \int_{-C_r} + \int_{L_2} = 0$$

i.e.,
$$\int_{C_R} \frac{e^{iz}}{z}\, dz + \int_{-R}^{-r} \frac{e^{ix}}{x}\, dx + \int_{-C_r} \frac{e^{iz}}{z}\, dz + \int_r^R \frac{e^{ix}}{x}\, dx = 0$$

Now, taking limit as $R \to \infty$ and $r \to 0$.

$$\lim_{R\to\infty} \int_{C_R} \frac{e^{ix}}{z}\, dz + \int_{-\infty}^{\infty} \frac{e^{ix}}{x}\, dx - \lim_{r\to 0} \int_{C_r} \frac{e^{iz}}{z}\, dz = 0$$

of which the first limit is zero, by Jordan's lemma, because

$$\text{Max. } 1/|z| = 1/R \to 0 \quad \text{as} \quad R \to \infty.$$

But e^{iz}/z has a simple pole at $z = 0$. Its residue there is 1. So, by (10.3.3),

$$\lim_{r\to 0} \int_{C_r} \frac{e^{iz}}{z}\, dz = i\,(1)\,(\pi - 0) = i\pi.$$

$$\therefore \quad \int_{-\infty}^{\infty} \frac{e^{ix}}{x}\, dx - i\pi = 0 \quad \text{or} \quad \int_{-\infty}^{\infty} \frac{e^{ix}}{x}\, dx = i\pi.$$

Thus, splitting into real and imaginary parts,

$$\int_{-\infty}^{\infty} \frac{\sin x}{x}\, dx = \pi.$$

Since $(\sin x)/x$ is an even function of x,

$$\int_0^{\infty} \frac{\sin x}{x}\, dx = \frac{\pi}{2}.$$

Illustration 2. We show that, if $0 < a < 1$, then

(*i*) $I = \displaystyle\int_0^{\infty} \frac{x^{a-1}}{1-x}\, dx = \pi \cos a\pi,$

(*ii*) $I' = \displaystyle\int_0^{\infty} \frac{x^{a-1}}{1+x}\, dx = \frac{\pi}{\sin a\pi},$ **(1978, 80, 81)**

Consider the function $f(z) = z^{a-1}/(1-z)$. Since $a - 1$ is negative, the singularities of $f(z)$ are $z = 0$ and $z = 1$.

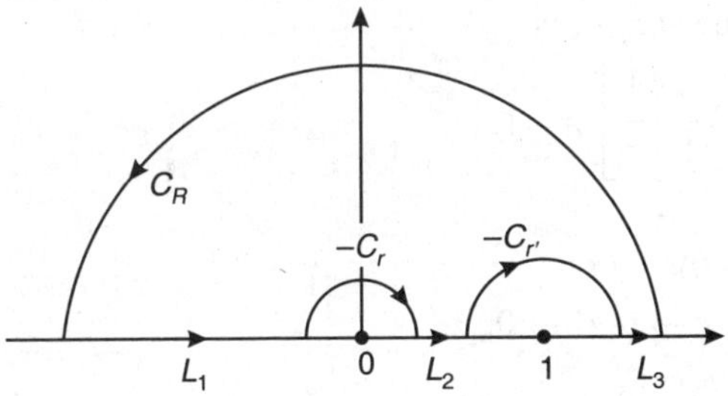

Fig. 10.3.4

Let C be the scro curve, as in figure 10.3.4, consisting of

(*i*)the positively oriented semicircle C_R specified by

$$|z| = R, \quad \textbf{Im } z \geq 0$$

(*ii*) the line segment $L_1 : [-R, -r]$

(*iii*) the negatively oriented semicircle $-C_r$ specified by

$$|z| = r, \quad \textbf{Im } z \geq 0$$

(*iv*) the line segment $L_2 : [r, 1-r']$

(*v*) the negatively oriented semicircle $-C_{r'}$ specified by

$$|z-1| = r', \quad \textbf{Im } z \geq 0$$

(*vi*) the line segment $L_3 : [1+r', R]$

For convenience we shall denote the integrals of $f(z)$ along these six bits of C by $I_1, I_2, I_3, I_4, I_5, I_6$ respectively. Then by Cauchy's theorem

$$\int_C f(z)\, dz = I_1 + I_2 + I_3 + I_4 + I_5 + I_6 = 0 \tag{1}$$

As $R \to \infty, r \to 0, r' \to 0$.

$L_1 + (L_2 + L_3) \to [-\infty\ 0] + \{[0, 1] + [1, \infty]\} = [-\infty, \infty]$

and
$$I_2 + (I_4 + I_6) = \int_{-\infty}^{0} \frac{x^{a-1}}{1-x}\, dx + \int_{0}^{\infty} \frac{x^{a-1}}{1-x}\, dx \tag{2}$$

Also
$$|I_1| \leq \left[\begin{matrix}\text{Max} \\ z \in C_R\end{matrix}\right] \left|\frac{z^{a-1}}{1-z}\right| l(C_R)$$

$$\leq \left[\begin{matrix}\text{Max} \\ z \in C_R\end{matrix} \frac{|z^{a-1}|}{|z|-1}\right] l(C_R) \text{ since } |z-1| \geq |z|-1$$

$$= \frac{R^{a-1}}{R-1}(\pi R) = \frac{\pi R^a}{R-1} \to 0 \text{ as } R \to \infty.$$

$\therefore\ I_1$ is 0 as $R \to \infty$.

In the lemma, $(z-a) f(z) \to k$ as $z \to a$. But in I_3

$$(z-0)\frac{z^{a-1}}{1-z} \to 0 \text{ as } z \to 0. \tag{3}$$

So, in I_3, as $r \to 0$, $z \to 0$, and consequently

$$I_3 \to i\,(0)\,[-(\pi - 0)] = 0 \qquad (4)$$

Since $z = 1$ is a simple pole of $f(z)$, its residue at $z = 1$ is

$$\lim_{z \to 1}\left[(z-1)\,\frac{z^{a-1}}{1-z}\right] = -1.$$

So, by the lemma, as $r' \to 0$,

$$I_5 \to i\,(-1)\,[-(\pi - 0)] = i\pi. \qquad (5)$$

Hence, from (1), (2), (3), (4), (5), as $R \to \infty$, $r \to 0$, $r' \to 0$,

$$\int_{-\infty}^{0} \frac{x^{a-1}}{1-x}\,dx + \int_{0}^{\infty} \frac{x^{a-1}}{1-x}\,dx + i\pi = 0$$

Making the substitution $X = -x$ in the first integral,

$$\int_{\infty}^{0} \frac{(-X)^{a-1}}{1+X}\,(-dX) + I + i\pi = 0$$

i.e., $$(-1)^{a-1}\,I' + I + i\pi = 0$$

i.e., $$(\cos \pi + i \sin \pi)^{a-1}\,I' + I + i\pi = 0$$

i.e., $$\{I' \cos (a-1)\,\pi + I\} + i\,\{I' \sin (a-1)\,\pi + \pi\} = 0$$

So $$I' \sin (a-1)\,\pi + \pi = 0, \qquad (6)$$

$$I' \cos (a-1)\,\pi + I = 0. \qquad (7)$$

(6) implies $I' \sin a\pi = \pi$ or $I' = \dfrac{\pi}{\sin a\pi}$

(7) implies $I = -I' \cos (a-1)\,\pi = I' \cos a\pi = \pi \cot a\pi$.

10.4. EXAMPLES

10.1 *Residue*

1. If a function $h(z)$ is analytic at $z = a$ and $h(a) \neq 0$, show that $z = a$ is a simple pole of the function $f(z) = h(z)/(z-a)$ and that the residue of $f(z)$ at $z = a$ is $h(a)$.

[**Hint:** $f(z) = \{h(a) + h'(a)(z-a) + ...\} / (z-a)$ by Taylor's expansion of the analytic function $h(z)$ about $z = a$]

2. If $f(z) = -\dfrac{1}{z-1} \times 2\,[1 + (z-1) + (z-1)^2 + ...]$, find the residue of $f(z)$ at $z = 1$.

[**Hint:** Residue of $f(z)$ at $z = 1$ equals Coef. of $\dfrac{1}{z-1}$ which is -2]

3. Find the residue of $f(z) = \dfrac{\cos z}{z}$ at its pole.

[**Hint:** Pole: $z = 0$. $f(z) = \dfrac{1}{z}\left(1 - \dfrac{z^2}{2!} + \dfrac{z^4}{4!} - ...\right)$. Residue = 1]

4. Solved sum. Find the residues at $z = 0$ of the functions

(i) $f(z) = e^{1/z}$

(ii) $f(z) = \dfrac{\sin z}{z^4}$

(iii) $f(z) = z \cos 1/z$

✍ The residues are the coefficients of $1/z$ in the Laurent's expansions of $f(z)$ about $z = 0$.

(i)
$$e^{1/z} = 1 + \frac{1}{1!}\frac{1}{z} + \frac{1}{2!}\frac{1}{z^2} + \frac{1}{3!}\frac{1}{z^3} + \ldots$$

$$\therefore \quad \left.\begin{matrix}\text{Res. of } f(z) \\ \text{at } z = 0\end{matrix}\right\} = \frac{1}{1!} = 1.$$

(ii)
$$\frac{\sin z}{z^4} = \frac{1}{z^4}\left[z - \frac{z^3}{3!} + \frac{z^5}{5!} - \ldots\right]$$

$$= \frac{1}{z^3} - \frac{1}{3!}\frac{1}{z} + \frac{z}{5!} - \ldots$$

$$\therefore \quad \left.\begin{matrix}\text{Res. of } f(z) \\ \text{at } z = 0\end{matrix}\right\} = -\frac{1}{3!} = -\frac{1}{6}.$$

(iii)
$$z \cos\frac{1}{z} = z\left[1 - \frac{1}{2!}\frac{1}{z^2} + \frac{1}{4!}\frac{1}{z^4} - \ldots\right]$$

$$= z - \frac{1}{2!}\frac{1}{z} + \frac{1}{4!}\frac{1}{z^3} - \ldots$$

$$\therefore \quad \left.\begin{matrix}\text{Res. of } f(z) \\ \text{at } z = 0\end{matrix}\right\} = -\frac{1}{2!} = -\frac{1}{2}.$$

5. Solved sum. Find the poles of the following functions and find the residues of the functions at them

(*i*) $\dfrac{1}{z^2 e^z}$ (*ii*) $\dfrac{z+1}{z^2 - 2z}$ **(1984)** (*iii*) $\dfrac{z^2 - 1}{(z^2+1)^2}$ (*iv*) $\dfrac{1}{z^3(z+4)}$.

(*i*)
$$\frac{1}{z^2 e^z} = \frac{1}{z^2}e^{-z} = \frac{1}{z^2}\left(1 - z + \frac{z^2}{2!} - \frac{z^2}{3!} + \ldots\right) \text{ in } 0 < |z| < \infty$$

$$= \frac{1}{z^2} - \frac{1}{z} + \frac{1}{2!} - \frac{z}{3!} + \ldots$$

So the function has a pole of order 2 at $z = 0$, the residue there, that is, the coefficient of $1/z$, being -1.

(*ii*) The function has simple poles at $z = 0$ and at $z = 2$.

$$\text{Residue at } z = 0 = \lim_{z \to 0}\left[z \cdot \frac{z+1}{z^2 - 2z}\right] = -\frac{1}{2}$$

$$\text{Residue at } z = 2 = \lim_{z \to 2}\left[(z-2)\frac{z+1}{z^2 - 2z}\right] = \frac{3}{2}.$$

(*iii*) The function has poles of order 2 at $z = -i$ and $z = i$.

$$\text{Residue at } z = -i = \frac{1}{(2-1)!}\frac{d}{dz}\left[(z+i)^2 \frac{z^2-1}{(z^2+1)^2}\right]\Bigg|_{z=-i}$$

$$= \frac{d}{dz}\frac{z^2-1}{(z-i)^2}\bigg|_{z=-i}$$

$$= \frac{(z-i)^2\,(2z)-(z^2-1)\,2\,(z-i)}{(z-i)^4}\bigg|_{z=-i}$$

$$= \frac{(-4)\,(-2i)-(-2)\,2\,(-2i)}{16} = 0\,.$$

$$\text{Residue at } z = i = \frac{1}{(2-1)!}\frac{d}{dz}\left[(z-i)^2 \frac{z^2-1}{(z^2+1)^2}\right]\Bigg|_{z=i}$$

$$= \frac{d}{dz}\frac{z^2-1}{(z+i)^2}\bigg|_{z=i}$$

$$= \frac{(z+i)^2\,2z-(z^2-1)\,2\,(z+i)}{(z+i)^4}\bigg|_{z=i}$$

$$= \frac{(-4)\,(2i)-(-2)\,2\,(2i)}{16} = 0.$$

(*iv*) The function has a pole of order 3 at $z = 0$ and a simple pole at $z = -4$.

$$\text{Residue at } z = 0 = \frac{1}{2!}\frac{d^2}{dz^2}\left[z^3 \frac{1}{z^3(z+4)}\right]\Bigg|_{z=0}$$

$$= \frac{1}{2}\frac{d^2}{dz^2}(z+4)^{-1}\bigg|_{z=0}$$

$$= \frac{1}{2}(-1)\,(-2)\,(4)^{-3} = \frac{1}{64}\,.$$

$$\text{Residue at } z = -4 = \lim_{z\to -4}\left[(z-4)\frac{1}{z^3(z-4)}\right] = -\frac{1}{64}\,.$$

6. Find the residues of the function $1/(z-1)(z-2)$ at the poles.

[**Hint:** Res. of $f(z)$ at $z = 1$

$$= \lim_{z\to 1}(z-1)\frac{1}{(z-1)\,(z-2)} = -1;$$

Res. of $f(z)$ at $z = 2$

$$= \lim_{z\to 2}(z-2)\frac{1}{(z-1)\,(z-2)} = 1]$$

7. Find the residues of the following functions at their poles :

(*i*) $\dfrac{z^2}{z^2+a^2}$ **(1978)** (*ii*) $\dfrac{z}{(z^2+1)(z-2)}$ **(1981)**

Ans. (*i*) Poles; $ai, -ai$, Res. : $\dfrac{1}{2}ai$, $-\dfrac{1}{2}ai$ (*ii*) Poles : $i, -i, 2$; Res. : $-(i+2)/10, (i-2)/10$, 2/5.

8. Solved sum. Find the residues of $f(z) = \dfrac{z^2}{z^2+a^2}$ at its singularities.

✍ The given function is

$$f(z) = \frac{z^2}{(z+ia)(z-ia)}.$$

So the singularities are simple poles $z = ia$ and $z = -ia$.

$$\left.\begin{array}{l}\text{Res. of } f(z)\\ \text{at } z = ia\end{array}\right\} = \lim_{z \to ia} (z-ia)\, f(z)$$

$$= \lim_{z \to ia} (z-ia) \frac{z^2}{(z+ia)(z-ia)}$$

$$= \lim_{z \to ia} \frac{z^2}{z+ia} = \frac{(ia)^2}{2(ia)} = \frac{ia}{2}$$

$$\left.\begin{array}{l}\text{Res. of } f(z)\\ \text{at } z = -ia\end{array}\right\} = \lim_{z \to -ia} (z+ia)\, f(z)$$

$$= \lim_{z \to -ia} (z+ia) \frac{z^2}{(z+ia)(z-ia)}$$

$$= \lim_{z \to -ia} \frac{z^2}{z-ia} = \frac{(-ia)^2}{2(-ia)} = -\frac{ia}{2}.$$

9. Solved sum. Find the residues of $f(z) = \dfrac{1}{(z^2+a^2)^2}$ at its singularities.

✍ The given function is

$$f(z) = \frac{1}{(z+ia)^2 (z-ia)^2}$$

So the singularities are the double poles $z = ia$ and $z = -ia$.

$$\left.\begin{array}{l}\text{Res. of } f(z)\\ \text{at } z = ia\end{array}\right\} = \frac{1}{(2-1)!}\left(\frac{d}{dz}[(z-ia)^2 f(z)]\right)_{z=ia}$$

$$= \frac{1}{(2-1)!}\left(\frac{d}{dz}(z-ia)^2 \frac{1}{(z+ia)^2 (z-ia)^2}\right)_{z=ia}$$

$$= \left[\frac{d}{dz}\frac{1}{(z+ia)^2}\right]_{z=ia} = \left[\frac{d}{dz}(z+ia)^{-2}\right]_{z=ia}$$

$$= \left[\frac{-2}{(z+ia)^3}\right]_{z=ia} = \frac{-2}{(2ia)^3} = \frac{-i}{4a^3}.$$

$$\left.\begin{array}{l}\text{Res. of } f(z)\\ \text{at } z=-ia\end{array}\right\} = \frac{1}{(2-1)!}\left[\frac{d}{dz}(z+ia)^2 f(z)\right]_{z=-ia}$$

$$= \frac{1}{(2-1)!}\left[\frac{d}{dz}(z+ia)^2 \frac{1}{(z+ia)^2(z-ia)^2}\right]_{z=-ia}$$

$$= \left[\frac{d}{dz}\frac{1}{(z-ia)^2}\right]_{z=-ia} = \left[\frac{d}{dz}(z-ia)^{-2}\right]_{z=-ia}$$

$$= \left[\frac{-2}{(z-ia)^3}\right]_{z=-ia} = \frac{-2}{(-2ia)^3} = \frac{i}{4a^3}.$$

10. Solved sum. Find the residues of

$$f(z) = \frac{z^4}{(z-1)^4(z-2)(z-3)}$$

at its singularities.

✍ The singularities of $f(z)$ are

(*i*) the simple pole $z = 2$

(*ii*) the simple pole $z = 3$

(*iii*) the pole $z = 1$ of order 4.

$$\left.\begin{array}{l}\text{Res. of } f(z)\\ \text{at } z=2\end{array}\right\} = \lim_{z\to 2}(z-2)f(z)$$

$$= \lim_{z\to 2}\frac{z^4}{(z-1)^4(z-3)} = \frac{16}{1(-1)} = -16$$

$$\left.\begin{array}{l}\text{Res. of } f(z)\\ \text{at } z=3\end{array}\right\} = \lim_{z\to 3}(z-3)f(z)$$

$$= \lim_{z\to 3}\frac{z^4}{(z-1)^4(z-2)} = \frac{81}{16(1)} = \frac{81}{16}.$$

$$\left.\begin{array}{l}\text{Res. of } f(z)\\ \text{at } z=1\end{array}\right\} = \frac{1}{3!}\left(\frac{d^3}{dz^3}[(z-1)^4 f(z)]\right)_{z=1}$$

$$= \frac{1}{3!}\left[\frac{d^3}{dz^3}\frac{z^4}{(z-2)(z-3)}\right]_{z=1}$$

Dividing z^4 by $z^2 - 5z + 6$, we get the quotient as $z^2 + 5z + 19$ and remainder as $65z - 114$. So

$$\frac{z^4}{(z-2)(z-3)} = z^2 + 5z + 19 + \frac{65z-114}{(z-2)(z-3)}$$

PO - Y FARUKI. I

SDTD

$$= z^2 + 5z + 19 - \frac{16}{z-2} + \frac{81}{z-3}$$

$$\left.\begin{array}{l}\text{Res. of } f(z)\\ \text{at } z=1\end{array}\right\} = \frac{1}{3!}\left[\frac{d^3}{dz^3}\left(z^2 + 5z + 19 - \frac{16}{z-2} + \frac{81}{z-3}\right)\right]_{z=1}$$

$$= \frac{1}{3!}\left[-16\frac{(-1)(-2)(-3)}{(z-2)^4} + 81\frac{(-1)(-2)(-3)}{(z-3)^4}\right]_{z=1}$$

$$= \left[\frac{16}{(z-2)^4} - \frac{81}{(z-3)^4}\right]_{z=1} = 16 - \frac{81}{16} = \frac{175}{16}.$$

11. Find the residues of the following functions at their poles:

(*i*) $\dfrac{z}{(z+1)^2 (z^2+1)}$ (*ii*) $\dfrac{z^2 - 2z}{(z+1)^2 (z^2+4)}$

(*iii*) $\dfrac{z^3}{(z-1)(z-2)(z-3)}$ (*iv*) $\dfrac{z^3}{(z-1)^4 (z-2)(z-3)}$

[**Ans.** (*i*) Poles: $i, -i, -1$; Residues: $-\dfrac{i}{4}, \dfrac{i}{4}, 0$

(*ii*) Poles: $2i, -2i, -1$; Residues: $\dfrac{7+i}{25}, \dfrac{7-i}{25}, -\dfrac{14}{25}$

(*iii*) Poles: 1, 2, 3; Residues: $\dfrac{1}{2}, -8, \dfrac{27}{2}$

(*iv*) Simple poles; 2, 3; Residues: -8, $27/16$. But the third pole $z = 1$ is of order 4. Residue at $z = 1$ is

$$\frac{1}{3!}\left[\frac{d^3}{dz^3}\left(z + 5 - \frac{8}{z-2} + \frac{27}{z-3}\right)\right]_{z=1} = \frac{101}{16}]$$

12. Solved sum. Find the residue of $f(z) = \dfrac{1}{(1+z^2)^4}$ at $z = i$

✍ Since $f(z) = \dfrac{1}{(z+i)^4 (z-i)^4}$, $z = i$ is a pole of order 4. Therefore

$$\left.\begin{array}{l}\text{Res. of } f(z)\\ \text{at } z=i\end{array}\right\} = \frac{1}{3!}\left[\frac{d^3}{dz^3}(z+i)^{-4}\right]_{z=i}$$

$$= \frac{1}{3!}[(-4)(-5)(-6)(z+i)^{-7}]_{z=i}$$

$$= -\frac{20}{(2i)^7} = -\frac{5i}{32}.$$

13. Solved sum. Find the residue of $f(z) = \dfrac{z \sin z}{(z-\pi)^3}$ at its singularity.

Now $z = \pi$ is the only finite singularity of $f(z)$ and its nature is not known. So we shall expand $f(z)$ in powers of $z - \pi$. Then the residue of $f(z)$ at $z = \pi$ is the coefficient of $(z-\pi)^{-1}$. Let

$$u = z - \pi, \quad z = u + \pi.$$

Then the function $f(z)$ in terms of u is

$$\frac{(u+\pi)\sin(u+\pi)}{u^3} = -\frac{(u+\pi)\sin u}{u^3}$$

$$= \frac{1}{u^3}(u+\pi)\left[-u + \frac{u^3}{3!} - \frac{u^5}{5!} + \ldots\right]$$

$$= \frac{1}{u^3}\left[-\pi u - u^2 + \frac{\pi u^3}{3!} + \frac{u^4}{3!} - \ldots\right],$$

where $u = z - \pi$. The residue at $z = \pi$ is the coefficient of u^{-1} which is -1.

14. Find the residue of $\dfrac{1+e^z}{\sin z + z\cos z}$ at the pole $z = 0$. **(1977)**

[**Hint:** $z = 0$ is a simple pole;

Method 1. $h(0) \neq 0$, $g(0) = 0$, $g'(0) \neq 0$; Res. : $\dfrac{h(0)}{g'(0)} = \dfrac{2}{2} = 1$.

Method 2. Res. : $\displaystyle\lim_{z \to 0} \frac{1+e^z}{(\sin z/z) + \cos z} = 1$]

15. Find the residues of $1/(z^2 + a^2)^2$ at ai and $-ai$. **(1978)**

[**Hint:** Residue at $z = ai = \dfrac{d}{dz}(z-ai)^2 f(z)\Big|_{z=ai}$

$$= \frac{d}{dz}(z+ai)^{-2}\Big|_{z=ai} = \frac{-i}{4a^3},$$

Residue at $z = -ai$ is $i/4a^3$]

16. Find residue of $1/(1+z^2)^n$ at $z = i$. **(1980)**

[**Hint:** $\displaystyle\frac{1}{(n-1)!}\lim_{z \to i}\frac{d^{n-1}}{dz^{n-1}}\frac{1}{(z+i)^n}$

$$= \frac{1}{(n-1)!}\left[(-1)^{n-1}\frac{(2n-2)!}{(n-1)!(2i)^{2n-1}}\right] = -i\frac{(2n-2)!}{2^{2n-1}[(n-1)!]^2}\ ]$$

17. Solved sum. Show that the residue of $1/z(e^z - 1)$ $z = 0$ is $-1/2$.

$z = 0$ is a double pole. The residue at $z = 0$ is

$$\frac{d}{dz}z^2 f(z)\Big|_{z=0} = \frac{d}{dz}\frac{z}{e^z - 1}\Big|_{z=0}$$

$$= \left. \frac{(e^z - 1) - ze^z}{(e^z - 1)^2} \right|_{z=0}$$

$$= -\left. \frac{z}{2(e^z - 1)} \right|_{z=0} = -\left. \frac{1}{2e^z} \right|_{z=0} = -\frac{1}{2} \text{ by Hospital's rule.}$$

18. Solved sum. Find the residue of $e^z/(\sin z)^2$ about where n is a positive or negative integer. Expanding e^z in powers of $z - n\pi$, say u,

$$e^z = e^{u+n\pi} = e^{n\pi}\left(1 + u + \frac{1}{2}u^2 + \ldots\right)$$

$$\frac{1}{(\sin z)^2} = \frac{1}{\sin^2(u + n\pi)} = \frac{1}{(\sin u)^2}$$

$$= \frac{1}{u^2}\left(\frac{1}{1!} - \frac{u^2}{3!} + \ldots\right)^{-2} = \frac{1}{u^2}\left(1 + \frac{u^2}{3} + \ldots\right)$$

$$\therefore \quad \frac{e^z}{(\sin z)^2} = \frac{e^{n\pi}}{u^2}\left(1 + u + \frac{u^2}{2} + \ldots\right)\left(1 + \frac{u^2}{3} + \ldots\right)$$

The residue at $z = n\pi$ is the coefficient of $1/u$ which is $e^{n\pi}$.

19. Solved sum. The only singularities of a single-valued function $f(z)$ are poles of orders 1 and 2 at $z = -1$ and $z = 2$ with residues 1 and 2 respectively at these points. If $f(0) = 7/4$ and $f(1) = 5/2$, determine the function and expand it in a Laurent's series valid in $1 < |z| < 2$. **(1975, 77, 80)**

Consider the sum of two functions

$$\phi(z) = \left(\frac{1}{z+1}\right) + \left(\frac{A}{(z-2)^2} + \frac{2}{z-2}\right).$$

The function $1/(z + 1)$ has a simple pole at $z = -1$ with a residue 1. But the function $A/(z - 2)^2 + 2/(z - 2)$ is analytic at $z = -1$. Therefore it can be expanded in a Taylor's series like $a_0 + a_1(z + 1) + \ldots$ in a neighbourhood of $z = -1$. Thus

$$\phi(z) = \frac{1}{z+1} + a_0 + a_1(z+1) + \ldots$$

Therefore $z = -1$ is a simple pole and the residue of $\phi(z)$ there is 1. Similarly, $\phi(z)$ has a double pole with residue 2. Therefore $f(z)$ is of the form $f(z) + a$ constant B because, if it contains a term z, then ∞ is its pole of order 1 and if it contains a term z^2, then ∞ is its pole of order 1 and if it contains a term z^2, then ∞ is its pole of order 2 and so on. But it is given that $f(z)$ has only two poles -1 and 2. Hence

$$f(z) = \frac{1}{z+1} + \frac{A}{(z-2)^2} + \frac{2}{z-2} + B.$$

But $f(0) = 7/4$ and $f(1) = 5/2$. Therefore

$$\left.\begin{array}{r} A + 4B = 7 \\ A + B = 4 \end{array}\right\} \quad \therefore \quad A = 3, B = 1$$

$$\therefore \qquad f(z) = \frac{1}{z+1} + \frac{3}{(z-2)^2} + \frac{2}{z-2} + 1.$$

In the annulus $1 < |z| < 2$, $|1/z| < 1$ and $|z/2| < 1$. Hence, for any z in the annulus,

$$f(z) = 1 + \frac{1}{z(1+1/z)} + \frac{3}{4}\frac{1}{(1-z/2)^2} + \frac{2}{(-2)(1-z/2)}$$

$$= 1 + \frac{1}{z}\left(1 - \frac{1}{z} + \frac{1}{z^2} - \frac{1}{z^3} + \ldots\right) + \frac{3}{4}\left(1 + 2\frac{z}{2} + 3\frac{z^2}{2^2} + 4\frac{z^3}{2^3} + \ldots\right) - \left(1 + \frac{z}{2} + \frac{z^2}{2^2} + \frac{z^3}{2^3} + \ldots\right).$$

RESIDUE THEREM

20. Solved sum. If C is the positively oriented circle $|z| = 2$, evaluate

$$I = \int_C \frac{5z-2}{z(z-1)}\, dz.$$

Denote the integrand by $f(z)$. The singular points of the integrand are $z = 0$ and $z = 1$. They are simple poles and they lie in C_i. The residues of the integrand at these points are respectively

$$\lim_{z \to 0}\left[z\frac{5z-2}{z(z-1)}\right], \lim_{z \to 1}\left[(z-1)\frac{5z-2}{z(z-1)}\right]$$

which equal 2 and 3. So, by the Residue theorem,

$$I = 2\pi i \text{ (sum of the Residues)} = 2\pi i\,(2+3) = 10\pi i.$$

21. If C is a scro curve encircling the origin, show that

(*i*) $\int_C \frac{2e^z}{z}\, dz = 4\pi i$ (*ii*) $\int_C \frac{3+z}{z}\, dz = 6\pi i$

(*iii*) $\int_C \frac{1}{z^2 e^z}\, dz = -2\pi i$

[**Hint:** Origin is the only singular point of the integrand in C_i].

22. If C is the positively oriented circle $|z| = 2$, show that

(*i*) $\int_C \frac{z-2}{z(z-1)}\, dz = 2\pi i$ (*ii*) $\int_C \frac{e^{2z}}{(z+1)^3}\, dz = \frac{4\pi i}{e^2}$

(*iii*) $\int_C \tan z\, dz = -4\pi i$.

[**Hint:** (*iii*) Singularities are $z = \pm\frac{\pi}{2}, \pm\frac{3\pi}{2}, \pm\frac{5\pi}{2} + \ldots$; $\tan z$ has simple poles at $z = -\frac{1}{2}\pi$, $z = \frac{1}{2}\pi$ in the interior of $|z| = 2$. Let $h(z) = \sin z$ and $g(z) = \cos z$.

Find $h\left(-\frac{1}{2}\pi\right)\Big/ g'\left(-\frac{1}{2}\pi\right)$ and $h\left(\frac{1}{2}\pi\right)\Big/ g'\left(\frac{1}{2}\pi\right)$].

23. If C is the positively oriented circle $|z| = 3$, show that

(*i*) $\int_C \frac{z+1}{z^2-2z}\, dz = 2\pi i$ (*ii*) $\int_C \frac{1}{z^3(z+4)}\, dz = \frac{\pi i}{32}$.

24. Solved sum. If C is the circle $|z-i|=2$, evaluate

$$\int_C \frac{z}{2z^2+5iz-2}\,dz.$$

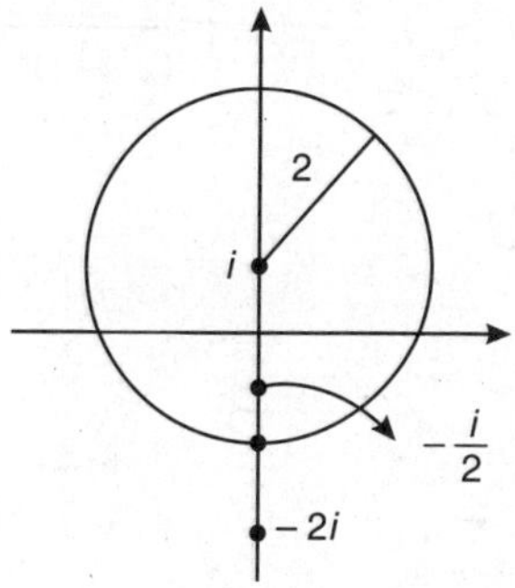

Denote the integrand by $f(z)$. Then

$$f(z) = \frac{z}{2z^2+5iz+2i^2} = \frac{z}{(2z+i)(z+2i)}$$

$$= \frac{z}{2(z+i/2)(z+2i)}.$$

So $f(z)$ has only two singularities

$$z=-i/2, \quad z=-2i$$

which are simple poles, of which only $z=-i/2$ lies inside C. Thus, by Residue theorem,

$$\int_C f(z)\,dz = 2\pi i \begin{Bmatrix}\text{Res. of } f(z) \\ \text{at } z=-i/2\end{Bmatrix} = 2\pi i\left[\lim_{z\to -i/2}(z+i/2)\,f(z)\right]$$

$$= 2\pi i\left[\frac{z}{2(z+2i)}\right]_{z=-i/2} = -\frac{\pi i}{3}.$$

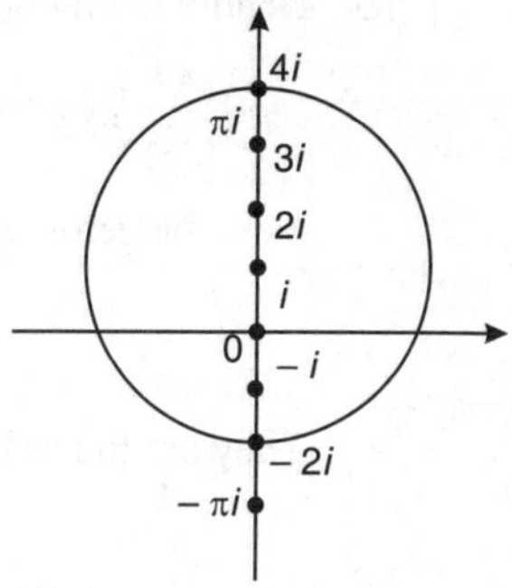

25. Solved sum. Evaluate $\int_C \frac{e^z}{z^2+\pi^2}\,dz$, where C is

$$|z-i|=3.$$

Denote the integrand by $f(z)$. Then

$$f(z) = \frac{e^z}{(z+\pi i)(z-\pi i)}.$$

Now $z=\pi i$ and $z=-\pi i$ are the only poles of $f(z)$. They are simple poles, of which only πi lies in C_i. So, by Residue theorem,

$$\int_C \frac{e^z}{z^2+\pi^2}\,dz = 2\pi i\,[\text{Res. of } f(z) \text{ at } z=\pi i]$$

$$= 2\pi i\left[\lim_{z\to \pi i}(z-\pi i)\,f(z)\right]$$

$$= 2\pi i\left[\frac{e^z}{z+\pi i}\right]_{z=\pi i} = 2\pi i\,\frac{e^{\pi i}}{2\pi i}$$

$$= \cos\pi + i\sin\pi = -1.$$

26. Solved sum. If C is $|z-1|=1$, evaluate

$$\int_C \frac{1}{(z^3-1)^2}\,dz.$$

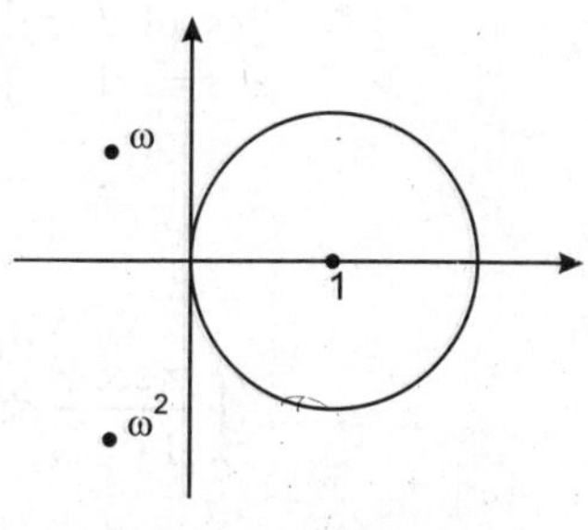

Denote the integrand by $f(z)$. Since

$$z^3-1=(z-1)(z-\omega)(z-\omega^2),$$

1, ω, ω^2 are the double poles of $f(z)$ of which only $z=1$ lies in C_i. This is the only singularily in C_i. Thus, by Residue theorem, we get

$$\int_C \frac{1}{(z^3-1)^2}\,dz = 2\pi i \left\{\begin{matrix}\text{Res. of } f(z)\\ \text{at } z=1\end{matrix}\right\} = 2\pi i \left[\frac{d}{dz}(z-1)^2 f(z)\right]_{z=1}$$

$$= 2\pi i \left[\frac{d}{dz}(z^2+z+1)^{-2}\right]_{z=1}$$

$$= 2\pi i \left[(-2)\frac{2z+1}{(z^2+z+1)^3}\right]_{z=1}$$

$$= -4\pi i\,\frac{3}{27} = -\frac{4\pi i}{9}.$$

27. Solved sum. Evaluate

$$\int_C \frac{z \sec z}{1-z^2}\,dz,$$

where C is the ellipse $4x^2 + 9y^2 = 9$.

✍ Denote the integrand by $f(z)$. Then

$$f(z) = \frac{z}{(1+z)(1-z)\cos z}.$$

Now the zeros of the denominator of $f(z)$ are

$$1, -1;\ \pm\frac{\pi}{2}, \pm\frac{3\pi}{2}, \ldots$$

They are the poles of $f(z)$. They are

$$1, -1;\ \pm\frac{3.1416}{2}, \pm\frac{9.4248}{2}, \ldots$$

all of which lie on the real axis. But the vertices of the ellipse are $\left(3/2, 0\right)$ and $\left(-3/2, 0\right)$

So only the simple poles $z = 1, z = -1$ lie inside C. Thus, by Residue theorem,

$$\int_C \frac{z \sec z}{1-z^2}\,dz = 2\pi i \left\{\begin{matrix}\text{Sum of the residues}\\ \text{of } f(z) \text{ at } z=1, z=-1\end{matrix}\right\}$$

$$\left.\begin{matrix}\text{Res. of } f(z)\\ \text{at } z=1\end{matrix}\right\} = \lim_{z\to 1}(z-1)f(z)$$

$$= \left[-\frac{z}{(1+z)\cos z}\right]_{z=1} = -\frac{1}{2\cos 1}$$

$$\left.\begin{matrix}\text{Res. of } f(z)\\ \text{at } z=-1\end{matrix}\right\} = \lim_{z\to -1}(z+1)f(z)$$

$$= \left[\frac{z}{(1-z)\cos z}\right]_{z=-1} = -\frac{1}{2\cos 1}.$$

$$\therefore \quad \int_C \frac{z \sec z}{1-z^2}\,dz = 2\pi i\left[-\frac{1}{2\cos 1} - \frac{1}{2\cos 1}\right]$$

$$= -2\pi i \sec 1.$$

28. If C is the circle $|z| = 3/2$, evaluate

$$\int_C \frac{4-3z}{z(z-1)(z-2)}\, dz.$$

[**Ans.** $2\pi i$]

29. If C is $|z| = 3$, evaluate

$$\int_C \frac{\sin \pi z^2 + \cos \pi z^2}{(z-1)(z-2)}\, dz.$$

[**Ans.** $4\pi i$]

30. If C is the circle $|z| = 1$, show that

$$\int_C \frac{e^{z^2}}{\cos \pi z}\, dz = 0.$$

[**Hint:** Simple poles: $z = \pm 1/2, \pm 3/2, \ldots$ of which $z = \pm 1/2$ lie in C_i.

Residues: $\frac{h(a)}{g'(a)} = \left[\frac{e^{z^2}}{-\pi \sin \pi z}\right]_{z=a}$, where $a = 1/2, -1/2$]

10.3 ***Definite integrals***

Type 1

31. Using contour integration along the unit circle, establish the following results:

(*i*) $\int_0^{2\pi} \frac{1}{a + b\cos\theta}\, d\theta = \frac{2\pi}{\sqrt{a^2 - b^2}}, \quad a > |b| > 0$ **(1982, 85)**

(*ii*) $\int_0^{2\pi} \frac{1}{1 + a\cos\theta}\, d\theta = \frac{2\pi}{\sqrt{1 - a^2}}, \quad |a| < 1$ **(1976)**

[**Hint:** (*i*) Solved already as an example for type 1 (*ii*) Set $a = 1$, $b = a$ in (*i*)]

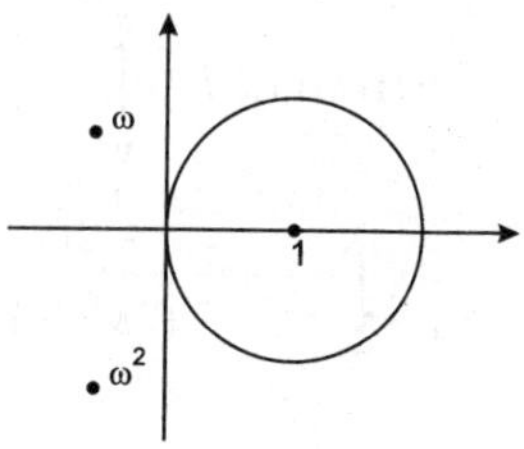

32. Solved sum. Using contour integration along the unit circle, establish the following results:

(*i*) $\int_0^{2\pi} \frac{1}{a + b\sin\theta}\, d\theta = \frac{2\pi}{\sqrt{a^2 - b^2}}, \quad a > |b| > 0$ **(1982)**

(*ii*) $\int_0^{2\pi} \frac{1}{(5/4) + \sin\theta}\, d\theta = \frac{8\pi}{3}$ **(1980)**

(*i*) On the unit circle C, $|z| = 1$, $z = e^{i\theta}$, $0 \le \theta < 2\pi$. So

$$\sin\theta = \frac{z - 1/z}{2i}, \quad d\theta = \frac{dz}{iz}.$$

$$\therefore \qquad I = \int_C \frac{2i}{2ai + b(z - 1/z)} \frac{dz}{iz} = \int_C \frac{2}{bz^2 + 2aiz - b}\, dz$$

The singularities of the integrand are the simple poles

$$z_1 = \frac{-ai + i\sqrt{a^2 - b^2}}{b}, \quad z_2 = \frac{-ai - i\sqrt{a^2 - b^2}}{b}.$$

But $z_1 z_2$ = product of the zeros $\left(\text{or } \frac{\text{constant term}}{\text{coef. of } z^2} \text{ of the quadratic equation}\right) = -1$,

$|z_1||z_2| = 1$, $(z_1)^2 < (z_2)^2$, $|z_1| < |z_2|$ and so z_1 lies inside the unit circle C and z_1 is a simple pole of the integrand. The residue of the integrand at $z = z_1$ is

$$\lim_{z \to z_1} (z - z_1) \frac{2}{bz^2 + 2aiz - b} = \lim_{z \to z_1} \frac{2}{2bz + 2ai} \text{ by L' Hospital's rule}$$

$$= \frac{1}{bz_1 + ai} = \frac{1}{i\sqrt{a^2 - b^2}}$$

$$\therefore \qquad I = 2\pi i \text{ (sum of the residues)} = \frac{2\pi}{\sqrt{a^2 - b^2}}.$$

(*ii*) Setting $a = 5/4$, $b = 1$, $I = 8\pi/3$.

33. By integration along C, the unit circle, establish the following results:

(*i*) $\int_0^{\pi} \frac{1}{a^2 + \cos^2\theta} d\theta = \frac{\pi}{a\sqrt{1 + a^2}}$ **(1979, 84)**

(*ii*) $\int_0^{\pi} \frac{1}{a^2 + \sin^2\theta} d\theta = \frac{\pi}{a\sqrt{1 + a^2}}$

(*iii*) $\int_0^{2\pi} \frac{1}{a^2 - 2a\cos\theta + 1} d\theta = \frac{2\pi}{1 - a^2}, 0 < a < 1.$

[**Hint:** (*i*) $I = \int_0^{\pi} \frac{2}{2a^2 + 1 + \cos 2\theta} d\theta = \int_0^{2\pi} \frac{d\theta'}{2a^2 + 1 + \cos\theta'}$, $(\theta' = 2\theta)$;

$$\int_C \frac{-2i}{z^2 + 2(2a^2 + 1)z + 1} dz = \int_C \frac{-2i}{(z - z_1)(z - z_2)} dz;$$

$|z_1||z_2| = 1$; Simple pole $z_1 = -(2a^2 + 1) + 2a\sqrt{1 + a^2}$ lies in C_i.

$$I = (-2i)(2\pi i)\left[\frac{1}{z - z_2}\right]_{z = z_1} = \frac{4\pi}{z_1 - z_2}$$

(*ii*) $I = \int_0^{\pi} \frac{2}{2a^2 + 1 - \cos 2\theta} d\theta = \int_0^{2\pi} \frac{1}{2a^2 + 1 - \cos\theta} d\theta$

$$= \int \frac{2i}{z^2 - 2(2a^2 + 1)z + 1} dz = \int_C \frac{2i}{(z - z_1)(z - z_2)} dz;$$

$|z_1||z_2|=1$; Simple pole $z_1 = (2a^2+1) - 2a\sqrt{1+a^2}$ lies in C_i

(*iii*) $-\dfrac{1}{ai}\displaystyle\int_C \dfrac{1}{z^2-(a+1/a)z+1}\,dz = -\dfrac{1}{ai}\int_C \dfrac{1}{(z-a)(z-1/a)}\,dz$.

Of the poles a and $1/a$, a lies inside the unit circle]

34. By integration along the unit circle C, establish the following results:

(*i*) $\displaystyle\int_0^{2\pi} \frac{1}{a+b\cos\theta+c\sin\theta}\,d\theta = \frac{2\pi}{\sqrt{a^2-(b^2+c^2)}},\ a>\sqrt{b^2+c^2}$

(*ii*) $\displaystyle\int_0^{2\pi} \frac{1}{3-2\cos\theta+\sin\theta}\,d\theta = \pi$ **(1975)**

[**Hint:** (*i*) $I = \dfrac{2}{i}\displaystyle\int_C \dfrac{1}{(b-ic)z^2+2az+(b+ic)}\,dz$.

The simple pole of the integrand, $\{-a+\sqrt{a^2-(b^2+c^2)}\}/(b-ic)$ lies in C_i because

$a-\sqrt{a^2-(b^2+c^2)} < \sqrt{b^2+c^2}$ which is true from $a^2 < \left(\sqrt{a^2-(b^2+c^2)}+\sqrt{b^2+c^2}\right)^2$

(*ii*) In (*i*) set $a=3$, $b=-2$, $c=1$]

35. By integration along the unit circle C, show that

(*i*) $\displaystyle\int_0^{2\pi} \frac{\cos 3\theta}{5-4\cos\theta}\,d\theta = \frac{\pi}{12}$ **(1981)**

(*ii*) $\displaystyle\int_0^{2\pi} \frac{\sin^2\theta}{a+b\cos\theta}\,d\theta = \frac{2\pi}{b^2}[a-\sqrt{a^2-b^2}],\ a>|b|>0,$ **(1980, 83)**

(*iii*) $\displaystyle\int_0^{2\pi} \frac{1+2\cos\theta}{5+4\cos\theta}\,d\theta = 0$ **(1983)**

(*iv*) $\displaystyle\int_0^{2\pi} \frac{\cos 2\theta}{1-2a\cos\theta+a^2}\,d\theta = \frac{2\pi a^2}{1-a^2},\ |a|<1$ **(1977)**

[**Hint:** (*i*) If $z=e^{i\theta}$, $5-4\left(\dfrac{1}{2}\right)(z+z^{-1})$ and dz/iz are real;

$$\therefore\quad \text{Re}\int_C \frac{z^3}{5-2(z+1/z)}\frac{dz}{iz} = \int_0^{2\pi}\frac{\cos 3\theta}{5-4\cos\theta}\,d\theta = I,\ \text{say}.$$

$$\therefore\quad I = \text{Re}\int_C \frac{\frac{1}{2}iz^3}{(z-\frac{1}{2})(z-2)}\,dz = \text{Re}\left\{(2\pi i)\lim_{z\to\frac{1}{2}}\frac{\frac{1}{2}iz^3}{z-2}\right\}$$

(*ii*) $$I = \text{Re} \int_C \frac{\frac{1}{2}(1-z^2)}{a + \frac{1}{2}b(z + 1/z)} \frac{dz}{iz} = \text{Re} \int_C \frac{(1-z^2)/i}{bz^2 + 2az + b} dz$$

$$= \text{Re}\left\{(2\pi) \lim_{z \to z_1} \frac{(z - z_1)(1-z^2)}{bz^2 + 2az + b}\right., \text{ where}$$

$$z_1 = \frac{-a + \sqrt{a^2 - b^2}}{b}$$

(*iii*) $$I = \text{Re} \int_0^{2\pi} \frac{1 + 2(\cos\theta + i \sin\theta)}{5 + 4\cos\theta} d\theta = \text{Re} \int_C \frac{1}{i(z+2)} dz$$

(*iv*) $$I = \text{Re} \int_C \frac{z^2}{1 - a(z + 1/z) + a^2} \frac{dz}{iz}$$

$$= \text{Re} \int_C \frac{iz^2/a}{(z-a)(z - 1/a)} dz$$

$$= \text{Re}\left\{(2\pi i)(i/a) \lim_{z \to a} \frac{z^2}{z - 1/a}\right\}$$

36. Solved sum. Using contour integration, evaluate

(*i*) $\displaystyle\int_0^{2\pi} \frac{1}{13 + 12\cos\theta} d\theta$ (*ii*) $\displaystyle\int_0^{2\pi} \frac{1}{2 + \cos\theta} d\theta$ **(1977)**

✍ (*i*) We know that, if C is the unit circle

$$z = e^{i\theta}, \quad 0 \le \theta < 2\pi,$$

then

$$\int_0^{2\pi} f(\cos\theta, \sin\theta)\, d\theta = \int_C f\left[\tfrac{1}{2}(z + 1/z), (1/2i)(z - 1/z)\right] \cdot \frac{1}{iz} dz$$

$$\therefore \int_0^{2\pi} \frac{1}{13 + 12\cos\theta} d\theta = \int_C \frac{1}{13 + 12\left(\frac{1}{2}\right)(z + 1/z)} \cdot \frac{1}{iz} dz$$

$$= \int_C \frac{1}{6z^2 + 13z + 6} \cdot \frac{1}{i} dz$$

$$= \frac{1}{i} \int_C \frac{1}{(3z+2)(2z+3)} dz$$

$$= \frac{1}{6i} \int_C \frac{1}{(z + 2/3)(z + 3/2)} dz$$

The denominator vanishes at $z_1 = -2/3$, $z_2 = -3/2$. Now z_1 lies inside C and z_2 lies outside C. Here z_1 is a simple pole of the integrand. Hence, by residue theorem,

$$I = \frac{1}{6i}[2\pi i \text{ (Residue of the integrand at } z = -2/3)]$$

$$= \frac{\pi}{3}\left[\lim_{z \to -2/3} (z + 2/3) \cdot \frac{1}{(z + 2/3)(z + 3/2)}\right]$$

$$= \frac{\pi}{3}\left[\lim_{z \to -2/3} \frac{1}{(z + 3/2)}\right] = \frac{\pi}{3}\left[\frac{1}{-2/3 + 3/2}\right]$$

$$= \frac{\pi}{3}\left[\frac{6}{5}\right] = \frac{2\pi}{5}.$$

Remark. The correctness of the result can be checked by substituting a = 13, b = 12, in (1) of the previous example,

(*ii*) Again, if C is the unit circle

$$z = e^{i\theta}, \quad 0 \le \theta < 2\pi,$$

$$I = \int_0^{2\pi} \frac{1}{2 + \cos\theta}\, d\theta = \int_C \frac{1}{2 + 1/2\,(z + 1/z)} \cdot \frac{1}{iz}\, dz$$

$$= \frac{2}{i}\int_C \frac{1}{z^2 + 4z + 1}\, dz = \frac{2}{i}\int_C \frac{1}{(z - z_1)(z - z_2)}\, dz,$$

where $z_1 = -2 + \sqrt{3}$, $z_2 = -2 - \sqrt{3}$ of which only z_1 lies inside C. Thus, by Residue theorem,

$$I = \frac{2}{i}\,[2\pi i\,(\text{Residue of the integrand at } z = -2 + \sqrt{3}\,)]$$

$$= 4\pi\left[\frac{1}{z + (2 + \sqrt{3})}\right]_{z = -2 + \sqrt{3}} = \frac{4\pi}{2\sqrt{3}} = \frac{2\pi}{\sqrt{3}}.$$

37. Solved sum. Evaluate

$$I = \int_0^{2\pi} \frac{1}{5 - 4\sin\theta}\, d\theta.$$

✍ We know that, if C is the unit circle

$$z = e^{i\theta}, \quad 0 \le \theta < 2\pi,$$

then

$$\int_0^{2\pi} f(\cos\theta, \sin\theta)\, d\theta = \int_C f\left[\left(1/2\right)(z + 1/z), \left(1/2i\right)(z - 1/z)\right] \cdot \frac{1}{iz}\, dz.$$

$$\therefore \quad \int_0^{2\pi} \frac{1}{5 - 4\sin\theta}\, d\theta = \int_C \frac{1}{5 - 4\left[\left(1/2i\right)(z - 1/z)\right]} \cdot \frac{1}{iz}\, dz$$

$$= \int_C \frac{1}{-2z^2 + 5iz + 2}\, dz = -\int_C \frac{1}{(z - 2i)(2z - i)}\, dz$$

$$= -\frac{1}{2}\int_C \frac{1}{(z - 2i)(z - i/2)}\, dz.$$

The denominator vanishes at $z_1 = i/2, z_2 = 2i$. But z_1 lies inside C and z_2 lies outside C. So z_1 is a simple pole of the integrand. Hence, by residue theorem,

$$I = -\frac{1}{2}\,[2\pi i\,(\text{Residue of the integrand at } z = i/2)]$$

$$= -\pi i\left[\lim_{z\to i/2}\left(z-\tfrac{i}{2}\right)\cdot\frac{1}{\left(z-\tfrac{i}{2}\right)(z-2i)}\right] = -\pi i\left[\lim_{z\to i/2}\frac{1}{z-2i}\right]$$

$$= -\pi i\,\frac{1}{(i/2)-2i} = -\pi i\left(\frac{2}{-3i}\right) = \frac{2\pi}{3}.$$

38. Solved sum. Using contour integration, evaluate

$$I = \int_0^{2\pi}\frac{1}{1-2a\sin\theta + a^2}\,d\theta,\quad 0<a<1$$

[**Hint:**

$$I = \int_C \frac{1}{1-a/i\left(z-\tfrac{1}{z}\right)+a^2}\cdot\frac{1}{iz}\,dz$$

$$= \int_C \frac{1}{-az^2 + i(1+a^2)\,z + a}\,dz$$

$$= -\frac{1}{a}\int_C \frac{1}{z^2 - i\left(a+\tfrac{1}{a}\right)z - 1}\,dz$$

$$= -\frac{1}{a}\int_C \frac{1}{(z-ai)\left(z-\tfrac{i}{a}\right)}\,dz.$$

So $z = ai$, $z = i/a$ are simple poles of the integrand of which $z = ai$ lies inside C since $0 < a < 1$. Thus, by Residue theorem,

$$I = -\frac{1}{a}\,(2\pi i)\,[\text{Residue of the integrand at } z = ai]$$

$$= -\frac{2\pi i}{a}\left[\frac{1}{z-\tfrac{i}{a}}\right]_{z=ai} = \frac{2\pi}{1-a^2}\,]$$

39. Solved sum. Using contour integration, evaluate

(*i*) $I = \displaystyle\int_0^{2\pi}\frac{\cos 3\theta}{5-4\cos\theta}\,d\theta$ (*ii*) $I = \displaystyle\int_0^{2\pi}\frac{\sin^2\theta}{a+b\cos\theta}\,d\theta.$

(*i*) If C is the unit circle $z = e^{i\theta}$, using the results

$$\int_0^{2\pi} f(\cos\theta, \sin\theta)\,d\theta = \int_C f,\left[\left(\tfrac{1}{2}\right)\left(z+\tfrac{1}{z}\right)\left(\tfrac{1}{2i}\right)\left(z-\tfrac{1}{z}\right)\right]\cdot\frac{1}{iz}\,dz$$

and $\cos 3\theta = \text{Re}\,(\cos\theta + i\sin\theta)^3 = \text{Re}\,z^3$, we get

$$I = \text{Re}\int_C \frac{z^3}{5-4\left[\left(\tfrac{1}{2}\right)\left(z+\tfrac{1}{z}\right)\right]}\frac{dz}{iz} = \text{Re}\,\frac{1}{i}\int_C\frac{z^3}{5z-2(z^2+1)}\,dz$$

$$= \text{Re}\left[-\frac{1}{i}\int_C\frac{z^3}{2z^2-5z+2}\,dz\right] = \text{Re}\left[i\int_C\frac{z^3}{(z-2)(2z-1)}\,dz\right]$$

$$= \text{Re}\left[\frac{i}{2}\int_C \frac{z^3}{(z-2)(z-\tfrac{1}{2})}\,dz\right].$$

So the integrand has only a simple pole $z = 1/2$ inside C. Thus, by Residue theorem,

$$I = \text{Re}\left[\frac{i}{2}(2\pi i)\left(\frac{z^3}{z-2}\right)_{z=1/2}\right] = (-\pi)\left[\frac{(1/2)^3}{(1/2)-2}\right] = \frac{\pi}{12}.$$

(*ii*) $$\int_0^{2\pi} \frac{\sin^2\theta}{a+b\cos\theta}\,d\theta = \int_0^{2\pi} \frac{(1/2)(1-\cos 2\theta)}{a+b\cos\theta}\,d\theta$$

If C is the unit circle $z = e^{i\theta}$, using the results

$$\int_0^{2\pi} f(\cos\theta, \sin\theta)\,d\theta = \int_C f\left[\left(\tfrac{1}{2}\right)\left(z+\tfrac{1}{z}\right), \left(\tfrac{1}{2i}\right)\left(z-\tfrac{1}{z}\right)\right]\frac{1}{iz}\,dz$$

and $1 - \cos 2\theta = \text{Re}\,(1 - z^2)$, we get

$$I = \frac{1}{2}\text{Re}\int_C \frac{1-z^2}{a+b(1/2)(z+\tfrac{1}{z})}\cdot\frac{1}{iz}\,dz$$

$$= \text{Re}\left[\frac{1}{i}\int_C \frac{1-z^2}{bz^2+2az+b}\,dz\right]$$

$$= \text{Re}\left[\frac{1}{i}\int_C \frac{1-z^2}{b(z-z_1)(z-z_2)}\,dz\right].$$

Then

$$z_1 z_2 = \frac{\text{Constant term}}{\text{Coefficient of } z^2} = \frac{b}{b} = 1.$$

$\therefore$ $$|z_1|\,|z_2| = 1.$$

So, of z_1 and z_2, one lies inside C and the other lies, outside C. Now

$$z_1 = \frac{-a+\sqrt{a^2-b^2}}{b}, \quad z_2 = \frac{-a-\sqrt{a^2-b^2}}{b}$$

are real and z_1 lies inside C and it is a simple pole of the integrand.

$$\therefore \quad I = \text{Re}\left[\frac{1}{i}(2\pi i)\,(\text{Residue of the integrand at } z = z_1)\right]$$

$$= \text{Re}\left[\frac{2\pi}{b}\left(\frac{1-z^2}{z-z_2}\right)_{z=z_1}\right] = \text{Re}\left[\frac{2\pi}{b}\left(\frac{1-z_1^2}{z_1-z_2}\right)\right]$$

$$= 2\pi\,\frac{1-\left(\dfrac{-a+\sqrt{a^2-b^2}}{b}\right)^2}{2\sqrt{a^2-b^2}}$$

$$= \frac{\pi}{b^2 \sqrt{a^2 - b^2}} [b^2 - (-a + \sqrt{a^2 - b^2})^2]$$

$$= \frac{\pi}{b^2 \sqrt{a^2 - b^2}} [b^2 - (a^2 - 2a\sqrt{a^2 - b^2} + a^2 - b^2)]$$

$$= \frac{\pi}{b^2 \sqrt{a^2 - b^2}} (2b^2 - 2a^2 + 2a\sqrt{a^2 - b^2})$$

$$= -\frac{2\pi}{b^2 \sqrt{a^2 - b^2}} (a^2 - b^2 - a\sqrt{a^2 - b^2})$$

$$= -\frac{2\pi}{b^2}\left(\sqrt{a^2 - b^2} - a\right) = \frac{2\pi}{b^2}(a - \sqrt{a^2 - b^2}).$$

40. Solved sum. Using contour integration, evaluate

(*i*) $\int_0^{2\pi} \frac{1}{(a + b\cos\theta)^2} d\theta, \quad a > |b| > 0,$ **(1979)**

(*ii*) $\int_0^{2\pi} \frac{1}{(a + b\sin\theta)^2} d\theta, \quad a > |b| > 0.$

✍ (*i*) If C is the unit circle $z = e^{i\theta}$, using the result

$$\int_0^{2\pi} f(\cos\theta, \sin\theta)\, d\theta = \int_C f\left[(1/2)\left(z + \tfrac{1}{z}\right), (1/2i)\left(z - \tfrac{1}{z}\right)\right] \cdot \frac{1}{iz} dz,$$

$$\int_0^{2\pi} \frac{1}{(a + b\cos\theta)^2} d\theta = \int_C \frac{1}{[a + b(1/2)(z + (1/z)]^2} \cdot \frac{1}{iz} dz$$

$$= \frac{1}{i}\int_C \frac{4z}{(bz^2 + 2az + b)^2} dz = \frac{4}{ib^2}\int_C \frac{z}{(z - z_1)^2 (z - z_2)^2} dz.$$

Then

$$z_1 z_2 = \frac{\text{Constant term}}{\text{Coefficient of } z^2} = \frac{b}{b} = 1.$$

$\therefore \quad |z_1||z_2| = 1.$

So, of z_1 and z_2, one lies inside C and other lies outside C. Now

$$z_1 = \frac{-a + \sqrt{a^2 - b^2}}{b}, \quad z_2 = \frac{-a - \sqrt{a^2 - b^2}}{b}$$

and z_1 lies inside C and it is a double pole of the integrand.

$$\therefore \quad I = \frac{4}{b^2 i}(2\pi i) \text{ [Residue of the integrand at } z = z_1]$$

$$= \frac{8\pi}{b^2}\left[\frac{1}{1!}\frac{d}{dz}(z - z_1)^2 f(z)\right]_{z = z_1} = \frac{8\pi}{b^2}\left[\frac{d}{dz}\frac{1}{(z - z_2)^2}\right]_{z = z_1}$$

$$= \frac{8\pi}{b^2}\left[\frac{(z-z_2)^2 - z\,(2)\,(z-z_2)}{(z-z_2)^4}\right]_{z=z_1}$$

$$= \frac{8\pi}{b^2}\left[\frac{z_1 - z_2 - 2z_1}{(z_1-z_2)^3}\right] = \frac{8\pi}{b^2}\left[\frac{-z_1 - z_2}{(z_1-z_2)^3}\right]$$

$$= \frac{8\pi}{b^2}\left[\frac{2a/b}{(2\sqrt{a^2-b^2}/b)^3}\right] = \frac{2\pi a}{(a^2-b^2)^{3/2}}.$$

(*ii*) If C is the unit circle $z = e^{i\theta}$, using the result

$$\int_0^{2\pi} f(\cos\theta, \sin\theta)\, d\theta = \int_C f[(1/2)(z + (1/z)), (1/2i)(z - (1/z)) \cdot \frac{1}{iz}\, dz,$$

$$\int_0^{2\pi} \frac{1}{(a + b\sin\theta)^2}\, d\theta = \int_C \frac{1}{[a + b\,(1/2i)\,(z - 1/z)]^2} \cdot \frac{1}{iz}\, dz$$

$$= -\frac{4}{i}\int_C \frac{z}{(bz^2 + 2aiz - b)^2}\, dz$$

$$= \frac{4i}{b^2}\int_C \frac{z}{(z-z_1)^2\,(z-z_2)^2}\, dz$$

Then

$$z_1 z_2 = \frac{\text{Constant term}}{\text{Coefficient of } z^2} = \frac{-b}{b} = -1.$$

$\therefore$ $|z_1||z_2| = 1.$

So, of z_1 and z_2, one lies inside C and the other lies outside C. Now

$$z_1 = \frac{-ai + i\sqrt{a^2-b^2}}{b}, \quad z_2 = \frac{-ai - i\sqrt{a^2-b^2}}{b}$$

and z_1 lies inside C and it is a double pole of the integrand.

$$\therefore \qquad I = \frac{4i}{b^2}\,(2\pi i)\,[\text{Residue of the integrand at } z = z_1]$$

$$= -\frac{8\pi}{b^2}\left[\frac{1}{1!}\frac{d}{dz}(z-z_1)^2 f(z)\right]_{z=z_1}$$

$$= -\frac{8\pi}{b^2}\left[\frac{d}{dz}\frac{z}{(z-z_2)^2}\right]_{z=z_1}$$

$$= -\frac{8\pi}{b^2}\left[\frac{(z-z_2)^2 - z\,(2)\,(z-z_2)}{(z-z_2)^4}\right]_{z=z_1}$$

$$= -\frac{8\pi}{b^2}\left[\frac{-z_1 - z_2}{(z_1-z_2)^3}\right] = -\frac{8\pi}{b^2}\left[\frac{2ai/b}{\left(2i\sqrt{a^2-b^2}\,/b\right)^3}\right]$$

$$= -\frac{8\pi}{b^2}\left[\frac{2ai}{b}\cdot\frac{b^3}{-8i\,(a^2-b^2)^{3/2}}\right] = \frac{2\pi a}{(a^2-b^2)^{3/2}}.$$

41. Solved sum. Using contour integration, evaluate

(*i*) $I = \int_0^{2\pi} \frac{1}{a + b\cos\theta + c\sin\theta}\,d\theta, \quad a > \sqrt{b^2+c^2}$

(*ii*) $I = \int_0^{2\pi} \frac{1}{3 - 2\cos\theta + \sin\theta}\,d\theta.$

✍ (*i*) If C is the unit circle $z = e^{i\theta}$, using the result

$$\int_0^{2\pi} f(\cos\theta, \sin\theta)\,d\theta = \int_C f\left[\left(\tfrac{1}{2}\right)\left(z+\tfrac{1}{z}\right), \left(\tfrac{1}{2i}\right)\left(z-\tfrac{1}{z}\right)\right]\cdot\frac{1}{iz}\,dz,$$

$$I = \int_C \frac{1}{a + b\left(\tfrac{1}{2}\right)\left(z+\tfrac{1}{z}\right) + c\left(\tfrac{1}{2i}\right)\left(z-\tfrac{1}{z}\right)}\cdot\frac{1}{iz}\,dz$$

$$= \frac{2}{i}\int_C \frac{1}{(b-ci)\,z^2 + 2az + (b+ic)}\,dz$$

$$= \frac{2}{i\,(b-ic)}\int_C \frac{1}{(z-z_1)(z-z_2)}\,dz.$$

Here

$$z_1 = \frac{-a + \sqrt{a^2-(b^2+c^2)}}{b-ic},$$

$$z_2 = \frac{-a - \sqrt{a^2-(b^2+c^2)}}{b-ic} \quad \text{and}$$

$$|z_1 z_2| = \left|\frac{b+ic}{b-ic}\right| = \frac{|b+ic|}{|b-ic|} = 1.$$

Therefore, of z_1 and z_2, one lies in C_i. Evidently z_1 lies in C_i. By Residue theorem,

$$I = \frac{2}{i\,(b-ic)}(2\pi i)\ [\text{Residue of the integrand at } z = z_1]$$

$$= \frac{2}{i\,(b-ic)}(2\pi i)\left[\frac{1}{z-z_2}\right]_{z=z_1} = \frac{4\pi}{(b-ic)}\left[\frac{1}{z_1-z_2}\right]$$

$$= \frac{4\pi}{b-ic}\cdot\frac{b-ic}{2\sqrt{a^2-(b^2+c^2)}} = \frac{2\pi}{\sqrt{a^2-(b^2+c^2)}}.$$

(*ii*) Set $a = 3$, $b = -2$, $c = 1$. Then $I = \pi$.

42. Using contour integration, show that $\int_0^{2\pi} \frac{1}{17 - 8\cos\theta}\,d\theta = \frac{2\pi}{15}.$

[**Hint:** S.P in C_i : 1/4; Res. : $1/15i$]

43. Using contour integration, show that $\int_0^{2\pi} \frac{1}{97 + 72\cos\theta}\,d\theta = \frac{2\pi}{65}.$

[**Hint:** S.P in C_i : – 4/9; Res. : 1/65i]

44. Using contour integration, show that $\int_0^{2\pi} \frac{1}{13 + 5\sin\theta}\, d\theta = \frac{\pi}{6}$.

[**Hint:** S.P in C_i : –i/5; Res. : 1/12i]

45. Using contour integration, show that $\int_0^{2\pi} \frac{\cos 2\theta}{5 + 4\cos\theta}\, d\theta = \frac{\pi}{6}$.

[**Hint:** S.P in C_i : –1/2; Res. : 1/12i]

46. Using contour integration, show that $\int_0^{2\pi} \frac{\cos 2\theta}{1 - 2a\cos\theta + a^2}\, d\theta = \frac{2\pi a^2}{1 - a^2}, \quad |a| < 1$.

[**Hint:** S.P in C_i : a ; Res. : $\frac{a^2}{i(1 - a^2)}$]

47. Using contour integration, show that $\int_0^{2\pi} \frac{\cos^2 3\theta}{5 - 4\cos 2\theta}\, d\theta = \frac{3\pi}{8}$.

[**Hint:** S.P's in C_i : $-\frac{1}{\sqrt{2}}, \frac{1}{\sqrt{2}}$; Res.: $\frac{3}{32i}, \frac{3}{32i}$]

48. Using contour integration, show that $\int_0^{2\pi} \frac{1}{(5 - 4\cos\theta)^2}\, d\theta = \frac{10\pi}{27}$.

[**Hint:** D.P in C_i : 1/2 ; Res. : 5/27i]

49. Using contour integration, show that $\int_0^{2\pi} \frac{1}{5 - 2\cos\theta - 3\sin\theta}\, d\theta = \frac{\pi}{\sqrt{3}}$.

[**Hint:** S.P in C_i : $\frac{-5 + \sqrt{12}}{-2 + 3i}$; Res.: $\frac{1}{i\sqrt{12}}$]

50. Using contour integration, show that $\int_0^{2\pi} \frac{1}{6 + 4\cos\theta - 2\sin\theta}\, d\theta = \frac{\pi}{2}$.

[**Hint:** S.P in C_i : – 1/(2 + i) ; Res. : 1/4i]

51. Using contour integration, show that $\int_0^{2\pi} \frac{1 + 2\cos\theta}{5 + 4\cos\theta}\, d\theta = 0$.

[**Hint:** S.P in C_i : – 1/2 ; Res. : 0]

52. By integration along the unit circle C show that $\int_0^{2\pi} \frac{1}{(5 - 3\sin\theta)^2}\, d\theta = \frac{5\pi}{32}$ **(1974)**

[**Hint:** Double pole in C_i ; $\frac{i}{3}$; Residue : $\frac{5}{64i}$

$$I = 4i \int_C \frac{z}{(3z^2 - 10iz - 3)^2}\, dz = \frac{4i}{9} \int_C \frac{z}{(z - i/3)^2 (z - 3i)^2}\, dz \,]$$

Type 2

53. Solved sum. Using contour integration, show that

(*i*) $\int_0^\infty \frac{1}{(x^2+1)^2}\,dx = \frac{\pi}{4}$ **(1973, 81)**

(*ii*) $\int_0^\infty \frac{1}{(x^2+1)^3}\,dx = \frac{3\pi}{16}$.

Let C be composed of the semicircle $\Gamma : |z| = R$, $\mathbf{Im}\, z \geq 0$ and L, the line joining $z = -R$ and $z = R$ (Fig. 10.3.1).

(*i*) Consider the contour integral

$$I' = \int_C \frac{1}{(z^2+1)^2}\,dz = 2\pi i \ \{\text{sum of the residues in } C_i\}.$$

Now $z = i$ and $z = -i$ are double poles of the integrand of which $z = i$ lies in C_i if $R > 1$. The residue of the integrand at $z = i$ is

$$\left[\frac{d}{dz}\frac{1}{(z+i)^2}\right]_{z=i} = -2\frac{1}{(2i)^3} = -\frac{i}{4} \tag{1}$$

$$I' = 2\pi i\,(-i/4) = \frac{1}{2}\pi. \tag{2}$$

$$\therefore \quad \int_\Gamma \frac{1}{(z^2+1)^2}\,dz + \int_{-R}^{R} \frac{1}{(x^2+1)^2} = \frac{\pi}{2}$$

since, along L, $z = x$. Now, taking limit as $R \to \infty$,

$$\int_{-\infty}^{\infty} \frac{1}{(x^2+1)^2}\,dx = \frac{\pi}{2} \quad \text{or} \quad \int_0^{\infty} \frac{1}{(x^2+1)^2}\,dx = \frac{\pi}{4}$$

if the integral along Γ vanishes in the limit. But, on Γ,

$$\left|\int_\Gamma\right| \leq \frac{1}{|z^2+1|^2}\, l(\Gamma) \quad \text{or} \quad \left|\int_\Gamma\right| \leq \frac{\pi R}{|z^2+1|^2}$$

i.e.,

$$\left|\int_\Gamma\right| \leq \frac{\pi R}{(|z^2|-1)^2} \quad (\because |z^2|-1 \leq |z^2-1|)$$

$$\therefore \quad \left|\int_\Gamma\right| \leq \frac{\pi R}{(R^2-1)^2}. \tag{3}$$

But $\pi R/(R^2-1)^2 \to 0$ as $R \to \infty$. This implies that the integral along Γ vanishes as $R \to \infty$.

(*ii*) $z = i$ and $z = -i$ are triple poles of which $z = i$ lies in C_i if $R > 1$. In this case the results (1), (2), (3) are respectively

$$\frac{1}{2!}\left[\frac{d^2}{dz^2}\frac{\pi R}{(z+i)^3}\right]_{z=i} = \frac{(-3)(-4)}{2!}\frac{1}{(2i)^5} = \frac{3}{16i},$$

$$I' = 2\pi i\left(\frac{3}{16i}\right) = \frac{3\pi}{8},$$

and $$\left|\int_\Gamma\right| \le \frac{R}{(R^2-1)^3}$$

54. Solved sum. Evaluate $\displaystyle\int_{-\infty}^{\infty} \frac{x^2-x+2}{x^4+10x^2+9}\,dx$

✍ Consider the integral

$$I' = \int_C \frac{z^2-z+2}{z^4+10z^2+9}\,dz = \int_C \frac{z^2-z+2}{(z^2+1)(z^2+9)}\,dz,$$

where C is the closed curve consisting the positively oriented circle

$$\Gamma := |z| = R, \quad \text{Im } z \ge 0,$$

and the line segment

$$L: \quad [-R, R]$$

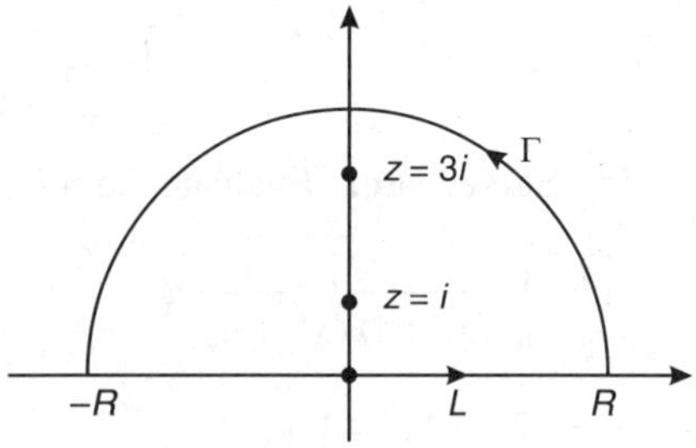

Now $i, -i, 3i, -3i$ are the simple poles of the integrand of which only

$$z = i, \quad z = 3i$$

lie inside C for an $R > 3$.

$$\left.\begin{array}{l}\text{Res. at}\\ z=i\end{array}\right\} = \left[\frac{z^2-z+2}{(z+i)(z^2+9)}\right]_{z=i} = \frac{-1-i+2}{2i(-1+9)} = \frac{1-i}{16i}.$$

$$\left.\begin{array}{l}\text{Res. at}\\ z=3i\end{array}\right\} = \left[\frac{z^2-z+2}{(z^2+1)(z+3i)}\right]_{z=3i} = \frac{-9-3i+2}{(-9+1)\,6i} = \frac{7+3i}{48i}$$

Therefore, by Residue theorem,

$$I' = 2\pi i\left[\frac{1-i}{16i} + \frac{7+3i}{48i}\right] = 2\pi\left[\frac{3-3i}{48} + \frac{7+3i}{48}\right] = \frac{5\pi}{12}.$$

Now, on L, $z = x$. Therefore

$$\int_\Gamma \frac{z^2-z+2}{z^4+10z^2+9}\,dz + \int_{-R}^{R} \frac{x^2-x+2}{x^4+10x^2+9}\,dx = \frac{5\pi}{12}. \qquad (1)$$

We shall show that the first integral vanishes as $R \to \infty$.

$$\left|\int_\Gamma \frac{z^2-z+2}{z^4+10z^2+9}\,dz\right| \le \left[\max_{\text{on }\Gamma}\left|\frac{z^2-z+2}{z^4+10z^2+9}\right|\right] l(\Gamma),$$

where $l(\Gamma) = \pi R$. By the result

$$|z_1+z_2+z_3| \le |z_1| + |z_2| + |z_3|,$$

now we have, for the numerator,

$$\begin{aligned}|z^2-z+2| &\le |z^2| + |(-z)| + |2|\\ &= R^2+R+2.\end{aligned}$$

By $\quad |z_1-z_2| \ge |z_1| - |z_2|$, we have, for the denominator,

$$\begin{aligned}|z^4+10z^2+9| &= |z^2+1| \times |z^2+9|\\ &= |z^2-(-1)| \times |z^2-(-9)|\\ &\ge [|z^2| - |(-1)|] \times [|z^2| - |(-9)|]\\ &= (R^2-1)(R^2-9) \quad \text{for } R > 3.\end{aligned}$$

Thus, for $R > 3$,

$$\left| \int_\Gamma \frac{z^2 - z + 2}{z^4 + 10z^2 + 9}\, dz \right| \le \frac{R^2 + R + 2}{(R^2 - 1)(R^2 - 9)} (\pi R)$$

which tends to 0 as $R \to \infty$ because of the higher power of R in the denominator. Thus, taking limit of (1) as $R \to \infty$,

$$0 + \int_{-\infty}^{\infty} \frac{x^2 - x + 2}{x^4 + 10x^2 + 9}\, dx = \frac{5\pi}{12}$$

$$\therefore \qquad \int_{-\infty}^{\infty} \frac{x^2 - x + 2}{x^4 + 10x^2 + 9}\, dx = \frac{5\pi}{12}.$$

55. Solved sum. Evaluate the following integrals:

(*i*) $\displaystyle\int_0^\infty \frac{x^2}{(x^2 + 1)(x^2 + 4)}\, dx$ **(1975, 77)**

(*ii*) $\displaystyle\int_0^\infty \frac{2x^2 - 1}{x^4 + 5x^2 + 4}\, dx$

✍ (*i*) Consider the integral

$$I' = \int_C \frac{z^2}{(z^2 + 1)(z^2 + 4)}\, dz,$$

where C is the closed curve consisting of the positively oriented circle

$\Gamma : |z| = R, \quad \text{Im}\, z \ge 0.$

and the line segment

$L : \quad [-R, R].$

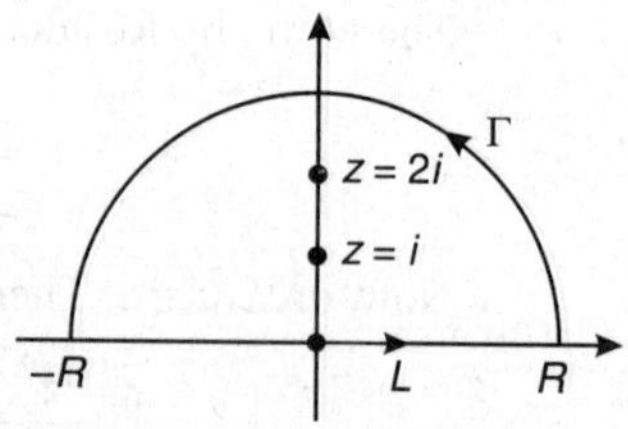

Now $i, -i, 2i, -2i$ are the simple poles of the integrand of which only

$$z = i, \quad z = 2i$$

lie in C_i for an $R > 2$.

$$\left.\begin{matrix}\text{Res. at} \\ z = i\end{matrix}\right\} = \left[\frac{z^2}{(z + i)(z^2 + 4)}\right]_{z = i} = \frac{-1}{2i(-1 + 4)} = -\frac{1}{6i}.$$

$$\left.\begin{matrix}\text{Res. at} \\ z = 2i\end{matrix}\right\} = \left[\frac{z^2}{(z^2 + 1)(z + 2i)}\right]_{z = 2i} = \frac{-4}{(-3)\, 4i} = \frac{1}{3i}.$$

Therefore, by Residue theorem,

$$I' = 2\pi i\left[-\frac{1}{6i} + \frac{1}{3i}\right] = 2\pi\left[-\frac{1}{6} + \frac{2}{6}\right] = \frac{\pi}{3}.$$

But, on L, $z = x$. Therefore

$$\int_\Gamma \frac{z^2}{(z^2 + 1)(z^2 + 4)}\, dz + \int_{-R}^{R} \frac{x^2}{(x^2 + 1)(x^2 + 4)}\, dx = \frac{\pi}{3}. \qquad (1)$$

Now we shall show that the first integral vanishes as $R \to \infty$. Considering its modulus,

$$\left|\int_{\Gamma} \frac{z^2}{(z^2+1)(z^2+4)}\,dz\right| \le \left[\begin{matrix}\max\\ \text{on }\Gamma\end{matrix}\left|\frac{z^2}{(z^2+1)(z^2+4)}\right|\right] l(\Gamma)$$

where $l(\Gamma) = \pi R$. By the result

$$|z_1 - z_2| \ge |z_1| - |z_2|,$$

we have, for the denominator,

$$\begin{aligned}|(z^2+1)(z^2+4)| &= |z^2+1| \times |z^2+4| \\ &= |z^2-(-1)| \times |z^2-(-4)| \\ &\ge [|z^2| - |(-1)|] \times [|z^2| - |(-4|] \\ &= (R^2-1)(R^2-4) \quad \text{for} \quad R > 2.\end{aligned}$$

$$\therefore \quad \left|\int_{\Gamma} \frac{z^2}{(z^2+1)(z^2+4)}\,dz\right| \le \frac{R^2}{(R^2-1)(R^2-4)}(\pi R)$$

which tends to 0 as $R \to \infty$ because of the higher power of R in the denominator. Thus, taking limit of (1) as $R \to \infty$,

$$0 + \int_{-\infty}^{\infty} \frac{x^2}{(x^2+1)(x^2+4)}\,dx = \frac{\pi}{3}.$$

Now the integrand is an even function. So

$$2\int_{0}^{\infty} \frac{x^2}{(x^2+1)(x^2+4)}\,dx = \frac{\pi}{3}. \quad \therefore \quad \int_{0}^{\infty} \frac{x^2}{(x^2+1)(x^2+4)}\,dx = \frac{\pi}{6}$$

(*ii*) Since $z^4 + 5z^2 + 4 = (z^2+1)(z^2+4)$, proceeding in the same way as in (*i*),

$$\left.\begin{matrix}\text{Res. at}\\ z = i\end{matrix}\right\} = \left[\frac{2z^2-1}{(z+i)(z^2+4)}\right]_{z=i} = \frac{-2-1}{6i} = \frac{-1}{2i}$$

$$\left.\begin{matrix}\text{Res. at}\\ z = 2i\end{matrix}\right\} = \left[\frac{2z^2-1}{(z^2+1)(z+2i)}\right]_{z=2i} = \frac{3}{4i}.$$

Therefore, by Residue theorem,

$$I' = 2\pi i\left[\frac{-1}{2i} + \frac{3}{4i}\right] = 2\pi\left[\frac{-2}{4} + \frac{3}{4}\right] = \frac{\pi}{2}.$$

But, on L, $z = x$. Therefore

$$\int_{\Gamma} \frac{2z^2-1}{(z^2+1)(z^2+4)}\,dz + \int_{-R}^{R} \frac{2x^2-1}{(x^2+1)(x^2+4)}\,dx = \frac{\pi}{2} \tag{1}$$

Now we shall show that the first integral vanishes as $R \to \infty$. Considering its modulus,

$$\left|\int_{\Gamma} \frac{2z^2-1}{(z^2+1)(z^2+4)}\,dz\right| \le \left[\begin{matrix}\max\\ \text{on }\Gamma\end{matrix}\left|\frac{2z^2-1}{(z^2+1)(z^2+4)}\right|\right] l(\Gamma)$$

where $l(\Gamma) = \pi R$. By the result,

$$|z_1 + z_2| \le |z_1| + |z_2|,$$

we have, for the numerator,

$$|2z^2 - 1| \le |2z^2| + |(-1)| = 2R^2 + 1.$$

By $\quad |z_1 - z_2| \geq |z_1| - |z_2|$, we have, for the denominator,

$$\begin{aligned} |(z^2+1)(z^2+4)| &= |z^2+1| \times |z^2+4| \\ &= |z^2-(-1)| \times |z^2-(-4)| \\ &\geq [|z^2| - |(-1)|] \times [|z^2| - |(-4)|] \\ &= (R^2-1)(R^2-4) \quad \text{for } R > 2. \end{aligned}$$

$$\therefore \qquad \left| \int_{\Gamma} \frac{2z^2-1}{(z^2+1)(z^2+4)}\, dz \right| \leq \frac{2R^2+1}{(R^2-1)(R^2-4)} (\pi R)$$

which tends to 0 as $R \to \infty$ because of the higher power of R in the denominator. Thus, taking limit of (1) as $R \to \infty$,

$$0 + \int_{-\infty}^{\infty} \frac{2x^2-1}{(x^2+1)(x^2+4)}\, dx = \frac{\pi}{2}.$$

Now the integrand is an even function. So

$$2\int_{0}^{\infty} \frac{2x^2-1}{(x^2+1)(x^2+4)}\, dx = \frac{\pi}{2}. \quad \therefore \quad \int_{0}^{\infty} \frac{2x^2-1}{(x^2+1)(x^2+4)}\, dx = \frac{\pi}{4}.$$

56. Solved sum. Evaluate the following integrals:

(*i*) $\displaystyle\int_0^{\infty} \frac{x^2}{(x^2+a^2)^2}\, dx$

(*ii*) $\displaystyle\int_0^{\infty} \frac{x^2}{(x^2+1)^2}\, dx.$ **(1982, 84)**

✍ (*i*) Consider the integral

$$I = \int_C \frac{z^2}{(z^2+a^2)^2}\, dz,$$

where C is the closed curve consisting the positively oriented circle

$\Gamma : |z| = R,$ **Im** $z \geq 0$, and the line segment

$L: \quad [-R, R].$

Now $ai, -ai$ are the double poles of the integrand, of which only

$$z = ai$$

lie in C_i for an $R > a$. Therefore, by Residue theorem,

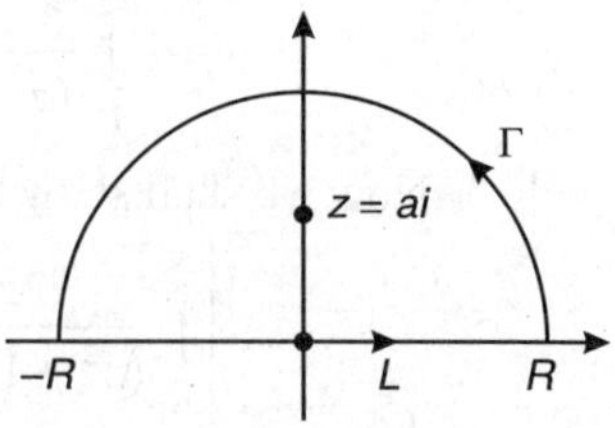

$$\begin{aligned} I &= (2\pi i)\left[\frac{1}{1!}\frac{d}{dz}(z-ai)^2 f(z)\right]_{z=ai} \\ &= (2\pi i)\left[\frac{d}{dz}\frac{z^2}{(z+ai)^2}\right]_{z=ai} \\ &= 2\pi i\left[\frac{(z+ai)^2\, 2z - z^2\,(2)(z+ai)}{(z+ai)^4}\right]_{z=ai} \\ &= 2\pi i\left[\frac{(z+ai)\,2z - 2z^2}{(z+ai)^3}\right]_{z=ai} = 2\pi i\left[\frac{2aiz}{(z+ai)^3}\right]_{z=ai} \end{aligned}$$

$$= 2\pi i \left[\frac{2ai\,(ai)}{(2ai)^3} \right] = \frac{\pi}{2a}$$

But, on L, $z = x$. Therefore

$$\int_\Gamma \frac{z^2}{(z^2+a^2)^2}\,dz + \int_{-R}^{R} \frac{x^2}{(x^2+a^2)^2}\,dx = \frac{\pi}{2a}. \tag{1}$$

Now we shall show that the first integral vanishes as $R \to \infty$. Considering its modulus,

$$\left| \frac{z^2}{(z^2+a^2)^2}\,dz \right| \le \left[\max_{\text{on } \Gamma} \left| \frac{z^2}{(z^2+a^2)^2}\,dz \right| \right] l\,(\Gamma),$$

where $l(\Gamma) = \pi R$. By the result

$$|\, z_1 - z_2 \,| \ge |\, z_1 \,| - |\, z_2 \,|,$$

we have, for the denominator,

$$|\,(z^2+a^2)^2\,| = |\, z^2 - (-a^2)\,|^2 \ge [\,|\, z^2\,| - |\,(-a^2)\,|\,]^2$$
$$= (R^2 - a^2)^2 \quad \text{for} \quad R > a$$

$$\therefore \qquad \left| \int_\Gamma \frac{z^2}{(z^2+a^2)^2}\,dz \right| \le \frac{R^2}{(R^2-a^2)^2}\,(\pi R)$$

which tends to 0 as $R \to \infty$ because of the higher power of R in the denominator. Thus, taking limit of (1) as $R \to \infty$,

$$0 + \int_{-\infty}^{\infty} \frac{x^2}{(x^2+a^2)^2}\,dx = \frac{\pi}{2a}$$

But the integrand is an even function. So

$$2\int_0^\infty \frac{x^2}{(x^2+a^2)^2}\,dx = \frac{\pi}{2a} \quad \text{or} \quad \int_0^\infty \frac{x^2}{(x^2+a^2)^2}\,dx = \frac{\pi}{4a}.$$

(*ii*) Setting $a = 1$. we get that the value of the integral as $\pi/4$.

57. In the following sums we take C as the closed curve formed by

(*i*) $\Gamma : |\, z\,| = R$, $\operatorname{Im} z \ge 0$; (*ii*) $L : [-R, R]$.

1. Using contour integration, show that

(*i*) $\displaystyle\int_0^\infty \frac{1}{x^4+a^4}\,dx = \frac{\pi}{2\sqrt{2}\,a^3}$ **(1974, 79)** (*ii*) $\displaystyle\int_0^\infty \frac{1}{x^4+1}\,dx = \frac{\pi}{2\sqrt{2}}$ **(1985)**

[**Hint:** If $a > 0$ S.P. in C_i : $ae^{i\pi}/4$, $ae^{i3\pi}/4$; Res. : $\dfrac{h\,(a)}{g'\,(a)}$.

Therefore

$$\left.\begin{matrix}\text{Res. at}\\ z = ae^{i\pi/4}\end{matrix}\right\} = -\frac{1}{4a^3}\left(\frac{1+i}{\sqrt{2}}\right); \quad \left.\begin{matrix}\text{Res. at}\\ z = ae^{i3\pi/4}\end{matrix}\right\} = \frac{1}{4a^3}\left(\frac{1-i}{\sqrt{2}}\right)$$

(*ii*) Set $a = 1$]

2. Using contour integration show that

$$\int_{-\infty}^{\infty} \frac{x^2}{(x^2+a^2)\,(x^2+b^2)}\,dx = \frac{\pi}{a+b}.$$

[**Hint:** S.P's in $C_i : a_i, b_i$; Res. : $\frac{a}{2i(a^2-b^2)}, \frac{-b}{2i(a^2-b^2)}$]

3. Using contour integration, show that

(*i*) $\int_0^\infty \frac{1}{(x^2+a^2)^2}\,dx = \frac{\pi}{4a^3}$ (*ii*) $\int_0^\infty \frac{1}{(x^2+1)^2}\,dx = \frac{\pi}{4}$

[**Hint:** (*i*) D.P. in $C_i : ai$; Res. : $\left[\frac{d}{dz}\frac{1}{(z+ai)^2}\right]_{z=ai} = \frac{1}{4ia^3}$. (*ii*) Set $a = 1$]

4. Using contour integration, show that

(*i*) $\int_0^\infty \frac{x^2}{(x^2+4)^2(x^2+9)}\,dx = \frac{\pi}{200}$, **(1979, 82)**

(*ii*) $\int_0^\infty \frac{1}{(x^2+1)(x^2+4)^2}\,dx = \frac{5\pi}{288}$.

[**Hint:** (*i*) S.P. $3i$ and D.P. $2i$ lie in C_i; Res. : $3i/50, -13i/200$.
(*ii*) S.P. i and D.P. $2i$ lie in C_i; Res. : $-i/18$, $11i/288$]

5. Using contour integration, show that

$$\int_0^\infty \frac{1}{x^4+x^2+1}\,dx = \frac{\pi\sqrt{3}}{6}.$$

[**Hint:** $(z^6-1) = (z^2-1)(z^4+z^2+1)$. Therefore the zeros of z^4+z^2+1 are the roots of $z^6 = 1$, namely,

$e^{i2\pi k/6}$, $k = 0, 1, 2, 3, 4, 5$ except $z = 1$ and $z = -1$.

So S.P's in C_i are $e^{i\pi/3}$, $e^{i2\pi}/3$; Res. : $= \frac{h(a)}{g'(a)}$ gives

$$\left.\begin{matrix}\text{Res. at}\\ z = e^{i\pi/3}\end{matrix}\right\} = \left[\frac{1}{4z^3+2z}\right]_{z=e^{i\pi/3}} = \frac{1}{-3+i\sqrt{3}}$$

$$\left.\begin{matrix}\text{Res. at}\\ z = e^{i2\pi/3}\end{matrix}\right\} = \left[\frac{1}{4z^3+2z}\right]_{z=e^{i2\pi/3}} = \frac{1}{3+i\sqrt{3}}$$

$$\left|\frac{1}{z^4+z^2+1}\right| = \frac{|z^2-1|}{|z^6-1|} < \frac{|z^2|+1}{|z^6|-|-1|}$$

$$= \frac{R^2+1}{R^6-1}\,]$$

6. Show that $\int_0^\infty \frac{x^2}{x^6+1}\,dx = \frac{\pi}{6}$.

[**Hint:** S.P's in C_i; $e^{i\pi/6}, e^{i3\pi/6}, e^{i5\pi/6}$; Now from $\frac{h(a)}{g'(a)} = \left[\frac{z^2}{6z^5}\right]_{z=a}$, we have

$$R_1 = \frac{e^{-i\pi/2}}{6}, R_2 = \frac{e^{-i3\pi/2}}{6}, R_3 = \frac{e^{-i5\pi/2}}{6} \text{ which are}$$

$$R_1 = \frac{e^{-i\pi/2}}{6}, R_2 = \frac{e^{-i\pi/2}}{6}, R_3 = \frac{e^{-i\pi/2}}{6} \text{ whose sum is } - i/6]$$

58. Using contour integration show that

$$\int_0^\infty \frac{1}{x^2+1}\, dx = \frac{\pi}{2}.$$

Type 3

59. Solved sum. By contour integration show that

(*i*) $\displaystyle\int_0^\infty \frac{\cos mx}{(x^2+a^2)^2}\, dx = \frac{\pi}{4a^3}(1+ma)\,e^{-ma}, m > 0, a > 0$ **(1981, 83)**

(*ii*) $\displaystyle\int_0^\infty \frac{\cos x}{(x^2+1)^2}\, dx = \frac{\pi}{2e}$ **(1977)**

(*i*) Let $C = \Gamma + L$, where Γ is the semicircle $|z| = R$, **Im** $z \geq 0$ and L is the line segment [R, – R]. Consider the integral

$$\int_C \frac{e^{imz}}{(z^2+a^2)^2}\, dz .$$

The integrand has a double pole ai lying in C_i if $R > a$. The residue of the integrand at $z = ai$ is

$$\left[\frac{d}{dz}\frac{(z-ai)^2\, e^{imz}}{(z-ai)^2\,(z+ai)^2}\right]_{z=ai} = -\frac{(1+ma)\, ie^{-ma}}{4a^3}$$

$$\therefore \qquad \int_C \frac{e^{imz}}{(a^2+z^2)^2}\, dz = \frac{\pi\,(1+ma)\,e^{-ma}}{2a^3}$$

As $R \to \infty$, by Jordan's lemma, the integral along Γ vanishes because $m > 0$ and, taking $R > a$,

$$\lim_{R\to\infty}\left[\max_{\text{on }\Gamma}\left|\frac{1}{(z^2+a^2)^2}\right|\right] \leq \lim_{R\to\infty}\left[\max_{\text{on }\Gamma}\frac{1}{(|z^2|-a^2)^2}\right]$$

$$= \lim_{R\to\infty}\frac{1}{(R^2-a^2)^2} = 0.$$

$$\therefore \quad \int_{-\infty}^\infty \frac{e^{imx}}{(a^2+x^2)^2}\, dx = \frac{\pi\,(1+ma)\,e^{-ma}}{2a^3} \qquad (\because \text{ on } L, z = x)$$

Equating the real parts and then using the fact that the integrand is an even function of x, we get the first result.

(*ii*) The second result is obtained from the first by setting $a = 1$ and $m = 1$.

60. Using contour integration, show that

(*i*) $$\int_0^{\infty} \frac{\cos x}{(x^2+a^2)(x^2+b^2)}\,dx = \frac{\pi}{2(a^2-b^2)}\left[\frac{e^{-b}}{b}-\frac{e^{-a}}{a}\right], \quad a>b>0$$

(*ii*) $$\int_0^{\infty} \frac{\cos ax}{(x^2+b^2)^2}\,dx = \frac{\pi}{4b^3}(1+ab)\,e^{-ab}, \quad a>0, b>0.$$

[**Hint:** (*i*) $\left|\frac{1}{(z^2+a^2)(z^2+b^2)}\right| \le \frac{1}{(|z|^2-a^2)(|z|^2-b^2)}$ if $R>a$. Consider

$\int_C \frac{e^{iz}}{(z^2+a^2)(z^2+b^2)}\,dz$, where $C=\Gamma+L$; S.P's $z=ai$ and $z=bi$ lie in C_i ;

Res. at $z=ai = \lim_{z\to ai} \frac{e^{iz}}{(z+ai)(z^2+b^2)}$]

61. Using contour integration, show that

(*i*) $$\int_0^{\infty} \frac{x\sin x}{x^2+a^2}\,dx = \frac{\pi}{2}e^{-a}, \quad a>0$$ **(1980, 81)**

(*ii*) $$\int_{-\infty}^{\infty} \frac{\sin x}{x^2+4x+5}\,dx = -\frac{\pi}{e}\sin 2$$ **(1982)**

[**Hint:** (*i*) Consider $\int_C \frac{ze^{iz}}{z^2+a^2}\,dz$, where $C=\Gamma+L$; $z=ai$ is the S.P. in C_i ;

$$\text{Res.} = \lim_{z\to ai} \frac{ze^{iz}}{z+ai} = \frac{e^{-a}}{2}$$

(*ii*) $\int_C \frac{e^{iz}}{z^2+4z+5}\,dz$; S.P. $-2+i$ lies in C_i ; For large R,

$$\left|\frac{1}{z^2+4z+5}\right| \le \frac{1}{|z^2|4z+5|} \le \frac{1}{|z^2|-(|4z|+5)}$$]

Type 4

62. Show that

(*i*) $$\int_0^{\infty} \frac{\sin mx}{x}\,dx = \frac{\pi}{2}, \quad \int_0^{\infty} \frac{\cos mx}{x}\,dx = 0, \quad m>0$$ **(1979, 82)**

(*ii*) $$\int_0^{\infty} \frac{\sin^2 x}{x^2}\,dx = \frac{\pi}{2}.$$

[**Hint:** (*i*) In illustration 1 under type 4, if the integrand is e^{imz}/z,

$$\int_C \frac{e^{imz}}{z}\,dz = 0\,; \quad \lim_{R\to\infty}\int_{C_R} \frac{e^{imz}}{z} = 0$$

by Jordan's lemma because $1/|z| \to 0$ as $R \to \infty$ and $m > 0$. The residue of e^{imz}/z at $z = 0$ is 1. Therefore

$$\int_{-\infty}^{\infty} \frac{e^{imx}}{x}\, dx - i\,(1)\,\pi = 0\,.$$

(*ii*) $2 \sin^2 z = 1 - \cos 2\theta = \mathbf{Re}\,(1 - e^{2iz})$;

$$\int_C \frac{1-e^{2iz}}{z^2}\, dz = 0\,;$$

$z = 0$ is a double pole of the integrand; Res. at $z = 0$ is $-2i$

$$\left|\int_{C_R} \frac{1-e^{2iz}}{z^2}\, dz\right| \le \frac{1+|-e^{2iz}|}{R^2}\,\pi R \to 0 \text{ as } R \to \infty]\,.$$

Other types

63. Solved sum. Using integration along the rectangle C whose sides lie on the lines $y = 0$, $x = R$, $y = 2\pi$, $x = -R$, show that

$$\int_{-\infty}^{\infty} \frac{e^{ax}}{1+e^x}\, dx = \frac{\pi}{\sin ax}, \quad 0 < a < 1\,.$$

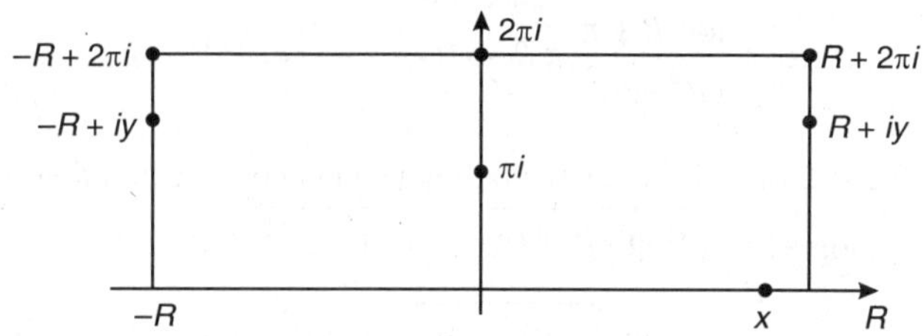

Fig. 10.4.1

Let the integrals of $e^{az}/(1 + e^z)$ along the four sides be I_1, I_2, I_3, I_4 respectively (Fig. 10.4.1). Since the solution of $e^z + 1 = 0$ is $z = (2n + 1)\,\pi i$ of which πi lies in C_i, the pole of $f(z)$ which lies in C_i is πi. the residue of $f(z)$ at πi is $-e^{a\pi i}$. So

$$I_1 + I_2 + I_3 + I_4 = 2\pi i\,(\text{Sum of Res.}) = -2\pi i\, e^{a\pi i}. \tag{1}$$

But
$$I_1 = \int_{-R}^{R} \frac{e^{ax}}{1+e^x}\, dx\,,$$

$$I_3 = \int_{R}^{-R} \frac{e^{a(x+2\pi i)}}{1+e^{x+2\pi i}}\, dx = -\,e^{2a\pi i} \int_{-R}^{+R} \frac{e^{ax}}{1+e^x}\, dx$$

On $x = R$,
$$|f(z)| = \left|\frac{e^{a(R+iy)}}{1+e^{R+iy}}\right| \le \frac{e^{aR}}{e^R - 1}$$

$$= \frac{e^{(a-1)R}}{1-e^{-R}}$$

On $x = -R$,
$$|f(z)| = \left|\frac{e^{a(-R+iy)}}{1+e^{-R+iy}}\right| \le \frac{e^{-aR}}{1-e^{-R}}$$

Therefore, as $R \to \infty$, I_2 and I_4 vanish. Thus (1) becomes

$$\int_{-\infty}^{\infty} \frac{e^{ax}}{1+e^x}\,dx - e^{2a\pi i}\int_{-\infty}^{\infty} \frac{e^{ax}}{1+e^x}\,dx = 2\pi i\,(-e^{a\pi i})$$

$$\therefore \quad \int_{-\infty}^{\infty} \frac{e^{ax}}{1+e^x}\,dx = \frac{-2\pi i\, e^{a\pi i}}{1-e^{2a\pi i}} = \frac{\pi}{(e^{a\pi i} - e^{-a\pi i})/2i} = \frac{\pi}{\sin a\pi}.$$

64. Solved sum. Show that $I = \int_0^{\infty} \frac{\log x}{(x^2+1)^2}\,dx = -\frac{\pi}{4}$.

Choose C as in figure 10.3.3. Then, for a small r and a large R, the pole of the integrand which lies in C_i is $z = i$ and the branch point $z = 0$ lies in C_e.

$$\therefore \quad \int_C \frac{\log z}{(z^2+1)^2}\,dz = 2\pi i\,(\text{Res. at } z = i) = 2\pi i\left[\frac{d}{dz}\frac{\log z}{(z+i)^2}\right]_{z=i}$$

$$= 2\pi i\left(\frac{\pi+2i}{8}\right) = \frac{\pi^2 i}{4} - \frac{\pi}{2}$$

Now, on C_R, $z = Re^{i\theta}$, $0 \le \theta \le \pi$, and

$$|\log z| = \sqrt{(\log R)^2 + \theta^2} \le \log R + \theta \le \log R + \pi$$

$$\therefore \quad \left|\int_{C_R} f(z)\,dz\right| < \frac{\log R + \pi}{(R^2-1)^2}\,\pi R \to 0 \text{ as } R \to \infty.$$

On C_r, $z = re^{i\theta}$, $0 \le \theta \le \pi$, and, for a small r, $\log(1/r)$ is positive and therefore

$$|\log z| = \sqrt{(\log r)^2 + \theta^2} = \sqrt{(-\log r)^2 + \theta^2}$$

$$= \sqrt{\{\log(1/r)\}^2 + \theta^2} < \log(1/r) + \pi$$

$$\therefore \quad \left|\int_{C_r} f(z)\,dz\right| < \frac{\log(1/r) + \pi}{(1-r^2)^2}\,\pi r \to 0 \text{ as } r \to 0$$

because, as $r \to 0$, $1/r \to \infty$ more rapidly than $\log(1/r)$. Thus, in conclusion, taking limit as $R \to \infty$ and $r \to 0$,

$$\lim_{\substack{r \to 0 \\ R \to \infty}} \left(\int_{-R}^{-r} \frac{\log x}{(x^2+1)^2}\,dx + \int_r^R \frac{\log x}{(x^2+1)^2}\,dx\right) = \frac{\pi^2 i}{4} - \frac{\pi}{2}$$

$$\therefore \quad \int_{-\infty}^{0} \frac{\log x}{(x^2+1)^2}\,dx + \int_0^{\infty} \frac{\log x}{(x^2+1)^2}\,dx = \frac{\pi^2 i}{4} - \frac{\pi}{2}$$

In the first integral x is negative, and so, on setting $x = -t$, the first integral becomes

$$\int_0^{\infty} \frac{\log(-t)}{(1+t^2)^2}\,dt = \int_0^{\infty} \log\frac{(e^{\pi i}\,t)}{(1+t^2)^2}\,dt = \pi i\int_0^{\infty} \frac{1}{(1+t^2)^2}\,dt + \int_0^{\infty} \frac{\log t}{(1+t^2)^2}\,dt$$

$$\therefore \quad \left(\pi i\int_0^{\infty} \frac{1}{(1+t^2)^2}\,dt + I\right) + I = \frac{\pi^2 i}{4} - \frac{\pi}{2}$$

Equating real parts we get that $I = -\pi/4$.

CHAPTER 11

MEROMORPHIC FUNCTIONS

11.1 Meromorphic Function

We know that an entire function is a function which is analytic in the whole of finite plane. Now we define another class of functions called meromorphic functions.

Meromorphic function. A function which has no singularities in a region other than a finite number of poles is said to be meromorphic in that region.

The following theorem reveals the number of zeros minus the number of poles of a meromorphic function inside a scro curve.

Theorem 11.1. Suppose

(*i*) D is a simply-connected region,

(*ii*) $f(z)$ is a meromorphic function in D,

(*iii*) C is a scro curve in D, not passing through a pole or zero of $f(z)$.

Then
$$\int_C \frac{f'(z)}{f(z)}\,dz = 2\pi i\,[n(Z,f) - n(P,f)],$$

where $n(Z,f)$ and $n(P,f)$ denote respectively the number of zeros and the number of poles of $f(z)$ in C_i, the zeros and poles being counted as many times as their orders.

Proof: Suppose $a_1, a_2, a_3, \ldots, a_m$
are the zeros of $f(z)$ in C_i of orders $h_1, h_2, h_3, \ldots, h_m$. Then $h_1 + h_2 + h_3 + \ldots + h_m = n(Z,f)$. Consider the zero a_j and let

$$f(z) = (z - a_j)^{h_j}\,\phi(z).$$

Then $\phi(a_j) \neq 0$ and $\phi(z)$ is analytic at $z = a_j$. So, finding the logarithmic differential of $f(z)$,

$$\frac{f'(z)}{f(z)} = \frac{h_j}{z - a_j} + \frac{\phi'(z)}{\phi(z)}$$

We know that the derivative of an analytic function is also analytic. So $\phi'(z)$ is analytic at $z = a_j$ and thus $\phi'(z)/\phi(z)$ is analytic there. Hence, expanding it about $z = a_j$ in Taylor's series,

$$\frac{f'(z)}{f(z)} = \frac{h_j}{z - a_j} + c_0 + c_1(z - a_j) + c_2(z - a_j)^2 + \ldots$$

Hence $z = a_j$ is a simple pole of $f'(z)/f(z)$ and the residue there is h_j. Similarly $a_1, a_2, \ldots$ are simple poles of $f'(z)/f(z)$ with residues $h_1, h_2, \ldots$

Also, suppose $b_1, b_2, b_3, \ldots b_n$ are the poles of $f(z)$ in C_i of order $k_1, k_2, k_3, \ldots, k_n$. Then $k_1 + k_2 + k_3 + \ldots + k_n = n(P,f)$.

Consider the pole b_j and let $f(z) = \dfrac{1}{(z - b_j)^{k_j}}\,\psi(z)$, where $\psi(z)$ is analytic at $z = b_j$ and its value there is not zero. So, finding the logarithmic differential, we get that $z = b_j$ is a simple pole of $f'(z)/f(z)$ and the residue there is $-k_j$. Thus, by the residue theorem,

$$\int_C \frac{f'(z)}{f(z)}\,dz = 2\pi i \,[\text{Sum of the residues of the poles of } f'(z)/f(z) \text{ in } C_i]$$

$$= 2\pi i \left[\sum_{j=1}^{m} h_j - \sum_{j=1}^{n} k_j\right]$$

$$= 2\pi i \,[n(Z,f) - n(P,f)].$$

Corollary. If $f(z)$ is analytic in D, then the number of poles in C_i is zero. Therefore

$$\int_C \frac{f'(z)}{f(z)}\,dz = 2\pi i \,\{\text{Number of zeros of } f(z) \text{ in } D\}.$$

Theorem 11.2. Principle of argument. Suppose

(*i*) D is a simply-connected region,

(*ii*) $f(z)$ is meromorphic in D,

(*iii*) C is a scro curve in D, not passing through a pole or zero of $f(z)$.

Then, as z travels along C once, the argument of $f(z)$ increases by

$$[n(Z,f) - n(P,f)] \text{ times } 2\pi$$

i.e., if $\Delta_C \arg f(z)$ denotes the change in the argument of $f(z)$ as z travels along C once, then

$$\Delta_C \arg f(z) = [n(Z,f) - n(P,f)]\,(2\pi)$$

Proof: Now, for any z on C,

$$\frac{d}{dz}\log f(z) = \frac{f'(z)}{f(z)}.$$

If A denotes a point $z = a$ on C and P denotes the point z on C (Fig. 11.1.1), then the integral of $f'(z)/f(z)$ along the arc AP is

$$\int_{AP} \frac{f'(z)}{f(z)}\,dz = [\log f(z)]_A^P = \log f(z) - \log f(a).$$

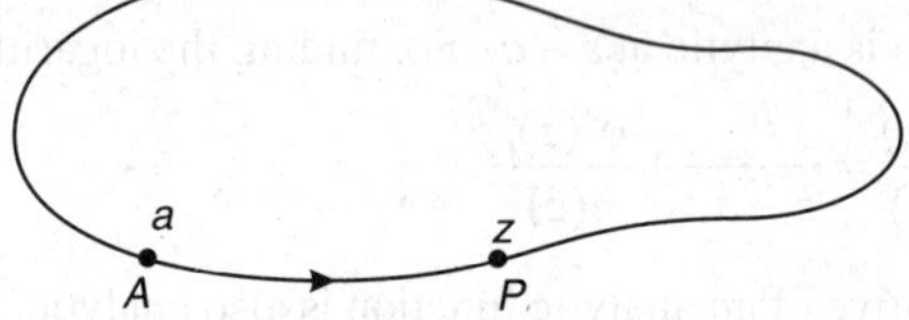

Fig. 11.1.1

Therefore, when P travels along C once (commencing from A),

$$\int_C \frac{f'(z)}{f(z)}\,dz = [\log f(z)]_C$$

$$= [\log|f(z)| + i \arg f(z)]_C.$$

But $\log|f(z)|$ is real and continuous. Therefore

$$[\log|f(z)|]_C = 0.$$

$$\therefore \quad \int_C \frac{f'(z)}{f(z)}\,dz = [i \arg f(z)]_C = i\,\Delta_C \arg f(z).$$

But $$\int_C \frac{f'(z)}{f(z)} = 2\pi i\,[n(Z,f) - n(P,f)] \quad \text{(Theorem 11.1)}$$

$\therefore$ $$\Delta_C \arg f(z) = 2\pi\,[n(Z,f) - n(P,f)].$$

Theorem 11.3. Rouche's theorem. Suppose

(*i*) D is a simply-connected region,

(*ii*) $f(z)$ is a meromorphic function in D,

(*iii*) C is a scro curve in D, not passing through a pole or zero of $f(z)$.

(*iv*) $f(z)$ is expressed as a sum of two functions $g(z)$ and $h(z)$ as

$$f(z) = g(z) + h(z),$$

where $g(z)$ is meromorphic in D and does not have a pole or zero on C as $f(z)$, and also

$$|g(z)| > |h(z)| \text{ on } C.$$

Then in C_i

$$n(Z,f) - n(P,f) = n(Z,g) - n(P,g).$$

Proof: From the principle of argument

$$\Delta_C \arg f(z) = 2\pi\,[n(Z,f) - n(P,f)].$$

$$\begin{aligned}\therefore\ 2\pi\,[n(Z,f) - n(P,f)] &= \Delta_C \arg f(z) = \Delta_C \arg [g(z) + h(z)] \\ &= \Delta_C \arg \left[g(z)\left\{1 + \frac{h(z)}{g(z)}\right\}\right] \\ &= \Delta_C \left[\arg g(z) + \arg\left\{1 + \frac{h(z)}{g(z)}\right\}\right] \\ &= \Delta_C \arg g(z) + \Delta_C \arg\left[1 + \frac{h(z)}{g(z)}\right] \\ &= 2\pi\,[n(Z,g) - n(P,g)] + \Delta_C \arg\left[1 + \frac{h(z)}{g(z)}\right]\end{aligned}$$

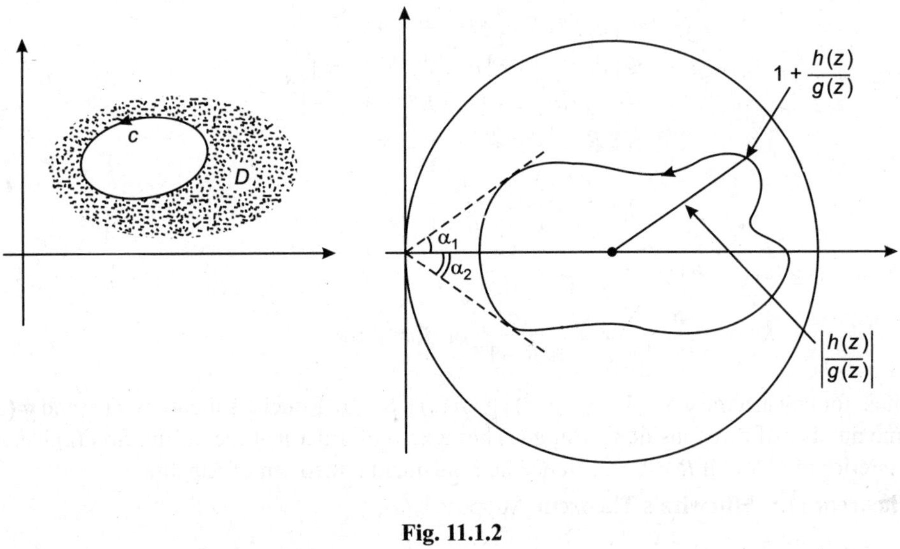

Fig. 11.1.2

But $|g(z)| > |h(z)|$ for all z on C. That is, $|h(z)/g(z)| < 1$ on C. Therefore the complex number

$$1+\frac{h(z)}{g(z)}$$

lies in the interior of the circle $|z-1|=1$ as shown in figure 11.1.2. So, for all z on C,

$$-\frac{\pi}{2} < -\alpha_2 \le \arg\left[1+\frac{h(z)}{g(z)}\right] \le \alpha_1 < \frac{\pi}{2}.$$

Thus, as z travels along C once

$$\arg\left[1+\frac{h(z)}{g(z)}\right]$$

increases from 0 to an acute angle α_1 and then decreases to an acute angle $-\alpha_2$ and finally increases to 0. Hence

$$\Delta_C \arg\left[1+\frac{h(z)}{g(z)}\right] = 0$$

$$\therefore \qquad n(Z, f) - n(P, f) = n(Z, g) - n(P, g).$$

Corollary. If $f(z)$ is analytic in D, then

$$n(Z, f) = n(Z, g).$$

The fundamental theorem of Algebra, already proved in theorem 8.15, can be proved by applying Rouche's theorem as follows:

Theorem 11.4. Fundamental theorem of Algebra. A polynomial of degree $n \ge 1$ has n zeros.

Proof: Consider the polynomial

$$f(z) = z^n + c_1 z^{n-1} + c_2 z^{n-2} + \dots + c_n$$

Let $f(z) = g(z) + h(z)$, where

$$g(z) = z^n,$$
$$h(z) = c_1 z^{n-1} + c_2 z^{n-2} + \dots + c_n$$

Then $f(z)$ and $g(z)$ are analytic in the finite plane. Let c be the maximum of $|c_1|, |c_2|, \dots, |c_n|$ and C, a circle $|z| = R$, $R > 1$. Then, on C,

$$|g(z)| = |z^n| = |z|^n = R^n$$

and

$$\begin{aligned} |h(z)| &= |c_1 z^{n-1} \; c_2 z^{n-2} + \dots + c_n| \\ &\le |c_1 z^{n-1}| + |c_2 z^{n-2}| + \dots + |c_n| \\ &= |c_1| R^{n-1} + |c_2| R^{n-2} + \dots + |c_n| \\ &\le c R^{n-1} + c R^{n-2} + \dots + c \\ &= c\frac{R^n - 1}{R-1} < c\frac{R^n}{R-1}. \end{aligned}$$

So, on C, $|g(z)| > |h(z)|$ if

$$R^n > c\frac{R^n}{R-1} \quad \text{or} \quad 1 > \frac{c}{R-1} \quad \text{or} \quad R > 1 + c.$$

Thus, for any arbitrary $R > 1 + c$, $|g(z)| > |h(z)|$. So, by Rouche's theorem, $f(z)$ and $g(z)$ have the same number of zeros inside C. But $g(z)$ has a zero of order n at the origin. So $f(z)$ has n zeros in the interior of C for all $R > 1 + c$. Hence he fundamental theorem of Algebra.

Theorem 11.5. Hurwitz's Theorem. Suppose

(*i*) D is simply-connected region,

(*ii*) $\{f_n\}$ is a sequence of analytic functions having no zeros in D and the sequence converges in D uniformly to a function $f(z)$, not identically zero.

Then $f(z)$ has no zeros in D.

Proof: Suppose $f(z)$ has zeros in D and $z = a$ is one of them. Let C be an scr curve in D encircling $z = a$ and not passing t through any zero such that no other zero of $f(z)$ lies either in C_i or on C. Let $\in$ be the minimum value of $|f(z)|$ on C. Then by hypothesis $\in > 0$. That is,

$$|f(z)| \geq \in, \quad \text{for all } z \text{ on } C. \tag{11.1.1}$$

But, due to the uniform convergency of the sequence, we can find, for all z on C, an n_0 such that,

$$|f_n(z) - f(z)| < \in, \quad \text{for all } n > n_0. \tag{11.1.2}$$

Dividing (11.1.2) by (11.1.1), we get for all z on C

$$\left|\frac{f_n(z) - f(z)}{f(z)}\right| < 1, \quad n > n_0$$

or

$$|f(z)| > |f_n(z) - f(z)|, \quad n > n_0$$

Therefore, writing $f_n(z)$ as the sum of two function as

$$f_n(z) = f(z) + \{f_n(z) - f(z)\}$$

and, using Rouche's theorem, we get that the functions

$$f_n(z), \quad n > n_0 \tag{11.1.3}$$

have the same number of zeros in C_i as $f(z)$. But $f(z)$ has one zero, $z = a$, in C_i. So $f_n(z)$ also has one zero in C_i, which contradicts the hypothesis. Therefore the assumption that $f(z)$ has zeros in D is wrong. This implies that $f(z)$ has no zeros in D.

Theorem 11.6. (Another form of Hurwitz's theorem). Suppose $f_n(z)$ is a sequence of analytic functions in a region D. Suppose this sequence converges uniformly in D to a function $f(z)$, not identically zero. If C is an scr curve in D not passing through any zero of $f(z)$, then there is an integer n_0 such that every function $f_n(z)$ with $n > n_0$ has the same number of zeros in C_i as the function $f(z)$.

Proof: Result (11.1.3) prove the theorem.

11.2 Function Meromorphic in the Extended Plane

We know that an analytic function in the extended plane is a constant (Corollary to Liouville's theorem). The following theorem reveals the nature of a function meromorphic in the extended plane.

Theorem 11.7. A function which is meromorphic in the extended plane is a rational function.

Proof: Let $f(z)$ be a function meromorphic in the extended plane. Let its poles be

$$a,\ b,\ \ldots,\ \infty.$$

Let their orders be $h, k, \ldots, t$. Then the principal parts of $f(z)$ at $a, b, \ldots, \infty$ are of the form

$$\frac{a_h}{(z-a)^h} + \frac{a_{h-1}}{(z-a)^{h-1}} + \ldots + \frac{a_1}{z-a},$$

$$\frac{b_k}{(z-b)^k} + \frac{b_{k-1}}{(z-b)^{k-1}} + \ldots + \frac{b_1}{z-b},$$

$$q_t z^t + q_{t-1} z^{t-1} + \ldots + q_1 z.$$

Let the sum of these principal parts be $P(z)$. Then $P(z)$ is a rational function and $f(z) - P(z)$ does not have any singularity in the extended plane. That is, $f(z) - P(z)$ is analytic in the extended plane. Hence, by the Corollary to Liouville's theorem, $f(z) - P(z)$ is constant, say C. Thus $f(z) = P(z) + C$. But $P(z)$ is a rational function, and so also is $f(z)$.

11.3 Examples

Rouche's Theorem

1. **Solved sum.** Find the number of zeros of the function $f(z) = z^7 - 4z^5 + z^2 - 1$ which lie inside the unit circle $C : |z| = 1$.

Of the coefficients 1, – 4, 1, – 1, the coefficient – 4 has the maximum absolute value. To use Rouche's theorem, choose the term $-4z^5$ as $g(z)$ so that

$$f(z) = g(z) + h(z),\ g(z) = -4z^5,\ h(z) = z^7 + z^2 - 1$$

Then, on $C : |z| = 1$,

$$|g(z)| = |-4z^5| = 4|z|^5 = 4(1)^5 = 4,$$
$$|h(z)| = |z^7 + z^2 - 1| \le |z^7| + |z^2| + |-1| = 1 + 1 + 1 = 3.$$

So $|g(z)| > |h(z)|$ and hence $f(z)$ and $g(z)$ have same number of zeros inside C. But $g(z)$ has $z = 0$ as a zero of order 5 and hence it has five zeros inside C. Therefore $f(z)$ also has five zeros inside C.

2. Find the number of roots of the following equations which lie inside the unit circle $|z| = 1$.

(*i*) $z^8 - 5z^5 - 2z + 1 = 0$ **(1976)**

(*ii*) $z^8 - 4z^4 + z^2 - 1 = 0$ **(1978)**

(*iii*) $2z^9 - 5z^5 + z^2 - 1 = 0$

(*iv*) $z^4 - 5z + 1 = 0$

(*v*) $z^6 + z^3 - 6z + 9 = 0$

[*Hint* : $g(z)$	$h(z)$	Answer
(*i*) $-5z^5$	$z^8 - 2z + 1$	5
(*ii*) $-4z^4$	$z^8 + z^2 - 1$	4
(*iii*) $-5z^5$	$2z^9 + z^2 - 1$	5
(*iv*) $-5z$	$z^4 + 1$	1
(*v*) 9	$z^6 + z^3 - 6z$	0]

3. **Solved sum.** Find the number of zeros of the function $f(z) = z^4 - 5z + 1$ which lie in the annulus $1 < |z| < 2$. **(1982)**

Let C_1 and C_2 be the circles $|z| = 1$ and $|z| = 2$. Then, if $g_1(z) = -5z$, $h_1(z) = z^4 + 1$, then on C_1,

$$|g_1(z)| = 5,\ |h(z)| \le |z|^4 + 1 = 2.$$

So $|g_1| > |h_1|$ and hence $f(z)$ and $g_1(z)$ have same number of zeros inside C_1. But $z = 0$ is the only zero of $g_1(z)$. Therefore $f(z)$ also has one zero inside C_1.

If $$g_2(z) = z^4,\ h_2(z) = -5z + 1, \text{ then on } C_2,$$
$$|g_2| = |z|^4 = 2^4 = 16;\ |h_2| = |-5z + 1| \le 5|z| + 1 = 11.$$

So $|g_2| > |h_2|$ and hence $f(z)$ and $g_2(z)$ have same number of zeros inside C_2. But $g_2(z)$ has one zero of order 4 inside C_2. Hence $f(z)$ also has four zeros inside C_2. Therefore the number of zeros in the given annulus is $4 - 1 = 3$.

4. Show that all the zeros of the following functions lie in the annulus regions shown against them:

(*i*) $16z^5 - z + 8$ $\quad \frac{1}{2} < |z| < 1$ **(1978, 80)**

(*ii*) $z^7 - 5z^3 + 12$ $\quad 1 < |z| < 2$ **(1980)**

(*iii*) $z^7 - 6z^3 + 12$ $\quad 1 < |z| < 2$ **(1981)**

[*Hint* : (*i*) $g_1(z) = 8$, $h_1(z) = 16z^5 - z$; $g_2(z) = 16z^5$, $h_2(z) = -z + 8$ (*ii*) $g_1(z) = 12$, $h_1(z) = z^7 - 5z^3$; $g_2(z) = z^7$, $h_2(z) = -5z^3 + 12$ (*iii*) $g_1(z) = 12$, $h_1(z) = z^7 - 6z^3$; $g_2(z) = z^7$, $h_2(z) = -6z^3 + 12$]

5. If $a > e$, show that $e^z - az^n$ has n zeros inside the unit circle $C : |z| = 1$.

[*Hint* : $g(z) = -az^n$, $h(z) = e^z$. On C, $|g(z)| = a > e$, $|h(z)| = |e^{\cos\theta + i\sin\theta}| = e^{\cos\theta} \leq e$; So $|g(z)| > |h(z)|$]

6. If $|a| > e^R/R^n$, prove that $e^z = az^n$ has n roots inside the circle $C : |z| = R$. **(1974)**

[*Hint* : $g(z) = -az^n$, $h(z) = e^z$; On C, $|g(z)| = |a| R^n > e^R$, $|h(z)| = |e^{R(\cos\theta + i\sin\theta)}| = e^{R\cos\theta}$ and $|g(z)| > |h(z)|$]

7. **Solved sum.** Show that the equation $e^{z-a} - z = 0$, $a > 1$, has just one root in the interior of the circle $C : |z| = 1$ and that it is real and positive.

Choose $g(z) = -z$, $h(z) = e^{z-a}$. Then, on C,

$$|g(z)| = 1,$$
$$|h(z)| = |e^{(\cos\theta - a) + i\sin\theta}| = e^{\cos\theta - a} = 1/e^{(a - \cos\theta)} < 1.$$

Therefore $f(z)$ has one root inside C as $g(z)$. Considering the real x and the real function $f(x) = e^{x-a} - x$,

$$f(0) = e^{-a} > 0$$
$$f(1) = e^{1-a} - 1 = (1/e^{a-1}) - 1 < 0.$$

So $f(x) = 0$ has a root say λ, between 0 and 1. This implies that $f(z)$ has a real root λ inside C.

8. Show that the equation $e^z = e^4 z$ has only one root inside the unit circle $|z| = 1$ and that it is positive.

Principle of Argument

9. **Solved sum.** Show that one root of $z^4 + z^3 + 1 = 0$ lies in the first quadrant. **(1979, 80, 83)**

First we show that no root lies on the right half of the real axis and on the upper half of the imaginary axis. Evidently 0 is not a root. Let α be real and positive. Then

$$\alpha^4 + \alpha^3 + 1$$

is positive for all α and so it does not vanish. Hence the equation does not have a root on the right half of the real axis. Also

$$(i\alpha)^4 + (i\alpha)^3 + 1 = \alpha^4 - i\alpha^3 + 1$$
$$= (\alpha^4 + 1) + i(-\alpha^3).$$

But no real value of α make $\alpha^4 + 1$ and α^3 vanish simultaneously. Therefore $i\alpha$ is not a root of the equation. Thus no root lies on the upper half of the imaginary axis.

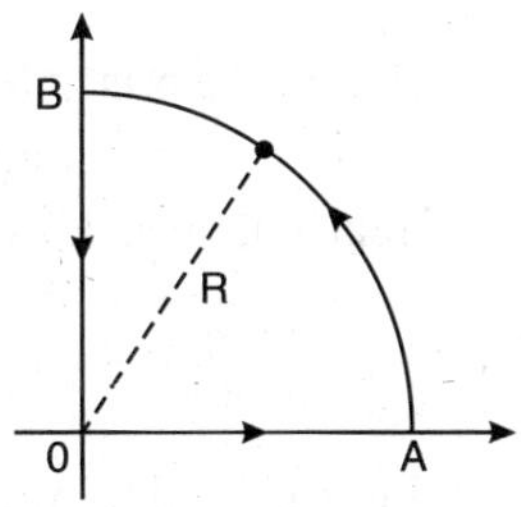

Fig. 11.3.1

Let C be a scro curve consisting of

(*i*) the line segment OA, $[0, R]$ (Fig. 11.3.1)

(*ii*) the circular $\widehat{AB}$, $|z| = R$, $0 \leq \arg z \leq \frac{1}{2}\pi$

(*iii*) the line segment BO, $[Ri, 0]$.

(*i*) Now the change in $\arg f(z)$ as z travels along the real axis from O to A, is

$$\Delta_{OA} \arg f(z) = [\arg f(\alpha)]_{\alpha=0}^{\alpha=R} = 0 \quad (1)$$

because $f(\alpha)$ is real and positve and as such its argument remains zero as α increases from 0 to R.

(*ii*) As z travels along $\widehat{AB}$, $z = Re^{i\theta}$, where θ varies from 0 to $\frac{1}{2}\pi$. Therefore

$$\Delta_{\widehat{AB}} \arg f(z) = \arg\left[R^4 e^{4i\theta} + R^3 e^{3i\theta} + 1\right]_{\theta=0}^{\theta=\frac{1}{2}\pi}$$

$$= \arg\left[R^4\left(e^{4i\theta} + \frac{R^3e^{3i\theta}+1}{R^4}\right)\right]_{\theta=0}^{\theta=\frac{1}{2}\pi}$$

$$= \left[\arg R^4 + \arg\left(e^{4i\theta} + \frac{R^3e^{3i\theta}+1}{R^4}\right)\right]_{\theta=0}^{\theta=\frac{1}{2}\pi}$$

$$= \left[0 + \arg\left(e^{4i\theta} + \frac{R^3e^{3i\theta}+1}{R^4}\right)\right]_{\theta=0}^{\theta=\frac{1}{2}\pi}$$

Taking limit as $R \to \infty$,

$$\lim_{R\to\infty} \Delta_{\widehat{AB}} \arg f(z) = \left[\arg e^{4i\theta}\right]_0^{\frac{1}{2}\pi} = \left[4\theta\right]_0^{\frac{1}{2}\pi} = 2\pi.$$

(*iii*) Next, for any positive α, $f(i\alpha) = \alpha^4 + 1 + i(-\alpha^3)$ is a point in the fourth quadrant and so its argument lies between $-\frac{1}{2}\pi$ and 0. Thus,

$$\Delta_{BO} \arg f(z) = \arg\left\{(\alpha^4+1) + i(-\alpha^3)\right\}_{\alpha=R}^{\alpha=0}$$

$$= \left[\tan^{-1}\frac{-\alpha^3}{\alpha^4+1}\right]_{\alpha=R}^{\alpha=0} = -\tan^{-1}\frac{-R^3}{R^4+1}.$$

$$\therefore \quad \lim_{R\to\infty} \Delta_{BO} \arg f(z) = -\tan^{-1} 0 = 0.$$

$$\therefore \quad \lim_{R\to\infty} \Delta_C \arg f(z) = 0 + 2\pi + 0. \qquad (1)$$

But, by the principle of argument,

$$\Delta_C \arg f(z) = (\text{Number of roots in } C_i)\, 2\pi.$$

Taking limit as $R \to \infty$,

$$\lim_{R\to\infty} \Delta_C \arg f(z) = \begin{pmatrix}\text{Number of roots in}\\ \text{the first quadrant}\end{pmatrix} 2\pi \qquad (2)$$

From (1) and (2) we get the number of roots in the first quadrant as 1.

Remark. The degree of the given equation is 4. Therefore it has four roots. We showed that there is no root on the right half of the real axis. We can also show that there is no root on the left half of the real axis. So all the four roots are imaginary. But the coefficients of the powers of z are real. Hence the imaginary roots occur in pairs such that in a pair one is the conjugate of the other. Now we have shown that only one root lies in the first quadrant. Its conjugate lies in the fourth quadrant. The remaining two roots lie on the left side of the imaginary axis, one in the second quadrant and the other in the third quadrant.

10. Show that $z^8 + 3z^3 + 7z + 5$ has two zeros in the first quadrant. **(1979, 80, 82)**

[*Hint* : Method: Same as in the previous sum. (*i*) $\Delta_{OA} = 0$. (*ii*) $\lim_{R\to\infty} \Delta_{\widehat{AB}} \arg f(z) =$

$\left[\arg e^{8i\theta}\right]_0^{\frac{1}{2}\pi} = 4\pi$ (*iii*) $f(i\alpha)$ is a point in the first or the fourth quadrant. So $\arg f(i\alpha)$ lies between $-\frac{1}{2}\pi$ and $\frac{1}{2}\pi$. $\lim_{R\to 0} \Delta_{BO} \arg f(z) = 0$].

11. Show that $z^4 - z^2 + 1 = 0$ has one root in the first quadrant. **(1975)**

12. **Solved sum.** Show that $z^3 + iz + 1 = 0$ has one root in the first quadrant.

Choose C as in sum 9. Let α be real and positive. Then

$$f(\alpha) = (\alpha^3 + 1) + i\alpha \neq 0 \text{ for any } \alpha,$$
$$f(i\alpha) = (1 - \alpha) + i(-\alpha^3) \neq 0 \text{ for any } \alpha.$$

So no root of the equation lies on the right half of the real axis and upper half of the imaginary axis.

(*i*) For any $z = \alpha$, on OA, $f(z)$ lies in the first quadrant. Therefore $\arg f(z)$ lies between 0 and $\frac{1}{2}\pi$. Thus

$$\Delta_{OA} \arg f(z) = \left[\tan^{-1}\frac{\alpha}{\alpha^3+1}\right]_{\alpha=0}^{\alpha=R} = \tan^{-1}\frac{R}{R^3+1}$$

$$\therefore \quad \lim_{R\to\infty} \Delta_{OA} \arg f(z) = 0.$$

(*ii*) $$\lim_{R\to\infty} \Delta_{\widehat{AB}} \arg f(z) = \lim_{R\to\infty}\left[\arg R^3 + \arg\left\{e^{i3\theta} + \frac{iRe^{i\theta}+1}{R^3}\right\}\right]_{\theta=0}^{\theta=\frac{1}{2}\pi}$$

$$= \left[\arg e^{i3\theta}\right]_0^{\frac{1}{2}\pi} = \left[3\theta\right]_0^{\frac{1}{2}\pi} = \frac{3\pi}{2}.$$

(*iii*) For any $z = i\alpha$, on BO, $f(z) = (1-\alpha) + i(-\alpha^3)$ of which the imaginary part is negative. So $f(z)$ lies below the real axis.

$$\Delta_{BO} \arg f(z) = \left[\tan^{-1}\frac{\alpha^3}{\alpha-1}\right]_{\alpha=R}^{\alpha=0}$$

$$\therefore \quad \lim_{R\to\infty} \Delta_{BO} \arg f(z) = \left[\tan^{-1}\frac{R^3}{R-1}\right]_{R=\infty}^{R=0}$$

$$= 0 - \left(-\frac{1}{2}\pi\right) = \frac{1}{2}\pi$$

Thus, by the principle of argument,

$$0 + \frac{3\pi}{2} + \frac{\pi}{2} = \begin{pmatrix}\text{Number of roots in}\\ \text{the first quadrant}\end{pmatrix} 2\pi$$

So the number of roots in the first quadrant is 1.

13. Show that $z^5 + z^3 + 2z + 3$ has one zero in the first quadrant. **(1980, 81, 85)**

[*Hint* : Method : Same as in sum 9 : zeros on the right half of the real axis and on the upper half of the imaginary axis: Nil; (*i*) $f(\alpha)$ is real. $\therefore \Delta_{OA} \arg f(z) = 0$; (*ii*) $\lim\limits_{R \to \infty} \Delta_{AB} \arg f(z) = [5\theta]_0^{\frac{1}{2}\pi} = \frac{5\pi}{2}$.

(*iii*) $f(i\alpha) = 3 + i\,(\alpha^5 - \alpha^3 + 2\alpha)$, where $\alpha\,\{\alpha^2\,(\alpha^2 - 1) + 2\}$ is positive, lies in the first quadrant.

$$\therefore \quad \lim_{R \to \infty} \Delta_{OB} \arg f(z) = \left[\tan^{-1} \frac{R\,(R^4 - R^2 + 2)}{3}\right]_{R=\infty}^{R=0} = -\frac{\pi}{2}\,]$$

14. Show that $z^4 + z^3 + 4z^2 + 2z + 3$ has two zeros each in the second and third quadrants. **(1979)**